ESSENTIAL LOGIC

BASIC REASONING SKILLS FOR
THE TWENTY-FIRST CENTURY

ESSENTIAL LOGIC

BASIC REASONING SKILLS FOR
THE TWENTY-FIRST CENTURY

Ronald C. Pine

Honolulu Community College

New York   Oxford
OXFORD UNIVERSITY PRESS

Oxford University Press

Oxford   New York
Auckland   Bangkok   Buenos Aires   Cape Town
Chennai   Dar es Salaam   Delhi   Hong Kong   Istanbul   Karachi
Kolkata   Kuala Lumpur   Madrid   Melbourne   Mexico City   Mumbai   Nairobi
São Paulo   Shanghai   Singapore   Taipei   Tokyo   Toronto

Published by Oxford University Press, Inc.
198 Madison Avenue, New York, New York 10016
http://www.oup-usa.org

Oxford is a registered trademark of Oxford University Press

ISBN 978-0-19-515505-1

Cover Painting: Chromatic Convergence Series, "Untitled," by Gloria Ross.
Represented by Select Art, Dallas, Texas.

Printing number: 9 8 7

Printed in the United States of America
on acid-free paper

TO DAVID, KYMBERLY, AND REYNA
AND OUR PUPUKEA DAYS

PREFACE

To the instructor:

Why should you consider yet another logic book?

I have taught a freshman logic course for almost 25 years at both a major university and a community college. I have watched as most students who started this course severely intimidated, gradually became confident as their natural critical thinking and formal reasoning abilities emerged. Over the years, several students have indicated that this course was a turning point in their lives; they said that they were able to build on the skills and attitudes learned in basic logic to tackle any course or problem that seemed at first too difficult. These students learned the essence of critical analysis: how not to be intimidated by a complex whole, how to break things into simpler pieces, how to remain disciplined and follow a reasoning trail, and how to stay focused so that the brain can do its natural pattern-recognition work.

In the past 40 years, thousands of new community colleges have drawn legions of nontraditional students, while universities have become increasingly multicultural. In general, I have found this nontraditional student pool, with its endless diversity, a joy. What they sometimes lack in academic preparation of the traditional sort, they make up for with rich experiential backgrounds and a seriousness of purpose. We teachers sometimes forget that students today are facing quite different challenges and opportunities than we faced as students. This, of course, is no reason for lack of rigor or a lowering of standards. But it should call us to rethink our approaches.

Here are the unique features this book provides.

1. A readable, dialogue-like, but challenging style more suitable for today's student.
2. Essential introductory skills, concepts, and attitudes: deductive and inductive, informal and formal logic placed within a philosophical perspective. A humble, but upbeat interpretation of the power of science and inductive reasoning, one that counters the polar evils of technological overconfidence and epistemological laziness. An integration of traditional introductory concepts with modern technological issues and the world view of modern science.
3. A unique structuring methodology and organization of informal fallacies that force students to interpret and argue, not just label. An exercise approach that forces students to write and argue by connecting criticism to key concepts.
4. Rigor and relevance in the symbolic logic chapters combined with a more student-centered sequencing of the learning steps, especially in Chapters 9 and 10.
5. Controversy. I deliberately take stands on some issues but encourage students to use the tools they have learned to critique these stands. I have also included a chapter on Fuzzy Logic and framed the presentation in terms of a debate between Western and Eastern philosophy.

Concerning (1), I try to talk directly to students in a narrative style. I explain some key terms at greater length than is customary (for instance, the treatment of validity in Chapter 1), attempt to add more texture to sequencing (the fallacy and propositional logic chapters), and then turn up the intellectual heat in places with either controversial digressions or advanced philosophical topics (value clarification, the nature of value disagreements, the nature of truth and issues in the philosophy of science, the role of reason in life, artificial intelligence, and the frontiers of logic.)

Concerning (2) and (5), I have tried to make this textbook interesting and intriguing by having a theme. Some instructors may be surprised by the digressions where I adopt controversial philosophical perspectives. But so-called more objective texts are not value- or position-neutral anyway, and I don't think I have hidden the controversial nature of my positions from student readers behind a cloak of authoritarianism. Furthermore, I have given students the format (exercises) to critique my positions. Those instructors who desire student exercises involving extended arguments should find this approach valuable.

Concerning (3), many logic books present fallacies as if their identification is purely a process of labelling and description; as if fallacies are simply anti-commandments rolled down from above, as absolute prohibitions in which no argument is needed. I assume that the interpretation and charge of fallacy require argumentation, and students need to learn how to make a case that arguments are weak. Thus, this book introduces a formalization strategy for fallacies that not only forces students to understand exactly how to critique each fallacy and connect the nuances of the critique to broader foundational concepts, but also serves as a transition to formal logic.

With the formalization schema introduced in Chapters 4 and 5, students are able to see the relevance of formalizing our thinking, of abstracting the essence of a thinking process so we only have to think about it once, so to speak, regardless of novel content. From here a smooth transition can be made to symbolic logic and mathematics. Students are also encouraged to see the positive nature of criticism: proper categorization and localization of the bad reasoning enables us to know what will make an argument better by identifying the type of evidence or support that will make a conclusion more reliable.

Concerning (4), in my experience with sympathetic teaching and sequencing, the average student with no more prerequisite than an ability to read at a college level can work on the more difficult problems presented in the text, such as those in Chapter 10. They may not solve each problem but they should be able to create significant, valid symbolic trails. And in the process they will gain a deep appreciation for an important aspect of our technological culture.

Hence, Chapters 9 and 10 present propositional logic without indirect or conditional proof. Because there is only so much that can be covered in a semester or quarter, and because there is a natural affinity between syllogistic and quantification logic, Chapter 11 provides students with an introduction to the first stages of quantification via a very brief exposure to Venn diagrams and syllogisms. This is a difficult chapter, but by this stage in the semester students should be able to master the basic concepts.

Given the amount of time in a semester or quarter, I have decided that although propositional logic should be covered thoroughly, just enough classical logic should be introduced to understand the debate in Chapter 12: As we enter the 21st century, Western bivalent logic has significant competition. Students should know that the crisp lines of a Venn diagram and the pleasing symmetry of a truth table represent a major philosophical assumption, and that different assumptions may literally be worth billions of dollars in new product development. Thus, I conclude the book with a balanced presentation of Fuzzy Logic consistent with my theme that these debates are not just irrelevant ivory tower discussions but can literally be "cashed in" to new technology.

A few final notes. I have kept the exercise sets shorter than is customary. It keeps the cost of the book down and acts as a confidence building measure for students. I have sought to provide quality in terms of conceptual understanding rather than mechanical mastery via quantity. For instance, I see no reason for students to construct a large number of truth tables, and especially any with more than four variables. In the propositional logic chapters I have used Copi's 19 rules, both for ease of student learning (it still is the best system I have tried) and for instructor convenience. There are only a few modifications. What Copi has called the rule of Tautology, I call Repetition, and I have replaced Copi's version of Implication, $(p \supset q) \equiv (\sim p \vee q)$, with $(\sim p \supset q) \equiv (p \vee q)$. The latter is easier for students to understand given English equivalents.

I would be happy to discuss any matter relevant to the material presented here via electronic mail. You can reach me via Internet at either **pine@uhunix.uhcc.Hawaii.edu** or **pine@pulua.hcc.Hawaii.edu**. Please feel free to have your students also correspond via e-mail.

ACKNOWLEDGEMENTS

In putting together any new textbook, an author receives assistance from many sources. I would especially like to thank the reviewers who examined this manuscript through various drafts in development. My thanks go to Bernard Allen, West Virginia University—Parkersburg; George Gale, University of Missouri—Kansas City; John Serembus, Widener University; Anne Morrissey, California State University—Chico; Christine Holmgren, Santa Monica College; Mark Cobb, Pensacola Junior College; Glenn A. Webster, University of Colorado—Boulder; and James Page, University of Kansas.

Closer to home, much gratitude is owed to my colleagues: Terry Haney, Cynthia Smith, Harvey Lacey, and Dan Petersen of Honolulu Community College; Ron Bontekoe and Larry Laudan of the University of Hawaii. Terry, Cynthia, and Harvey provided encouragement in many ways related to our instructional mission. Dan's passion in promoting the value of critical thinking and the relevance of a well-sequenced symbolic logic component for introductory logic constantly kept me focused. Ron's many conceptual and editing suggestions were of significant help in the last two chapters. Larry's assistance has been indirect, but monumental. Anyone who knows of his work will see its influence in my development of the concept of a reliable belief and the nonfoundational approach to rationality and evidence.

Most important, however, is the gratitude that I owe some very special students. In many ways, the following helped me remember who the book is really for: Brian Gonzaga, Kapiolani Iseke, Aloha Kekoolani, Tana-Lee Rebhan-Kang, Maureen Liu-Brower, Mary Martz, Karl H. Klaassen, Susan O'Neal, Al Tuvera, Sharon Rasos, Jennifer Schiereck, and Gregory Tom.

TABLE OF CONTENTS

INTRODUCTION

When I was in the fourth grade I had a teacher by the name of Mrs. Vagnus. Although she was probably a very good teacher, to a precocious little boy she seemed just like Mrs. Wormwood does to the imaginative little rascal Calvin, in the comic strip Calvin and Hobbes. Like Calvin I got into trouble a lot in Mrs. Vagnus's class. I can remember quite vividly Mrs. Vagnus drawing a small circle with a piece of chalk on the blackboard and exclaiming in a shrill voice, "Ronnie Pine, get up here right this minute and put your nose against this spot for the rest of the period!" It was very effective punishment. For a little fourth grade boy it was not only excruciatingly confining to be still with my face claustrophobically plastered into the dusty blackboard, but also very embarrassing with my rear facing the class, and my friends giggling and even occasionally firing a spit wad at me when Mrs. Vagnus was not looking.

Calvin and Hobbes   by **Bill Watterson**

1

Calvin and Hobbes

by Bill Watterson

I tell this story because you are about to work through a logic course and probably at least ten chapters of this textbook. Many students are intimidated by logic and mathematics. I am often intimidated by art. We did a lot of art work in Mrs. Vagnus's class. Needless to say, I cannot draw very well. To this day when I am required to draw the simplest picture I tremble. A small child could do better. My artistic friends claim that anyone can learn to draw, that the basic ability is in all of us, and my problem is not lack of talent but that I just need some "therapy"— I need to exorcise Mrs. Vagnus from my mind!

One of my reasons for writing this textbook is to convince students who are intimidated by logic and mathematics that all they may need is some therapy. I hope to show you that once the game of analysis is understood, anyone should be able to see that it is just a matter of staying calm enough to see that a complex whole is really just a bunch of simple parts in disguise. In my opinion, logical and mathematical analyses are much easier than writing papers, say in a humanities class where you must synthesize multiple perspectives and a wealth of information, or like this Introduction where there are so many interrelated thoughts I want to convey to you. As part of this therapeutic process, you should also be motivated by realizing that logic and math are very practical tools that simply extend our common sense for the most part to make life easier and more successful. People are not born logical or illogical. We not only all have the basic ability, but we all need this ability because critical thinking is crucial to the success of a democratic society.

Another reason for writing this book is that most logic books use a much different style than I have used. They write in a very objective, neutral, and noncommittal manner and avoid discussing issues that are obviously very important to us. By just offering the student a disconnected "recipe" of logical techniques, they do not put these techniques into a big picture, a cultural or philosophical context. They seem to assume that everyone just knows that being logical is good and that each student can figure out alone how each technique is relevant.

This book will be more personal and take some risks in terms of taking positions on controversial topics. However, the book not only attempts to provide you with some basic critical thinking tools, but also invites you to turn those tools against my own presentation of these tools. In other words, you should feel free to criticize me.

Up front, here are some of my beliefs that you will find directing much of the presentation in this book: I believe that being logical is good. I also believe that modern science has done much more for us than provide high technology and sophisticated gadgets for entertainment. It has provided us with a picture, or "world-view" (to use the terminology of philosophers), that is startling, humbling, and exhilarating. A world view that in my opinion we all ought to pay attention to and learn. Science teaches us that our earth is but a fragile, biologically precious grain of sand in a vast and very old universe. It also teaches us that evolution is true, and that we cannot ignore this anymore than we can ignore the law of gravity.

However, both logic and science are primarily products of Western culture, and if one has any faith in people and their ability over time to create worthwhile ways of life, it follows that other cultural perspectives ought to be of value also. This implies that there may be many limitations to logic, science, and technology. One of the most interesting questions of our time involves how to achieve a balance between unity (one world) and diversity (many cultures), and what should be the relationship of the many cultures in the world today with our scientific-technological culture. As you will see in Chapter 12, logic itself is not a static, finished discipline, and the philosophies of different cultures may play a role in its growth.

In this book, I have made every effort to present the basic material of introductory logic with a constant link to relevant context and the big picture. What is the relationship between logic and technology? How is logic related to the way a computer works? Do computers think? What is the relationship between logic and values? What does it mean to be a reasonable person? Are there alternate rationalities, different logics depending on one's culture? Is being reasonable simply a matter of what you believe? Can it be shown that some beliefs are better than others? Should everyone accept the results of science? What is the relationship between logic and emotion? Between logic and creativity? Imagination? A happy life?

From these questions it ought to be obvious that this book is not only for students who are intimidated by logic and mathematics. It is also for the science or applied technology major who needs to have not only the underlying logical principles of our Western scientific-technological culture systematically presented, but also needs a bridge to the humanities and the questions that concern us all as human beings.

There is also another reason for attempting such an interconnected approach. Students are confronted daily with an avalanche of perspectives, persuasive appeals, and virtual data bases of information. There is so much going on that it is easy to perceive today's student as "unprepared," as a member of a video generation who needs pictures for everything and cannot read or think. I believe that some of my professor colleagues misread the contemporary college student. They compare how prepared they were in communication skills and the so-called classics of culture and rhetoric with students in their present classrooms and they are often appalled. What they see as lack of preparation, I prefer to see as "unfocused potential," as for the most part highly intelligent, often street-wise young human beings, networked with and buffeted by enumerable competing subcultures, attempting to assimilate much more than I ever had to at their ages. Thus, a disconnected recipe of thinking skills is no more likely to be successful at producing a critical thinker than a bunch of disconnected historical facts and dates are likely to produce an informed individual.

We live in an increasingly interconnected world; we have no choice but to try to present disciplines as part of this big picture.

So we are going to undertake a process of communication, collaboration, and critical thinking together. It will not be perfect. There will be times when you will struggle. There will no doubt be times when you will be angry at me when you confront a difficult concept or procedure. However, with effort and a sympathetic teacher you should do fine. In my opinion, the most important concepts that you learn in college may not be something that you understand right away, and they will frequently require struggle and effort. If my own students are any guide to your probable success, we should be able to predict that this book will change you: You will not just learn fancy techniques, but a discipline of mind, a "tightness" of focus, "thinking surgery"—to use the terminology of some of my former students— a valuable tool to turn on given the appropriate context.

A few suggestions:

1. **Don't miss your class if at all possible**. Learning logic, like mathematics, is a sequential activity. It is a *step-by-step process*, and, as noted above, wholes only look hard if you can't see the simple pieces. As they say in mathematics, "By an inch it's a cinch, by a yard it's hard." Your instructor will make the material as clear as possible, but he or she will not be able to help you make connections and build up technical proficiency if you miss too many steps.

2. **Put in the time**. To do well in any college course of this nature, for every one hour you spend in class, you should spend at least two hours outside of class reading, doing exercises, reading again, thinking, and discussing the concepts with other students. As noted above, there are many concepts in this book that you will not understand right away. But if you put in at least six hours a week outside of class—a normal commitment for a college course—connections will result.

3. **Read slowly and sequentially**. Don't jump around looking for a fast way to finish each chapter. The first five chapters have concept summaries and lists of key terminology. Use them. If after reading a chapter you do not understand a key concept, check it off, and then be bold in class. Ask your instructor to explain the term again and give different examples from those of the text. For the symbolic chapters, read very slowly. Study the examples carefully and constantly go back to them when doing the exercises.

Good luck. If you have access to a network e-mail account at your college, please feel free to share your thoughts with me or ask questions. My Internet mail addresses are:

pine@uhunix.uhcc.Hawaii.edu or **pine@pulua.hcc.Hawaii.edu**

Chapter 1

WHY STUDY LOGIC?

You can fool some of the people all of the time, and all of the people some of the time, but you cannot fool all of the people all of the time.
—Abraham Lincoln

Logic as a Defensive Tool

Why study logic? It is often taken for granted that being logical is a good thing and that studying logic will make us better persons. But will being logical help us be more ethical in our behavior? Will it help us be nicer to people in our families, to our friends, to strangers? Will it help us be more objective and more tolerant of other ways of life? Will it make us more intelligent, wiser, more likely to make the right decisions in a complex world? Will it make us happier?

As promised in the Introduction, this book will be more than a standard introduction to logic and will address the above philosophical issues. But for now consider that your study of logic has a more focused mission. Logic can be viewed as a **defensive tool**, a tool that allows us to defend ourselves against the onslaught of powerful persuasive appeals that bombard us daily. As an example of this, bear with me and take the brief critical thinking test below.

Items 1–4 require an evaluation of commonly used advertisements. Select the option that you think is the best answer.

1. "No other digital video recorder retailer can beat Sam Kung's prices."
   If the above statement is true, which one of the following must also be true?
   a. Only Sam Kung has low prices for digital video recorders.
   b. Sam Kung's digital video recorders are the least expensive of all video recorders.
   c. The best buys in digital video recorders are at Sam Kung's.
   d. Sam Kung's digital video recorders are priced at least as low as other digital recorders sold at other retail shops.

2. "Now you can save up to 50% and more on many famous brand items at PayLess."
For this claim to be true, which one of the following must also be true? What are the minimum conditions necessary for this claim to be true?
   a. At least one famous brand item must be more than 50 percent off.
   b. At least some of the famous brand items must be 50 percent off, and a few must be more than 50 percent off.
   c. At least half the store's items must be 50 percent off and a few famous brand items must be more than 50 percent off.
   d. At least half of the famous items must be more than 50 percent off.

3. "Only Chrysler offers 11.9% financing on new cars."
Even if this statement is true, which one of the following could be true?
   a. Some of Chrysler's more expensive cars could have a higher financing rate.
   b. Another car maker could have a lower financing rate.
   c. Many other car makers could have a lower financing rate.
   d. a, b, and c.

4. "Dunlop's SP-4 radial tire had the highest rating in *Car and Driver's* tire test."
The persuasiveness of the above claim is weakened by its ambiguous or vague use of the word(s)
   a. test.   d. highest.
   b. rating.   e. a, b, and c.
   c. tire   f. a, b, c, and d.

Did you choose a, b, or c in number 1? If you did, then you did exactly what the creators of this advertisement hoped you would—you were **psychologically tricked** by the language into believing that Sam Kung has the lowest or best prices for the digital video recorders sold at this store compared to the competition. Note also that this trick does not involve lying or misrepresenting the truth, rather you were tricked into making an **illogical inference** from a statement we said was true to a statement that we do not know to be true. By an inference logicians mean that you were lead to a particular belief based on another belief. However, if you read the advertisement carefully—and the creators of this advertisement are hoping that you will not—you can see that all this appeal is saying is that no one can be lower, which is very different from saying that Sam Kung's prices are the lowest.

Let's think of a hypothetical scenario by which we can demonstrate that this is so—that even if the advertisement is true, we would not be entitled to infer that Sam Kung's store has the lowest prices. Suppose upon reading this advertisement we went to shop at Sam Kung's. (Already the advertisement would have served one of its purposes—it "hooked" us into going to the store where a sales person can work on us.) Suppose we find the latest high-end Sony digital video recorder at a price of $499.99.[1] Suppose next door at Kaneshiro's TV Sales and Service we find

---

[1] It should be obvious to you now that advertisers use .99, as in $499.99, for psychological appeal. We all know that $499.99 is for all intents and purposes really $500. But it feels better to say that we paid about $400. This is a case where advertisers know that they will not be deceiving us, but that we will want to deceive ourselves.

that the same Sony recorder is priced at $499.99. The advertisement for Sam Kung's would still be true. Kaneshiro's did not "beat" Sam Kung's price, but Sam Kung's price was not the lowest or the best price. All the original advertisement is saying is that Sam Kung's is committed to having the same prices as their competitors, but how many of us would rush down to Sam Kung's if we saw an advertisement that read, "Come to Sam Kung's, we have the same prices as everyone else"?

It is surprising how many people fall for this trick. It is no accident that this same type of appeal is made in many other advertisements, such as "No one can beat United to Chicago," "No one beats Midas," "No regular aspirin product reduces fever better than Bufferin," or the Safeway supermarket slogan, "Nobody does it better." Also consider, "Nothing outlasts an Energizer (battery)," and the Duracell slogan, "You can't top the Copper Top." The last two examples show the empty nature of this general appeal: similar slogans can be used for competing products! These advertisements work, however. It is a sad commentary on the state of human affairs that a great deal of human creativity involves one group of people devising ways to deceive another group of people.[1] These advertisements are good examples of how a creative writer can create what seems a highly positive statement about a product without really saying anything at all.

However, it is also encouraging to see how many people quickly see the trick in these ads after a little careful reading and common sense reflection. As you progress through the more difficult portions of this book, it will be important to remember for motivational purposes that logic is simply *organized common sense*. Most logic books define logic as the study of principles used in distinguishing correct from incorrect reasoning. For now, from our point of view, all this means is that we will be slowing our thinking down and reflecting carefully on reasoning trails, using our common sense to see if these trails are tricks or not and if we ought to follow them or not. Hence, everyone should be good at logic; it is a mistake to think that logic is something that only a special type of person is good at. Some people may be faster at learning it, just as some people learn languages better than others. But just as in learning a language, we all have the capability if we immerse ourselves in a situation where such a tool is needed all the time. Anyone could learn to speak Russian if they lived in Russia long enough.

Logicians make a living by formalizing reasoning, by identifying the simplest acts of good reasoning and organizing them into principles of inference and analysis, such that more complex presentations of thought can be judged. Thus, logic provides a defensive discipline for staying calm and organized when confronted by the buzzing, confusing voices of all those who want our attention, support, and, most often, money. In Chapter 2 we will discuss this in more detail, but as an early example of this formalization process we can indicate how our first advertising

---

[1] In his book, *The Biology of Moral Systems*, Harvard naturalist Richard Alexander argues that we cooperate and identify with a group only to compete with another group of human beings. According to Alexander, because of our intelligence we no longer have to worry as much about the hostile forces of nature. So, "Apparently, no other species has accomplished this peculiar evolutionary feat, which has led to an unprecedented level of group-against-group *within species* competition. It is this competition that draws us toward strange and ominous consequences." This may mean that deception is inevitable, but see Chapter 6, Logic and Hope.

example could be formalized into various arguments—three bad arguments if *a, b,* or *c* are taken as inferences from the original advertisement, and one good or "valid" argument, if *d* is taken as the inference.

Unlike its meaning in everyday speech of a verbal fight or quarrel, an **argument** is a technical term in logic that refers to a persuasive appeal involving a chain or a trail of reasoning with two or more statements, where there is a **conclusion** offered and reasons or **premises** given to support or lead to the conclusion. Also, unlike the meaning in everyday speech, as when someone refers to a concept, idea, or belief being valid, **validity** in logic is a technical concept that applies only to the reasoning of arguments, not to the premises or conclusions. Hence, statements, premises, or conclusions will be referred to as true or false. Arguments will be referred to as valid or invalid. In logic there is no such thing as a "valid premise" or a "true argument." Sometimes we do say that a conclusion is valid; then we mean that the conclusion follows validly from the premises.

So, if the original advertisement in the first example is taken as a premise, and either the *a, b,* or *c* options are considered the conclusion, then we can see that they would make bad or **invalid** arguments: in each case, even if the premise is true, we would not have good grounds for inferring or believing each conclusion true. On the other hand, if option d is taken as the conclusion, then the argument would be valid. If the ad is true, then we would know that no one is lower than Sam Kung on digital video recorders. The reasoning trail is solid. If the ad is true (premise), then option (*d*) (conclusion) is true. We have inferred (believed correctly) the right conclusion from the ad. Much of this book will consist in formalizing and analyzing arguments to see if they are valid or invalid.

For now, let's return to our brief test above and take a look at number 2. Although most thoughtful people will pick *b*, I will argue that the correct answer is *a*.

In a real advertising situation, such as in a newspaper insert, the sizing of the lettering is often significant. The **"50% OFF"** is usually written very large relative to the almost microscopic "up to." This variation of the "fine-print" trick accents the "50% OFF" so that this will stick in our minds and create the impression that a major sale is taking place. The average reader is so busy that he or she most often skims articles, headlines, and advertisements. It is no accident that advertisers take advantage of this and carefully think about how to structure a layout. Particular items of text will be accented for maximum effect, and others will be buried— stated to perhaps "cover" the advertisers, but hidden nevertheless. But what does the "up to" mean? Clearly it implies some sort of range from a minimum to 50 percent. But what is the minimum? If I said, "Members of the UCLA basketball team are up to seven feet tall," the context implies that the members of the basketball team range from some unspecified minimum height to seven feet tall. That is, because no human being is zero feet tall, and most college basketball players are more than six feet tall, the range must be from some number greater than zero to seven. However, if we are just referring to the bare, mathematical numbering system, and apply it to "up to 50%," then we mean anywhere from zero to fifty, because zero is a number. When a commercial says that "many items are up to 50% OFF" technically the meaning of the "up to" cancels the "many" in the sense that not everything is on sale, and many items may be 0 percent off or the same regular price. Hence, for the phrase "up to 50% and more on many" to be true, only one item would need to be more than 50 percent off its regular price. Everything else could be 0 percent off!

Most people who choose b for number 2 are no doubt correct in the sense that the context implies that at least a number of items would be for sale within the range of some minimum positive number and 50 percent off, and a few would even be more than 50% off. Stores would quickly go out of business if on repeated occasions of using an ad like the above, only one item was on sale. We may often be gullible, but eventually we realize when we are being tricked. Recall Abraham Lincoln's famous statement, "You can fool some of the people all the time, and all the people some of the time, but you can't fool all of the people all of the time." However, when we contemplate the absolute minimum conditions for the advertisement to escape misrepresenting the truth, we see that, like our first advertisement, this advertisement has been constructed to give maximum flexibility to the store by being *vague*, while simultaneously creating a maximum positive psychological impression. Such ads get us in the store, and once we are there, even though we may find that what we wanted to buy is not on sale, we often buy something that is on sale anyway. Our time is precious, and we don't want to waste it or admit that we are wasting it by coming to the store for no reason. The managers of the store are well aware of this, and many commercials are devised as "hooks" just to get us into the store.

Let's look at number 3. This advertisement for Chrysler occurred in the late 1970s with a background of high inflation and high interest rates. When this pitch was made on TV, complete with a marching band, pretty girls with batons whirling in the air, and a sports figure announcing this interest rate breakthrough, it was not uncommon for car interest rates obtained from banks to be around 15 percent. The buyer would pay more than $15,000 for a $10,000 car on a six-year loan—more than one half the price of the car extra! So Chrysler's offer seemed like a very good deal. However, note again the vagueness of the ad. The ad leaves open the question of whether this applies to all of Chrysler's cars or is only a special deal for some of Chrysler's cars. It also does not state that this is the lowest interest rate in the industry. For the ad to be true, Chrysler must be the only one to offer this particular rate (note the particular number: not 12, not 11.8, but 11.9). But this does not rule out the possibility of another car company offering a lower rate (how about 11.89!), or many other car manufacturers competing with banks and offering a lower rate. Hence, the answer to number 3 is *d*.

While we are on the subject of car buying, consider the negotiating situation you often face at your local car dealership. First of all, like the above, the advertisements on TV and in the newspaper are designed to get you onto the car lot. During the late 1980s and early 1990s you would often see ads for Honda Accords for $10,000–$12,000, but on the lot the sticker price for these cars was often near $21,000. The phrase "sticker shock" became common. Today the disparity between the ad and the car lot sticker price is about the same, and if you look closely at the advertisement, you will often see a stock number (again it will be very small, almost undetectable). Or, sometimes in fine print you will see the notes, "Cars are subject to prior sale. Vehicle not exactly as shown. Financing rate for selected models only." What this means, of course, is that the price and deal quoted do not refer to Hondas in general, but to some specific car that may or may not even be available any longer—it may have already been sold.[1] The advertisement is structured so that the newspaper reader or TV viewer will hastily generalize and conclude that these prices represent a typical price for Hondas. So the real fun begins when you get on the lot and find that the price is much more than you had planned to pay.

Fear not, the car salesperson will come to your "rescue" and negotiate a deal closer to what you can afford, even though it will be a price that the company can "barely" live with in terms of profit margin and will hurt the commission of the poor hardworking car salesperson. Here is how it works. Figure 1-1 shows an example of a typical car sticker price, in this case a 1990 Honda Accord LX. Notice that it is divided vertically into two parts. First we see a basic description of what the car contains and at the bottom is the "suggested retail price." The next part shows the so-called dealer-added equipment and services with a total at the bottom— your local car dealership's offering price for the car. In reading this or negotiating with a salesperson, you are given the impression that the car manufacturer has charged the car dealership selling the car $15,645, and the $21,829 price includes the car dealership's profit and cost of the equipment they have added. So, you start negotiating with the impression that any price that you can get under $21,829 is a good deal. After a long day of driving the car around and haggling with perhaps several sales people—when you try to leave to shop around you invariably get turned over to another salesperson, who is supposedly in a better mood and can give you a better deal—if a weary-looking manager finally comes out of some secret back room and offers $3,000 for your rusted 1965 Volkswagen Bug, you jump at the offer and leave thinking that you have won a major battle of wits. Little do you realize that in this situation the car dealership would make a $4,529 profit off you!

Here is how. First of all, the suggested retail price is known in the car-selling trade as the "Maronie Price." In virtually all cases, this price already includes a profit for the car dealership, regardless of the equipment added. In this particular case, the car dealership was charged $14,300 by the car manufacturer. The so-called adjusted market value is pure negotiating hype. But what about the added equipment? Wouldn't it be fair to pay for this in addition to the amount needed for the car dealership to make a profit and the salesperson to make a living? These charges add up to $1189, so, with the $15,645 Maronie price, wouldn't a fair price be $16,834? No, because the added equipment did not cost the car dealership $1189; they mass handle these costs, and the actual cost was about $300! Each item is marked up anywhere from 300 to 600 percent. For instance, the car dealership's actual cost of rust proofing was not $295, but $76. And the actual cost of the stripes was nearer $20 than $125.

---

[1] Sometimes the car will actually be on the lot, but it will be a dirty, stripped down, less-expensive and less-popular version of the car, in the back of the lot. To get to it you will probably have to walk by a fully loaded, immaculate higher-end version of the car.

FIGURE 1-1   A Typical Price Sticker From a New Car Lists The Various Charges The Buyer Incurs.

Model: 1990 Accord 4DR LX **Color:** Gray		**Interior:** Gray	Manufact. Suggested
Vehicle Identification No: 1HCGD7650MA132151		**Eng. No.** F 22A2-2495212	Retail Price
			$15,420.00

2156CC 4 Cylinder 16-Valve Engine w/Programmed Fuel Injection	STD
4-Speed Electronic Automatic Transmission w/sport mode & lock-up torque con/tach	STD
4-Wheel Independent Double Wishbone Suspension	STD
Front Stabilizer Bar/Ventilated Front Disc Brakes	STD
Variable-assist Rack and Pinion Power Steering	STD
P185/70 R14 Michelin LX1 Radial Tires/Full Wheel Covers	STD
Power Windows/Power Doors/Dual Body-Color Power Door Mirrors	STD
AM/FM Electronics Full Logic 12.5W X 4 Stereo Cassette W/12 Radio Presets, Four Speakers, and Trunk Mounted Automatic Power Antenna	STD
Soft Touch Heat and Vent Controls w/5-speed Fan Operation	STD
Air Conditioning/ Cruise Control/ Illuminated Vanity Mirror	STD
Etc.	
Etc.	
Etc.	
Destination and Handling Charge	$225.00
Total (Includes Pre-Delivery Service)	**$15,645.00**

ADD-A-TAG   PROTECTING THE CONSUMER*

Model: 1990 Accord 4DR LX	
Vehicle Identification No: 1HCGD7650MA132151	
DEALER-ADDED EQUIPMENT SERVICES	
ADJUSTED MARKET VALUE	4,995
RUST PROOFING	295
PAINT SEALANT	295
CAR COAT INTERIOR	195
SIDE MOLDING	145
STRIPES	125
CARPET MATS	85
REAR MUD GUARDS	49
TOTAL PRICE	**$21,829.00**

*This protecting-the-consumer label is not an official factory or government sticker. This label has been affixed to this vehicle by this dealer.

This is not all, however. Someone who pays the full list price of $21,829 would be given a "full pop" by the car dealership. (This result, by the way, is more likely to happen if the buyer shows a lot of emotional attachment to the car before negotiations begin.) But buyers can be "full popped" through the "front end" or the "back end," and here is where our interest rate example is relevant again. Back end charges can include paying $600 for an extended warranty that costs the car dealership $200; paying a dealer-arranged financing charge much higher than one arranged through a credit union or a bank equity loan; or purchasing an unnecessary life, accident, and health insurance policy. The latter is unnecessary because it automatically comes with a bank loan!

A lesson that emerges from the planned vagueness of these examples is that when we decide to turn on the critical ability we are all born with, we should give some attention to what an advertisement leaves out. These examples make no explicit mention that the products have the lowest price or interest rate. If the goal of advertising is to state positively everything possible about a product, one should get suspicious when the words "lowest," "highest," or "best" are not there in black and white. Unfortunately, a trick can be involved even when these words are used. Let's look at our last example. My answer is *f.*

Number 4 was given to me by a female student several years ago who was offended initially not so much by the trick involved but by the sexism in the full ad. The original magazine spot showed this statement in bold, dark print above a very pretty girl in a white bikini with her arm wrapped suggestively around a nice, new, black Dunlop tire. Because she was one of the first females in a traditionally all-male automotive technology program at my college, the student asked me if she could do a little research on this claim. She had learned enough through her own interest in automobiles and the critical analysis of advertisements in my class to know that the words "test," "rating," and "tire" were vague and could mean just about anything. What kind of test was conducted? With what other tires did Dunlop compete? Was a **Brand-X approach** used, in which a radial tire (the Dunlop) was compared to a well-known inferior type of tire? And what was rated? For radial tires, the appropriate traits to test are longevity of tire wear and ability to grip the road in adverse conditions. For all we know from the advertisement, tires were dropped from the Empire State Building and Dunlop's bounced the highest upon hitting the ground!

So my student found the original tire-test publication (she was a little surprised that there actually was one), and initially was disappointed to find that none of her suspicions appeared confirmed. Fifteen radial tires were tested, at least four of which were well-known major competitors to the Dunlop radial: Michelin, Bridgestone, Toyo, and Goodyear. The test appeared objective, and all the pertinent performance characteristics of radial tires were examined. Points were assigned for each characteristic tested, and different tire experts were used to assign points in cases where judgment calls were needed. All tire identification markings were removed to prevent the experts knowing the brand of each tire. An average was then taken from the points assigned by the experts and totalled with other objective point totals. Categories were assigned based on point spreads. For instance, to receive an excellent rating a tire needed to receive 300–350 points, and to receive an above average rating a tire needed 250–299 points. Average was 200–249, and below average was 150–199. This publication was complicated with numerous charts, difficult text, and distracting qualifying footnotes.

The trick involved in Dunlop's use of this tire test was not easy to see. But after considerable reflection, my student finally saw that the trick was in the apparently clear phrase *the highest rating*. To receive the highest rating of excellent in this test a tire had to get at least 300 points, but could rate as high as 350. More than one tire could receive this rating, and receiving this rating did not eliminate the possibility of other tires receiving a much higher rating within the category. So, guess which tire had the lowest rating within this category? Dunlop rated lower than its four major competitors—the above-named tires all had a higher rating within the category of excellent—but with a little creative use of an ambiguity on the phrase *the highest*, Dunlop was able to advertise that their product had achieved the highest rating in this test, which was true from one perspective—it did make enough points to be rated excellent, while actually being the lowest rated among other major competitors![1]

It is an important theme of this book that logic and critical analysis are not ways of life, but tools to be used in appropriate situations. There are times in which we desire to clear our minds of details and problems, times in which it is healthy and appropriate to avoid cutting life into little pieces of analysis and instead let our thoughts bubble away until we are perhaps rewarded by quiet insight. However, it should be clear from these examples that logic and critical analysis are also valuable tools, because it is very easy to use language to "set us up" for mistaken inferences. As individuals we are up against an enormous external power pounding on our doors daily. Billions of dollars are spent on commercial and political advertising every year. Even the names that corporations give themselves are often the result of a large investment intended to produce maximum psychological effect.

For instance, in the 1960s what we now call the Exxon corporation consisted of several regional oil and gas companies, such as Esso and Enco. The directors of this company were smart enough to realize that the United States would soon be competing in a global market economy and that the company needed a unifying logo to take up this challenge. What could be involved in simply coming up with a name? Well, quite a lot. Would you believe $100 million? First of all, the company wanted a name that was culturally sensitive to all the major countries in the world in which they might compete. Enco meant "stalled car" in Japanese! Hardly a hallmark for a company that wants to sell its oil products in Japan. Second, they wanted a name that provided for maximum name recognition in all the cultures of the world, a name that went beyond the particulars of culture to tap something deep within human nature. They also wanted a name that was distinctive and not used by anyone else. So they hired computer experts, psychologists, linguists, and

---

[1] Many of my students get very cynical after covering these examples, concluding that all advertisements must be out to "get us." But this is a hasty conclusion. Not all advertisements are bad, and some impart very important information. For instance, for years in Honolulu a tire retailer by the name of Lex Brodie published no-nonsense ads in the daily newspapers, listing the failure rates of tires that his staff had kept track of over the years. He showed that the most expensive radial tires had the lowest failure rate. He did not make a hard-sale pitch for everyone to buy the most expensive tire, but just presented the information, allowing people to make up their own minds based on the kind of driving they did, the price they could afford, and the risk they wanted to take.

other research specialists on human nature and spent $100 million to arrive at the name "Exxon." Then they raised the price of gas a little to pay for it or deducted this expense from their income tax so that as much as possible of the $6 billion of company annual profit stayed in company hands.

As one final example of our defensive-tool theme, consider the language tricks politicians use to paint favorable pictures of themselves. In 1982, President Reagan was not running for reelection, but the Democrats decided that they had a major issue with which to challenge the Republicans during the congressional races. The Reagan administration had been busy the last two years attempting to gradually eliminate the Federal Department of Education, believing that almost all matters of governance should be decentralized with the exception of defense. By 1982, a middle-class backlash was developing because the programs in jeopardy involved grants and services for middle-class college students. There were also numerous negative reports on the state of U.S. education and warnings of our eventual noncompetiveness in a global arena. The Democrats thought that they could make some political gains from emphasizing the massive cuts in education proposed by the Reagan administration. While campaigning for Republican congressional candidates, Reagan responded by claiming that the Democratic charges that his administration was anti-education were not true by noting that,

"We will propose $3 billion more for (the budget in) education this year than last year."

The trick here is, of course, in the word *propose*. In 1981, the budget for the Department of Education was $15 billion. For the 1982 budget, the Reagan administration proposed a budget of $9 billion! After compromising with panicked members of Congress who would have to face angry constituents, the actual budget for 1982 was $12 billion. When for 1983, the Reagan administration planned to propose $3 billion more than the $9 billion they "proposed" the year before, the actual budget would be the same as the year before, or less when inflation is factored in. Again the message is the same. Reagan's claim did not involve lying or misstatement of the facts; instead we were cleverly offered an opportunity to mistakenly *infer* that the Reagan administration was increasing the education budget, when actually they were proposing to lower it.[1]

Deductive Reasoning

One of the major purposes of this book is to help you avoid mistaken inferences. To do this, we need not only to practice recognizing inferences or arguments, but we also need to study criteria that separate good inferences from bad ones. The most important concept in the study of logic is what logicians call a valid deductive argument. Many of the practical tasks offered in this book, as well as the wider

---

[1] Reagan also said during this time that we spend more on education than on defense. Yet the average budget for defense during the Reagan years was $300 billion. When asked about this, Reagan said he meant education spending throughout the entire United States, including at the state and local levels. But this fact, although true, was misleading related to the issue of whether we are spending enough of federal tax money on education.

philosophical message as to the nature of logic and its relationship to science and technology, will be better understood if you can fully grasp the concept of a valid deductive argument. In preparation for a more technical discussion of this concept in the next section, let's take another little test. This time we are going to work on a little brain teaser. Its solution is much simpler than it first appears, and it provides not only a nice example for the technical concepts discussed in the next section, but also a thoughtful analogy for many aspects of life and reasoning.

Here is the situation. Imagine that we know of three prisoners who share a particular jail cell. A particular security guard is somewhat of a friend with these three men because they are all quite intelligent and they often have philosophical discussions and play games such as chess together. Let's refer to these three men simply as prisoners x, y, and z. Imagine further that prisoner x has perfect vision, prisoner y has sight in only one eye, and prisoner z is totally blind. One day, our security guard brings in a little basket with five small hats in it. Because the blind man can't see, the security guard announces that there are three white hats and two red hats as he explains the intent of bringing the basket of hats with him this day. He has decided, he says, to offer freedom to the first one who can solve a little logic riddle. He intends, he says, to blindfold each of the men who can see and then mix up the hats and randomly place a hat on each of their heads. Then he will remove the remaining hats and basket and place them in a nearby closet. Upon removing the blindfolds, the first man who can tell what color hat he has on will go free. (Because the hats are small—like flat berets—no man can see the color of his own hat.) All three prisoners agree to the challenge, but after being surprised how readily the blind man agrees given his distinct disadvantage, the two prisoners with sight demand that all of them be blindfolded. Although they have known the blind man for years, and are almost positive that he is indeed blind and will not have the advantage of seeing the color of the hats of their opponents, they decide to take no chances.

So the security guard complies with their request. He blindfolds all three men, mixes the hats up, places one on each of the prisoners' heads, removes the basket and the remaining hats to the closet, and then removes the blindfolds. The two prisoners with sight begin to look around quickly at their colleagues, and just before both begin to give their answer based on what they see, the security guard stops them with a warning. He explains that there is one more little item that must be a condition for their answer. This condition, he says, is a result of prior consultation with the blind man. The blind man agreed to participate in this riddle if and only if each answer was based on **deductive** reasoning; that is, **inductive** reasoning and answers based on probability would *not* be allowed. We will discuss inductive reasoning in Chapter 3. For now, think of good inductive reasoning as the inferring of a reasonable, but not completely certain, conclusion from available evidence. For instance, upon the removal of his blindfold if prisoner x saw two white hats, one on prisoner y and one on prisoner z, because there would remain only one white hat but two red hats, the probability would be greater that x would have one of the red hats on, and it would be reasonable, although not conclusive, for him to "guess" red.

Originally 3 white, 2 red—if one white and two red left, then

| x | y | z |
| (*probably red*) | (white hat) | (white hat) |

This would be an inductive inference by prisoner x. But it will not be allowed. To ensure that all answers will be deductive, the security guard has decided to impose a severe penalty for a wrong answer. If one of the prisoners answers without being deductively sure what color hat he has on and is then wrong, that prisoner will be executed! After all, it would be reasonable for prisoner x to pick red in the above situation based on inductive reasoning, but it would still be possible for him to have the one remaining white hat and be wrong and dead. (It will be important later when we discuss scientific method and inductive reasoning in Chapter 3 to remember that an inference can be reasonable even though it could be wrong in terms of the conclusion drawn.)

So here is the situation now. No prisoner will use inductive reasoning. Each prisoner will use only deductive reasoning. With deductive reasoning, if the reasoning is good, there is no doubt about the conclusion, given the truth of the initial information. For instance, assuming we know no hats were already in the closet and there were only five hats initially, we know that with three on the prisoner's heads, there remain only two in the closet. This is an elementary deduction; we are not guessing that there are only two, we are not assuming that it is probable that there are only two, we know that there are only two hats in the closet. So, each prisoner will either answer "no" to avoid being executed, if he is not certain of what color hat he has on, or "yes" when he is certain what color hat he has on.

Here is what now happens: Prisoner x looks at his fellow prisoners and without much thought looks discouraged and says "no," he can not tell what color hat he has on with deductive certainty. Prisoner y, with a very thoughtful look, at first stares straight ahead, neither looking at x nor z. Then after an "aha!" expression of insight, without looking at x at all, he turns his head to look at z with great expectation. Then a look of discouragement appears, and he says "no," he also cannot tell with deductive certainty what color hat he has on his head. Upon y's failure, the security guard and prisoners x and y begin to depart and go about their business. Upon hearing them leaving, the blind man asks casually if he is not going to be given an opportunity to give his answer. Surprised at this request given his disadvantaged state, all three men remind him that he must answer deductively; he must know with certainty what color hat he has and cannot just guess; that he may have a three out of five chance of having a white hat on, but he risks execution if he chooses white and in fact has a red hat on. The blind man replies, "I do not need to have my sight, from what my friends with eyes have said I clearly see that my hat is ___?___." He has the right answer and he does not use induction or probability, nor does he simply venture a guess. Based on the information he is given, he knows what color hat he has on, just as we know that there must be only two hats in the closet.

Take a break here from your reading and see if you can recreate the blind man's reasoning. What color hat does he have on, and how does he figure it out? What trail of reasoning does he use to arrive correctly at the right deductive conclusion?

Valid, Invalid, and Sound Arguments

Before we see how close you came to reconstructing the blind man's reasoning, note that your problem is similar to many daily situations. First, I have flooded your mind with a lot of information. Is it all relevant to the problem to be solved? You not only have to draw the correct inference from the information, you also have to select the right starting place. Is it relevant to solving the problem that one man has perfect vision and another has sight in only one eye? Second, do we know that all the information is true? Does the blind man know for sure that there are three white hats and two red hats? After all, he is blind. And finally, are there assumptions we need to make in constructing the correct reasoning trail? And how reasonable are these assumptions?

Many people after a little thought are able to see the first step in how the blind man figures out his hat color. When the man with perfect vision (prisoner x) answers first, we can at least know what he does not see in order for him to say "no." If there are only two red hats, then it is not possible for him to see two red hats (one each on the other two prisoners), for if he did, he would know conclusively that he had on one of the remaining white hats, and the game would be over. Because he said that he could not tell with deductive certainty what color hat he had, we can conclude that he did not see simultaneously a red hat on y and a red hat on z. This is an elementary deduction, just like the case where we knew there were only two hats in the closet.

x y z

(would have to be white)   (red hat)   (red hat)

(not possible)

With this simple insight we have started a trail of reasoning. The next step—and there is only one more step—is just as easy. If we have been able to figure out that the prisoner x could not see two red hats, then so too would the man with one eye (y) and our blind man (z). This is an assumption on our part, but given the information that these men are intelligent, it is a reasonable assumption. A similar little assumption was made even when we concluded that prisoner x could not see two red hats. We must assume that he is at least smart enough to remember that there were only two red hats to begin with. If he is mentally retarded or has a brain disease that impairs his memory, he may have forgotten that information. So, given the reasonable assumptions that x is intelligent, that nothing is wrong with his memory, and that he really wants to go free, we can conclude conclusively that he did not see two red hats. Similarly, we can conclude that y and z know as soon as x says "no" that x did not see two red hats.

So, if y knows that x did not see a red hat on his (y's) head at the same time he saw one on z's head, then what is y thinking when he looks quickly and expectantly at z? What is he hoping to see? He is hoping to see a red hat on z. If he does, he would know conclusively that he must have a white hat on, and he would be the one to go free. If both hats can't be red, then at least one is white. But he also said "no," implying (and this is the last step) that he did not see a red hat on z, and if he did not see a red hat he had to have seen a white hat. Because two reds are not possible (one on y and one on z), it is also not possible for there to be a red on z, otherwise y would know that he had a white hat on.

x   y   z

(would have to be white)   (red hat)   (red hat)

(not possible)

x   y   z

(would have to be white)   (red hat)

(also not possible, because the above case is not possible)

So, we know, as the blind man knows, that there cannot be a red hat on the blind man. He must have a white hat. It is crucial now to reflect to what extent we know this, to reflect on the strength of this knowledge. The blind man is able to construct what logicians call a valid deductive argument. It is not possible for all the information he uses to construct his trail of reasoning to be true and his conclusion that he has on a white hat be false. Unfortunately for the blind man, he does not know conclusively whether all the information he has been given is true. For all he knows the security guard and the other two prisoners planned an elaborate hoax to lie to him about the color of the hats. There could be two white hats and three red hats, or the hats could be blue and yellow and when the blind man answers white, he could still be executed! This possibility does not change the nature of the blind man's logical reasoning; it is still impeccable. It only shows that a good reasoning trail is a very neutral thing, and that the content of a reasoning trail (the starting place) and the reasoning trail itself (the steps given the starting place) are two different things. The blind man's starting place, the information he was given, could have been incorrect, but his reasoning steps were perfect.

It is very important to understand this to fully understand the nature of logic and its role in our modern technological society. Computers, for instance, are marvelous logic machines. But not only do human beings construct the logic trails for them to follow, but also, and more importantly, we give them (input) the premises. This means even the most sophisticated computer analysis (economic, military, and such) is a fallible process; if the information put into the computer to analyze is wrong, the conclusion could be wrong.

Logicians define a **valid deductive argument** as one in which **it is not possible to have true premises and a false conclusion**. Notice some important implications that follow from this definition. It is not true, as some students interpret this to mean, that given any argument with true premises and a true conclusion that argument must be valid. Furthermore, to say that in a valid argument, it is impossible to have true premises and a false conclusion is not the same as saying that valid arguments always must have true premises and a true conclusion. At this point it is natural to be a little confused, because this definition implies we can have valid arguments with false premises and even a false conclusion, and that we can have arguments with true premises and a true conclusion and the reasoning not be valid. But stay calm. Let's look at a simple example.

You don't know me, so you don't know if what I am about to tell you is true. Suppose I tell you that in my garage I have a normal, twentieth-century automobile that runs on gasoline. I believe my car has no gasoline, and I tell you this. Because you know that for my car to start it must have at least some gasoline, you conclude that my car will not start. Here is a formalization of your reasoning:

EXAMPLE 1-1	(Premise)	In order for my car to start, it must have at least some gasoline.
	(Premise)	My car does not have any gasoline.
	(Conclusion)	Therefore, my car will not start.

Based on our definition, this example is a valid argument. A good reasoning trail is created. If it were really true that I have a normal twentieth-century gasoline car, and that car needs at least some gas to start, and if it were really true that it has none, then my car will not start. If the premises of this argument are true, then the conclusion is true. Remember, however, you do not know me. For all you know I could go out in my garage right now and start my car, which happens to be a new solar-powered car or electric vehicle (EV) running on a hydrogen fuel cell or batteries, and drive happily away. In other words, the conclusion of the above example could be false. But this would not change the *form* of the reasoning of the above example; it would only mean that not all of the **content** of the argument is true. (If I have a solar-powered car or EV, the first premise would be false, the second premise would be true, and the conclusion could be true or false. But the reasoning would still be valid.)

Let's look at another example. Several years ago I decided with great resolve that I would end the decades-long frustration of trying to get every student to understand the full implications of a valid deductive argument. I thought I had an example that was so clear that anyone living in the state of Hawaii, where I teach, would understand. To my great surprise, the example totally bombed, and I had at first an even harder time making this concept clear. Here is the example I gave: Because Sandy Beach is north of Kona, and Haleiwa is north of Sandy Beach, and because Hanalei is north of Haleiwa, Hanalei must be north of Kona. Here is the formalization:

EXAMPLE 1-2	(Premise)	Sandy Beach is north of Kona.
	(Premise)	Haleiwa is north of Sandy Beach.
	(Premise)	Hanalei is north of Haleiwa.
	(Conclusion)	So, Hanalei must be north of Kona.

I had thought that we could start with an example that students could easily relate to, one that used familiar places in the island chain. This way they would not be distracted by unfamiliar content and would have a comfortable example of an argument with a valid reasoning trail. The example bombed not because the students were unfamiliar with the places mentioned, but because to my surprise many local people living on Oahu (the island where I teach) don't use the directions north, south, east, and west. Instead they give directions by referring to mauka (in the direction of the mountains), makai (in the direction of the ocean), Diamond Head (in the direction of the major Waikiki landmark), or Ewa (a leeward or westward part of the island). Because they didn't understand yet the difference between the reasoning trail and the content of that trail, they were distracted by their lack of understanding of the phrase "north of" and were hesitant to agree that this was a valid argument.

As you read this you may be in a similar situation. If you don't know the places mentioned above, you can't judge whether these statements are true. Like the blind man, you and some of my students are blind as to what the truth is, but also like the blind man, you should be able to judge the reasoning trail to be valid anyway. In other words, it does not matter if you know the premises to be true or not; all you need to understand is the meaning of the statements. Let's show this by representing what these statements claim to be true, by representing the argument with a picture. We will represent "north of" as "up" and see what the premises imply.[1]

Hanalei   (north of) above Haleiwa, and hence (north of) above Kona

Haleiwa   (north of) above Sandy Beach

Sandy Beach   (north of) above Kona

Kona

Pictured this way, we see that there is no doubt that *if* Sandy Beach is really (north of) above Kona, and *if* Haleiwa is really (north of) above Sandy Beach, and *if* Hanalei is really (north of) above Haleiwa, then Hanalei must also be above (north of) Kona. We can judge this argument to be valid absolutely, even if the truth of each statement is uncertain. We can illustrate this point further by showing that we can create a valid argument that contains statements that contradict the statements in the above example, such as: Because Sandy Beach is south of Kona, and Haleiwa is south of Sandy Beach, and because Hanalei is south of Haleiwa, Hanalei (conclusion) must be south of Kona. We can again see that this argument is valid by representing the claims within the argument with a picture, now using "down" for "south of."

Kona

Sandy Beach   (south of) below Kona

Haleiwa   (south of) below Sandy Beach

Hanalei   (south of) below Haleiwa, and hence (south of) Kona

Notice that in both examples the conclusion is already contained in the premises; that once we represent the premises we are "locked into" the conclusion, because in a sense we have already stated the conclusion. Our method of representation

---

[1] As a North American I reflect here the North American bias that north is "up." From space we know there is no up and down, and thus we could just as well make all our globes with the South Pole facing upward rather than the North Pole. I wonder what long-term political and cultural ramifications would ensue if children in South America grew up with globes in their classrooms with the South Pole on top?

shows that stating the conclusion is only making explicit what was already implicit. This feature is the hallmark of all valid deductive arguments.

Two very important points about logic and rationality follow from these examples. *Being logical does not guarantee truth* or being right all the time, and *rational people can still disagree*—two people can disagree over a belief but each have logically valid reasons for holding their respective beliefs. Consider these two examples.

EXAMPLE 1-3   (Premise)   If the Constitution implies that each person has a right to self-determination concerning matters of a person's physical body, then abortion should be legal.
(Premise)   The Constitution does imply this.
(Conclusion)   Therefore, abortion should be legal.

EXAMPLE 1-4   (Premise)   If the Constitution grants every person the right to self-determination, then the involuntary termination of life is wrong.
(Premise)   If the involuntary termination of life is wrong, then abortion should not be legal.
(Premise)   The Constitution does grant every person the right to self-determination.
(Conclusion)   So, abortion should not be legal.

Both of these arguments are valid. The reasoning is not the cause of the disagreement shown in the conclusions. The problem is in the premises and the disagreement over the implications of the Constitution in the case of abortion—whether an absolute right of self-determination should be given to a woman for control of what takes place in her body or whether this normal right should be limited because the fetus is also a person with a right to self-determination. The acceptance or rejection of these premises is not an easy matter, but at least both arguments follow a valid reasoning trail.

In the north and south examples, however, the truth is known—the first example (**1-2**) contains true premises, and thus, also contains a true conclusion. However, as the abortion examples show, in many situations that deal with major issues of life the beliefs that serve as premises are either not agreed upon or are not known to be true. Thus, being logical alone will not suddenly make everyone in the world agree. However, as we shall see, being logical will go a long way toward encouraging us to objectively discuss and test our beliefs.

Example **1-2** illustrates what logicians call a *sound argument*, an argument that is *valid and has true premises*. The islands of Hawaii line up in a southeast to northwest direction. They were created by a Pacific plate that is passing over what geologists call a "hot spot"—essentially a hole in the Earth through which molten material oozes from deep within the Earth. This plate has been moving in a northwest direction for millions of years, and as it passes over the hot spot islands are formed and then eroded to make nice beaches for tourists. Kona is on what local people call the "Big Island," the island of Hawaii, the southernmost island in the island chain. It is also not far from South Point, a location on the Big Island that has the distinction of being the southernmost point in the entire United States, and a place that the state government would like to develop for a rocket launching facility. (Being closer to the equator than any other place in the United States allows for the best possible satellite orbits.) Sandy Beach is on the island of Oahu, the fourth island "up," or northwest, of the Big Island. This beach is on what local people call the "south shore" of Oahu. Haleiwa is also located on Oahu and is part of what the local people call the "north shore." Finally, Hanalei is on Kauai, the next island up the chain about ninety miles northeast of Oahu, and Hanalei is on the "north shore" of Kauai.

We will see that truth is a much more controversial and difficult matter to establish than logical validity. Do you know whether all the things I just told you about the Hawaiian islands are true? Should you just take my word for it? You could look at a map, but how do you know the map is correct? Almost everywhere these days— in advertisements, in textbooks, on TV—we see a picture of our fragile little biological paradise floating in space. It is spherical and possesses north and south poles. But how do we know these pictures are not fake and part of a massive conspiracy? A group of about four thousand people who call themselves the Flat Earth Research Society believe not only that the Earth is flat, but also that astronauts from the United States never went to the moon. They also assert that the accepted scientific belief that the Earth is a spinning globe is the result of a massive government conspiracy and successful brainwashing campaign that starts in elementary school. Can you attest absolutely that they are wrong? If you think they are wrong, do you "know" this just on the basis of testimony? Are you taking someone else's word for it? Are you accepting the belief that the Earth is spherical on the basis of authority—because they say it is true? Who are "they," and what evidence do they have?

These questions are more difficult than they first seem, and they involve some complex issues in what philosophers call epistemology, or theory of knowledge acquisition. But, for what it is worth, I believe that not only is the world round but that we "knew"[1] this in the third century BC, not in 1492 from Columbus's expeditions as my generation was told in elementary school. We will discuss this in more depth shortly, but for now consider that even Flat Earth supporters will agree on what is a valid argument.

So, let's take one thing at a time. We claim that logical validity gives us at least a framework of rationality, a way to test our beliefs. Let's see why by looking at an example of an *invalid argument*. Logicians define invalid arguments as arguments where it is ***possible to have true premises and a false conclusion***. Suppose we changed our gasoline car example as follows:

EXAMPLE 1-5   (Premise)   In order for my car to start, it must have at least some gasoline.

(Premise)   My car has at least some gasoline.
(Conclusion)   My car will start.

---

[1] In the sense that at least one person knew this and that anyone who followed his evidence and reasoning would also know. See the Eratosthenes' example in Chapter 3.

Clearly this argument is invalid, because even if it is true that I have a normal twentieth-century gasoline car and I have just looked at the gas gauge and see that it has some gas, as we have all been made painfully aware at one time or another, I am not guaranteed that my car will start. Unfortunately, many things could go wrong with the car, even though in this case being out of gas is not one of them. The conclusion *could be* false, even if the premises are true. An invalid argument does not follow a good reasoning trail; it does not lock us into the conclusion even when the premises are true. Argument (**1-5**) makes the same mistake as

EXAMPLE 1-5A   (Premise)   If it is raining, the streets are wet.
                    (Premise)   The streets are wet.
                    (Conclusion) It is raining.

The first premise is not stating that streets only get wet when it rains. It is essentially stating that being rained on is one way the streets get wet. Thus, because the streets could be wet for other reasons, we don't know that it is raining, even if the premises are true. Recall we also said above that an argument that has true premises and a true conclusion is not necessarily valid. Invalid arguments can also have true premises and a true conclusion. So, in considering the reasoning of **1-5**, could my car still start anyway? Of course it could, and I reason every day (unconsciously) that because my car has gas it will start. Only sometimes am I disappointed. However, this is the point: I am not guaranteed a true conclusion when I use invalid reasoning. Even though it is possible for the above arguments to have true premises and true conclusions, having true premises does not guarantee a true conclusion when we use invalid reasoning. This shows that invalid reasoning does not guarantee a false conclusion either. All we mean by calling an argument invalid is that it is a **weak** argument; its conclusion may be true, but the premises do not provide good deductive reasons to believe this is so. The main problem with an invalid argument is that if its premises are true, we know nothing about its conclusion—the conclusion could be true or false based on the premises. If we always knew that the conclusions of invalid arguments were false, this would be valuable information—we would at least know what not to believe. But we don't even know this.

Valid arguments preserve truth. They don't guarantee truth unless the premises are true; when the premises are true, we know the conclusion is true. We also know something about the premises if the conclusion is false. At least one premise will be false, if the conclusion is false, because valid arguments do not allow for true premises and a false conclusion. Invalid arguments do not preserve truth; even if the premises are true the conclusion could be false. The conclusions of invalid arguments could be true, but even if the premises are also true, the connection to the conclusion is such that it is not the premises that are locking us into the truth of the conclusion. The conclusion would be true based on independent reasons not stated in the argument. In the above case (**1-5**), I am lucky, in a sense. I think that because my car has gas, it will start. But my reasoning does not guarantee that it will start, even if my premises are true. Compare the above case to this valid argument:

EXAMPLE 1-6   (Premise)   In order for my car to start, it must have at least some gasoline.
(Premise)   My car starts.
(Conclusion)   My car must have at least some gasoline.

Like the example of an invalid argument (**1–5**), the conclusion of this example could be false. I could have a solar-powered car, or my car could really be completely out of gasoline. But if the conclusion is false, I would also know that at least one of the premises must be false. It is not possible for the premises of this argument to be true and the conclusion false. If the premises are true, it is not a matter of luck that the conclusion is true. The premises lock us into the conclusion.

You may be experiencing the same confusion that most students experience at this stage. Valid arguments can have false conclusions; invalid arguments can have true conclusions. Aren't valid arguments suppose to be the "good guys"? Unfortunately, it is not a black-and-white world, and uncertainty and tentativeness prevail in many important parts of our lives. But such a world is not inconsistent with rationality. To understand what logic is and what it means to follow a logical trail correctly (a valid deductive argument), you must be able to distinguish between the **form** (reasoning) of an argument and its **content** (whether you believe or know the individual statements—the premises and conclusion—to be true or false). Consider the following argument given to me some time ago by a student, and before you read further make a judgment whether it is valid or invalid.

EXAMPLE 1-7   (Premise)   Only people who believe in the Christian God are moral.
(Premise)   John Smith is moral.
(Conclusion)   Therefore, John Smith must believe in the Christian God.

As a teacher in Hawaii, I am often faced with a multicultural, mixed-background student population. I teach many "local" students from the surrounding community, but also (because Hawaii is such a popular place to visit) many transfer students from all over the U.S. mainland, Europe, and Asia-Pacific region. One particular semester, I had a transfer student from a very conservative, Christian fundamentalist southern Bible college. He felt it was his mission in life to persuade as many people as possible to convert to his version of the Christian religion, and he would bring his Bible wherever he went and read it to others at every opportunity. When I asked the class to give me an example of a valid argument, he volunteered the above argument and proceeded to read to us Biblical passages that in his mind demonstrated the truth of the first premise. His constant conversion techniques did not sit well with many of the students in my class, many of whom held Buddhist, Islamic, or other religious beliefs. In general, most people in Hawaii have culturally tolerant beliefs. When I asked the class whether this was an example of a valid argument, most of the students responded that it was invalid. But they were wrong! Did you make the same mistake? This argument is valid. If it is true that only people who believe in the Christian God are moral, that is, if it is really true that all people who are good, kind, don't steal, and so on must believe in the Christian God, and John is good, kind, and so on, then it must be also true that John believes in the Christian God. Essentially this argument says that if you are moral, you must believe in the Christian God; John is moral, so he must believe in the Christian God.

What my students should have said is that this argument is indeed valid, but based on their beliefs it is not sound. It is an objective logical fact that the reasoning is valid, but the content is another matter. Most of my students did not agree that to be moral one had to be a Christian; they believed that people can be good to each other and have other religious beliefs, or none at all for that matter.

However, you may still be puzzled, as most of my students usually are, as to why we should be logical if this is the nature of logic: rational people can disagree, arguments with offensive premises can be valid, valid arguments can have false conclusions, and invalid arguments can have true premises and true conclusions. If you can be wrong after being logical, and right after being illogical, why be logical? The answer to this important question is very long. In fact, this entire book is an answer to this question. For now, however, in addition to the defensive-tool points made at the beginning of this chapter, consider the following little story as my answer.

Logic and Belief Testing

Several years ago, I had the good fortune to be an assistant coach on a Little League baseball team. My team's district had a long tradition of support for Little League baseball. The families were also mostly ethnic Japanese, many of which still had strong ties to relatives or friends in Japan. They loved their baseball, and each summer the games were quite a social event for the whole family—elaborate "pupu" (appetizer) parties followed every game, fundraisers were conducted yearly for an all-star team to be sent to play teams in Japan, and families would host visiting teams from Japan. At the beginning of each spring, tryouts would be held for new players, usually eight-year-olds who would begin an important four years of socialization. We would line up each new child opposite another and let them throw a baseball to each other. We were determining primarily if they were ready to play and would not get hurt, whether they watched the ball when they tried to catch it. If it looked like they were not experienced enough, we asked the parents to keep the child out one more year and practice a little with them. Only in rare cases was this necessary, however. Almost all the children who wanted to play were allowed to play at the eight-year-old level. If they were boys, that is! I can still remember my father shouting at me when I didn't watch the ball, "Don't catch like a girl!"

One particular year, a little eight-year-old girl calmly and confidently walked out and took up a position in a line and began playing catch with a boy in the opposite line. Everyone knew who she was. She was the all-star coach's daughter and the younger sister of the coach's twelve-year-old son, one of the best players in the league. Evidently, she had watched how much attention her brother had received during the last four years, and how much of the family's life revolved around Little League baseball. She had decided, without consulting anyone, that it was her turn regardless of the unstated, but accepted, restriction that Little League baseball was only for boys. When we realized what was going on, that she was not just playing or trying to help out by playing catch with a boy trying out, one of the coaches ran over to the far side of the field where her father was supervising the advanced players. When he heard what she was doing the father rushed over in an agitated mood. He grabbed his daughter's arm and pulled her aside, and an argument ensued. The next thing we knew, the little girl was walking home alone crying.

The father had given his daughter the following argument for why she could not play Little League baseball. He pointed out to her that her grandfather had been one of the first Little League baseball players in this district. Then, years later, he (the girl's father) had followed in his father's footsteps and also played in this district. Now her brother was following the family tradition. "Don't you see," he said, "it's always been this way, only boys play Little League baseball." Later in this book we will examine a common informal fallacy called *Traditional Wisdom*. The father had committed this fallacy, but it is worth our while to contemplate exactly what is wrong with this argument independent of the fallacy specifics. When I ask my students what was the real tragedy of the father's argument, some will usually refer to some vague principle prohibiting discrimination against women in our society. This may be true from a constitutional or ethical perspective, but I tell them that from a logical perspective this is only an indirect concern. Why is discrimination wrong? Why was discrimination wrong in this case? What was the father really worried about?

From a logical point of view, the tragedy in this situation was that the real issue was not tested. The real issue was whether the little girl had the potential skill to play at the eight-year-old level without getting hurt. If the father's reasoning was followed, we would never find out. In discussions of gender and racial discrimination issues this point often seems forgotten. Discrimination against a race or a sex is not just bad for those being discriminated against, it is bad for all of us. In a democracy we assume that whenever people congregate in a social situation, that situation— be it in a game, a business, or a country—is most good if individuals have the freedom to develop their full potential. We all lose if a valuable potential is stifled. Little League baseball and the father both lose if the daughter is not tested. She might be good. The father loved his daughter and feared that she would get hurt or embarrass herself. The real tragedy of the father's reasoning, however, was that it shifted attention away from what was relevant and shielded him from testing his beliefs. And this is the reason for my little story: Much of the *bad reasoning* identified in this book is just like this example; *an excuse not to test our beliefs*. Most often, bad reasoning is no more than an excuse not to think about what we should think about, to not test our beliefs against the world of experience.

We are not always right when we use good reasoning. However, this is the point. Because valid arguments cannot have true premises and a false conclusion, when the conclusion is false—when we realize that we are wrong even though we have followed a good reasoning trail—we know something must be wrong with our premises and we must think about them. With invalid reasoning, because the conclusion can be false even if the premises are true, having a false conclusion tests nothing, the premises can be true or false. Because we carry with us a web of beliefs that often act as premises for decisions, if we do not infer valid implications from our beliefs, we will not be testing our beliefs even if we are often successful in using them.

An additional virtue of good reasoning is that it helps us localize what we disagree about. Too often many arguments between people are just arguments in the sense of fighting or quarreling, where egos are at stake and the goal is not to understand each other but just to disagree in an unfocused way for its own sake. If two people disagree on whether a conclusion is true or false, but the reasoning that led to that conclusion is valid, then these two people know that they must also disagree over at least one of the premises. (The abortion examples, (1-3) and (1-4), are examples of this.) Both the localization of disagreement and the knowledge that our premises or beliefs must be reexamined if they lead to false conclusions constitute at least progress, if we assume as we do in a democracy that the critical discussion and examination of freely expressed beliefs is the way better ideas are found.

As a final aside, you might be interested how my little story turned out. The next weekend the little girl was back trying out again for one of the eight-year-old teams, under the watchful eye of a now sheepish-looking father. Evidently, the mother of the little girl—a very quiet, traditional wife in many respects—stepped in during the week and demanded that the daughter be given a chance. So the little girl played, and guess what? She was very good. Evidently for the past few years she had to shag balls for her brother and his friends when they were practicing, so she had developed (for an eight-year-old) a very powerful and accurate throwing arm. At any level, but especially at the eight-year-old level, this is a major asset. Few children could make an accurate throw from the short-stop position, the position she eventually played. She was so good that she made the district all-star team that year, and guess who was the proudest parent at the playoffs?! Lucky for him, the father was forced to test his beliefs.

However, our story is not over, and it is worth finishing to illustrate some additional points that we will be discussing in detail later. The next year, after the little girl's success, many girls in the district, perhaps at the urging of their parents, hastily concluded that they too could compete. Unfortunately, most of these girls were acting more on a whim or fad and had little real motivation or previous baseball experience. The results were disastrous. Some got hit in the head, and almost all were embarrassed and soon quit. More surprising, the little girl who had been so successful the year before also quit! Between seasons the father had begun to work with his daughter earnestly, probably dreaming that his daughter would be the first woman to play in the major leagues. However, the little girl wanted no part of the next level that the father was preparing her for. At the next level, the pitching was overhand and much more difficult and dangerous. Just before the season started, she quit, much to the dismay of her father. She had proved her point and went on to be a successful soccer player.

The world has a way of breaking through our most cherished belief-shields regardless of how hard we try to insulate ourselves from its sometimes painful touch. But, as Harvard scientist Melvin Konner has commented, "The truth may not always be helpful, but the concealment of it cannot be." Confronting the world and getting a clear response from it is what we call learning. Following correct reasoning trails keeps us on track and forces us to discuss the relevant issues related to testing our beliefs. Like an adventure, good reasoning forces us to take risks with the world of experience, but also like an adventure we expand our horizons in the process.[1]

A number of important philosophical consequences follow from an understanding of good reasoning. Good reasoning is not a panacea. Although good reasoning is a neutral tool, reasonable people can still disagree. We will be studying numerous objective processes that are available to check for valid reasoning. Two people can agree on the validity of the reasoning, yet come to different conclusions by disagreeing on the truth status of the information in the premises. This does not mean that logic is useless, rather it shows that often the world is complicated and that arriving at truth or consensus is difficult. People often have different initial opinions, but this does not necessarily imply that truth is subjective and only a matter of opinion. Most philosophical descriptions of democracy assume that truth is objective but difficult to find, and so the more points of view available, the more likely truth will be known. As the philosopher Socrates stated many centuries ago, we will be "better, braver, and less idle" if we believe there are answers to our questions and solutions to our disagreements, than if we believe there are no answers or solutions and that we need not even try. In this regard, logic can be a valuable tool in localizing genuine disagreement and showing where further investigation is needed.

It also follows from the above understanding of good reasoning that a person can reason impeccably and still be dead wrong. The conclusions of the greatest economic experts, the most sophisticated computer studies, the most eminent scientists can all be incorrect regardless of the precision or sophistication of the reasoning involved. All that is needed is a little false information somewhere in their premises. This is all the more reason why as individuals we should not rely solely on experts and authorities, and learn to think for ourselves.

Key Terminology

Arguments · Valid arguments
Premise · Invalid arguments
Conclusion · Sound arguments
Deductive reasoning · Form of an argument
Inductive reasoning · Content of an argument

Concept Summary

Logical thinking can be seen as slow, disciplined, defensive thinking. Often we can be tricked into believing something, not because we have been lied to, but because

[1] As you contemplate this paragraph, keep in mind that it expresses a set of values—primarily Western—inherited from the ancient Greeks. The ancient Greek culture (seventh century B.C. to second century B.C.), the historical source of modern science and technology, had a philosophy that assumed it was good to test beliefs, to think critically, to challenge opposing viewpoints, so that in time better beliefs would be found.

we have accepted an inference that we ought not to accept. As a critical tool, logic can be used to avoid being tricked into accepting an inference that we ought not to accept. Advertisements and political claims are often deliberately vague or misleading in such a way that they invite incorrect inferences.

Logicians have learned to practice this slow, critical thinking by analyzing persuasive appeals in terms of arguments—series of statements that contain premises (evidence) and a conclusion (an inference from the evidence). The goal of this analysis is to be able to separate good inferences from bad ones.

The most important concept in this regard is that of a **valid deductive argument**. Valid deductive arguments allow for all possible inferences but one—they do not allow for an inference from true premises to a false conclusion. Valid arguments make no claim about the specific truth content of the premises and the conclusion; we are assured only that truth will be preserved by using this mode of reasoning, that if the premises are true, the conclusion will be true. Another way of understanding this is that of a reasoning trail that stays on track; each step in the reasoning leads firmly to the next. Valid reasoning trails stay on track, even though we can't be assured whether we started our trail in the right place. The blind man stayed on track; he correctly carried out the implications of the information he was given. But he didn't know if his starting place—three white hats and two red hats—was absolutely guaranteed. Valid arguments may have false premises and false conclusions, but if the conclusion of a valid reasoning trail is false, then we know at least one of the premises is also false. Hence, reasonable people can disagree and being reasonable does not guarantee being successful. Valid arguments, however, encourage us to test our beliefs.

Invalid deductive arguments do not stay on track. They allow for all possible inference situations, including the very important one of having all true premises and a false conclusion. Because the conclusion of an invalid argument can be true or false, even if the premises are all true, we do not test any of the content when we use this form of reasoning. When we use invalid reasoning the truth or falsity of the conclusion of our inference is a matter of luck or some other factor, rather than the work of our inference. Unlike a valid argument, if the conclusion of an invalid argument is false, the premises could be true or false. Because testing our beliefs is desirable and we want the effort of our inferences to mean something, we should value valid reasoning. Ideally, we should strive for **sound arguments**—arguments that are valid and have premises that are known to be true.

We live in a complicated, uncertain world, where there are many conflicting opinions about what is true and false, wise and unwise. Logical thinking will not remedy this situation completely, nor is it intended to. To get along, to reach agreement on what is true and what is false, or right and wrong, to establish consensus on courses of action, all these require more human abilities than just being logical. To be open to new ideas, to be tolerant of other opinions and ways of life, and to understand our fallible nature are important practical virtues. However, having the skill and understanding the nature of logical thinking not only aid the implementation of these virtues by allowing for focused, disciplined discussion, but help us understand the practical value of having these virtues in the first place. If logical thinking gave us more of a guarantee than that provided by a valid argument, we would not need to be tolerant of other opinions. Truth would be too easy to establish.

EXERCISE I

Indicate whether the following are true or false.

1. The conclusion of a computer analysis will always be true.

*2. Arguments that have all true premises and a true conclusion will always be valid.

3. Arguments that have all false premises and a false conclusion will always be invalid.

4. The blind man knows that he has on a white hat, if the information he is given is true.

5. If an argument is valid, but the conclusion is known to be false, then at least one of the premises is false.

6. If an argument is invalid, and the conclusion is known to be false, then at least one of the premises must be false.

7. Invalid arguments always have false conclusions.

*8. The conclusions of experts who use valid reasoning are always true.

9. People who reason illogically are always wrong about the truth of the conclusions they infer.

10. Sound arguments always have true conclusions.

11. If two people disagree on the truth of a conclusion, then they must also disagree on the validity of the reasoning.

12. In logic, correct usage of terminology requires that we refer to arguments as valid or invalid and statements as true or false. In logic, we do not refer to arguments being "true" and statements being "valid."

EXERCISE II

Identify vague phrases and explain what each of the following statements imply in terms of the minimum conditions needed for each statement to be true.

1. Sale—up to 30% OFF! Now you can save between 10 and 30% on our most popular models at Shelley cars.

*2. When it comes to first class treatment, no one can beat American Airlines' air fares to Hawaii.

3. Johnsons' Hotels have consistently had one of the highest ratings in customer satisfaction in the past 10 years.

4. Full page JC Penney advertisement:

Up to 30% to 50% OFF
THOUSANDS OF ITEMS
THROUGHOUT OUR STORES!

Very fine print in lower left corner: "Regular prices are offering prices only. Sales may or may not have been made at regular prices. Percentages off represent savings on regular prices. Intermediate markdowns may have been taken."

EXERCISE III

Read the introduction to each of the following deductive arguments and then write out a short essay response explaining why each of the arguments is valid or invalid. Be sure to answer the question asked at the end of each argument. Be sure to develop and explain your answer to the question consistent with your explanation on the validity or invalidity of the argument. (The last statement in each argument is the conclusion, and the other statements are premises.)

*1. Suppose that the police chief in San Francisco is running for reelection. He wants to demonstrate that under his leadership the police department is more efficient in protecting the citizens of San Francisco. Suppose he offered the following argument:

We all know that if police departments do a better job crime will decrease.
Crime has decreased in San Francisco.
Therefore, the police department in San Francisco must be doing a better job.

Valid or invalid? If the premises of the above argument are true, what would be known about the conclusion?

2. A coach of a high school football team has a rule that having short hair is a necessary condition for being on the team. Eric, who knows this rule, overhears another student saying that the new boy Greg is not on the football team. Although he has not met Greg yet, he reasons as follows that Greg does not have short hair:

Everyone on the football team must have short hair.
Greg is not on the football team.
So, Greg does not have short hair.

Valid or invalid? Because the first and second premises are true, what would we know about the conclusion?

3. At the zoo, a father is attempting to demonstrate to his children his knowledge of zoology by stating that a particular bird they are viewing must be a crow. He gives the following argument.

Only crows are black.
This bird is black.
Therefore, this bird is a crow.

Valid or invalid? Suppose we know that the conclusion is false, that a zoo attendant overhears the father and tells him that the bird in question is a rare black honeycreeper. How would this affect the validity of this argument? What would this imply about the first premise? Would knowing that the conclusion is false and the second premise true constitute a test of the first premise?

4. Suppose John is explaining to his girlfriend why he has no money to take her out on the upcoming weekend. He explains that he has just had to install safety belts in his recently purchased used 1969 car because he received a traffic citation on the way home from the purchase. All U.S.-made cars by law must now have safety belts, whether or not they were originally so equipped. His girlfriend is suspicious, because although John's car is very old, she believes that all cars have always had safety belts installed at the factory. John gives his girlfriend the following argument:

No U.S.-manufactured car built before 1970 was equipped with safety belts at the factory.
John has a U.S.-manufactured car built in 1969.
So, John's car was not equipped with safety belts at the factory.

Valid or invalid? If the premises of the above argument are true, what would we know about the conclusion? Suppose John's girlfriend finds out that Congress passed a law in 1969 mandating factory installed safety belts for all U. S.-manufactured cars by 1970. Should she now be convinced that John's conclusion is true? (Hint: Think of the different meanings of (1) "No U.S.-manufactured car built before 1970 was equipped with safety belts at the factory" and (2) "All U.S.-manufactured cars built from 1970 on had safety belts installed at the factory.")

*5. In general, political liberals were against Ronald Reagan's plan for a missile defense system in space, nicknamed "Star Wars." They argued that it would cost trillions of dollars by the time it was in place, probably not even work, establish a dangerous precedent by placing nuclear weapons in space, and in general make us all less secure rather than protect us. Suppose you are a newspaper reporter and you are trying to estimate the outcome of a congressional vote allocating funds for the continuation of this program. Although you have not checked his voting record, you are told that John Dunn, a senator from New York, is a political liberal. You reason as follows:

All liberals are in favor of discontinuing funding for the Star Wars program.
John Dunn is a liberal.
Therefore, John Dunn must be in favor of discontinuing funding for the Star Wars program.

Valid or invalid? Suppose, when the actual vote is taken you find out that the conclusion is false, that Mr. Dunn is definitely in favor of funding the Star Wars program. What would we know about the premises?

6. During the second term of the Reagan presidency it was discovered that our government sold arms to the Iranian government in hopes of getting American hostages released from Lebanon. This revelation was very embarrassing, because the Reagan policy at the time was to never negotiate with terrorists, that hostage taking was an illegal international action, and that negotiation was therefore totally inappropriate. Negotiation was also argued to be counterproductive as it would encourage more hostage taking, not to mention that Reagan beat Jimmy Carter for the presidency in 1980 in part because he promised to be a stronger president and not let us be kicked around by radical countries such as Iran. After the news story broke, Reagan admitted that he had lied to the American people and that he did know about the arms sale to Iran. It also became known at this time that the money that we received from the Iranians was then given to the Contras who were fighting a guerrilla war against the communist Sandinistas in Nicaragua. The arms sales to Iran were not illegal, just embarrassing. But the diversion of the proceeds to the Contras was illegal, because at the time Congress had passed a law specifically prohibiting our government from helping the Contras militarily. Hence, if Reagan did know about this, he would have been impeached. Reagan never admitted to knowing anything about the Contra diversion. Analyze the following argument, which claims that we know that Reagan must have lied about the Contra diversion:

If President Reagan lied about the Contra money, then he would also lie about the Iranian arms for hostage deal.
It is clear (because he has admitted this to the American people) that he did lie about the Iranian arms for hostage deal.
So, President Reagan lied also about the Contra money.

Valid or invalid? Because we know that the second premise is true, what do we know about the conclusion?

7. Suppose a history professor tells her class on the first day of a new semester that a necessary condition for passing her course

is to pass the final exam. Suppose a student who is retaking the course reasons that he should be able to pass the final without coming to class the whole semester; he has had most of the material before and should be able to pass the final exam by just reviewing his notes from the previous semester. A key part of the student's reasoning is the following:

A necessary condition for passing this history course is that I pass the final exam.
So, if I pass the final exam, I will be guaranteed to pass the course.

Valid or invalid? Assuming that the professor was true to her word, would it be possible for a student to pass the final exam and still not pass the course?

8. Suppose John is having a party at his apartment. His hidden agenda for having the party is that he wants to get to know Delia, a new girl at school. He knows that Virginia and Betty do everything together and that if he invites Virginia, Betty will also come. This is important because Delia has recently become close friends with Betty. He reasons as follows:

If Virginia attends John's party, Betty will definitely also attend.
If Betty attends the party, Delia will also come.
So, if Virginia attends John's party, Delia will also come to the party.

Valid or invalid? Suppose Virginia attends John's party, but Delia does not come as expected. From this, would we be able to infer anything about the premises? If so, what?

9. Suppose John also wants to make his party special by making sure that Ken, the captain of the football team, attends. He reasons as follows:

If Virginia attends John's party, Sam will definitely attend.
If Sam does not attend the party, then Ken will not attend the party.
So, if Virginia attends, we can be sure that Ken will attend.

Valid or invalid? What if Virginia attends, but Ken does not attend as expected? Suppose also that we know that the second premise is true. Would this result (false conclusion and true second premise) constitute a test of the first premise?

10. Suppose someone is trying to explain to a friend that Bill Clinton is not just another liberal Democrat, but a new type of Democrat who has many things in common with moderate Republicans. She reasons as follows:

All liberal Democrats oppose the GATT free trade agreement, which will make trade between countries easier.
Bill Clinton does not oppose the GATT free trade agreement which will make trade between countries easier.
Therefore, Bill Clinton is not a liberal Democrat.

Valid or invalid? Explain. If the premises of the above argument are true, what would we know about the conclusion? Explain.

11. Consider the difference between number 10 and the following argument.

All liberal Democrats oppose the GATT free trade agreement, which will make trade between countries easier.
Bill Clinton is not a liberal Democrat.
So, Bill Clinton is not opposed to the GATT free trade agreement, which will make trade between countries easier.

Valid or invalid? Explain. How is this argument different from number 10? If the premises of the above argument are true, what would we know about the conclusion? Explain.

12. Suppose a peace advocate at a U.S. university argues that the cause of U.S. domestic problems is the tendency of the United States to get involved in wars. He argues as follows:

Stopping inflation is a necessary condition for the solution of our domestic problems.
If we are going to stop inflation, we must stop periodically getting involved in wars such as the Persian Gulf and Vietnam wars.
Hence, if we stop getting involved in wars such as the Persian Gulf and Vietnam wars, then we will solve our domestic problems.

Valid or invalid? Suppose most economists dispute the conclusion of this argument, arguing that the link asserted here between war and the United States' domestic problems is too strong. From a logical point of view, would these same economists also necessarily dispute one or both of the premises? Would they necessarily believe that at least one of the premises is false?

EXERCISE IV

Essay questions: Write a short essay response to each of the following. Answer yes or no to the following questions and then explain your answer.

1. Do valid arguments always have true conclusions? Do invalid arguments always have false conclusions? Explain by comparing the difference between validity, invalidity, and soundness.

2. If valid arguments allow for all possible inferences except ones from true premises to a false conclusion, what happens when we

know that at least one of the premises of a valid argument is false? Do we know that the conclusion must be false? Look at **1-1** again and explain what we would know about the conclusion if the first premise is false. Assume that I have a solar-powered car, so it is false that my car must have gasoline to start.

3. If an argument is valid and has all false premises, do we know that the conclusion must also be false? Consider **1-1** again, then consider that I have a very special car. It is a regular gasoline car with a specially fitted solar-powered system, such that it can operate either in the normal way with gasoline or by solar power during very sunny days. This would make the first premise of **1-1** false: I do not have to have gasoline for my dual-operated car to start. Suppose at the time we are thinking of **1-1** my car does have gasoline, making the second and hence both premises false. Would we be sure that the conclusion is also false? Would we be sure that the conclusion is true?

4. Explain why invalid reasoning does not allow us to test our beliefs. Explain why valid reasoning does allow us to test our beliefs.

ANSWERS TO SELECTED STARRED EXERCISES

I.

2. False. Invalid arguments can also have true premises and a true conclusion.

8. False. The conclusion of any valid argument can be false, if any of the premises are false.

II.

2. The phrase *first-class treatment* is vague. It could refer to first-class air fare or the way people are treated in general on American Airlines given the amount of money they pay. Either way, the *no one can beat* phrase implies that no one is better and hence does not claim that American is the best. So this ad does not claim that American has the best prices for first-class air fare or that it treats people better than any other airline. It only claims minimally that no one is better, that American is at least as good as any other airline.

III.

1. Here is the kind of complete answer you should give for the remaining arguments in part III.

This argument is invalid. Invalid arguments do not lock us into or guarantee the truth of the conclusion. In other words, even if the premises are true, the conclusion could be false. In this argument, to say, as the first premise does, that if police departments do a better job, then crime will decrease is not the same as saying that "only if" police departments do a better job, crime will decrease. The first premise, even if true, leaves open the possibility that there are other ways for crime to decrease. Crime could go down for, say, economic reasons or low unemployment. So, if crime does decrease, as stated in the second premise, it could be for other reasons, and we would not be sure that the police department was the cause. Even if the premises of this argument are true, we would know nothing about the conclusion. It could be true or false.

Incidently, appeals such as this are very common during election campaigns. Incumbents will always take credit for anything positive that has happened during their watch, and usually blame their predecessors for anything negative that happened. President Reagan blamed Jimmy Carter for the poor state of the economy for several years into his presidency. When the economy finally turned around, he took the credit. President Clinton also took credit when the economy improved during his administration.

5. This argument is valid. Valid arguments guarantee a true conclusion, if the premises are true. In the above case, if it is really true that being a liberal assures us that one is also against Star War funding, and John Dunn is a liberal, then he must also be against Star War funding. However, valid arguments do not guarantee having true premises, nor do they guarantee true conclusions unless all the premises are true. So, if we discover that this conclusion is false, then at least one of the premises must be false, because valid arguments do not allow for true premises and a false conclusion. Although the argument is valid regardless—if the premises are true, the conclusion must be true—the first premise could be false, some liberals may be in favor of Star War funding. Likewise, the second premise could be false: John Dunn may not be a liberal. In fact, both premises and the conclusion could be false. But this would not change the logical fact that *if* the premises were true, the conclusion would have to be true.

Chapter 2

ARGUMENTS AND LANGUAGE

However elegant and memorable, brevity can never, in the nature of things, do justice to all the facts of a complex situation. . . . Abbreviation (though) is a necessary evil, and the abbreviator's business is to make the best of the job which, though intrinsically bad, is still better than nothing. He must learn to simplify, but not to the point of falsification. He must learn to concentrate upon the essentials of a situation, but without ignoring too many of reality's qualifying side issues. In this way he may be able to tell, not . . . the whole truth . . . but considerably more than the dangerous quarter-truths and half-truths which have always been the current coin of thought.

—Aldous Huxley, *Brave New World Revisited*

Recognizing Arguments

In Chapter 1 we saw how logicians use the term *argument* in a technical way to refer to a formalized presentation of a reasoning trail. In everyday life, people argue about all sorts of things, and what they argue about is usually much messier and more involved than the neatly packaged formalizations found in logic texts. To conduct logical analysis we must be abbreviators in the sense described by Huxley above. However, this does not mean that formalizations are useless. We also saw in Chapter 1 that when something is very important to us, it is useful to break up a complex appeal into pieces, to analyze a perplexing situation into simple parts to see if they fit together correctly.

This implies that we are not just interested in analyzing quarrels or arguments as in "John and Kym got into an argument." A logical argument is not limited to disagreements between one person and another, or between one group and another. Arguments can also be objective, formal presentations of reasoning trails that are studied by people who already accept the conclusions, as in science and mathematics, where the goal may be shorter, tighter, more elegant presentations of systems of belief. In mathematics and logic, proofs or arguments can also be created like an artist creates a statue, to try to say something clearly, to bring out a perspective, or to make a discovery for the appreciation of others.

Analytic thinking is often contrasted with synthetic or holistic thinking. When we analyze, we take a whole and break into its parts (some would say we destroy it). In synthetic thinking, we take or are given parts and details and (if we are lucky, brilliant, or insightful) create a perspective that brings the parts into a new light. Synthetic thinking is often associated with so-called artistic thinking, and some philosophers have worried that too much analytic thinking ruins one's ability for synthetic thinking. This may be so, and we will adopt the view that both of these ways of thinking are valuable tools that are part of our potential as human beings, and that like any tool, using one for every task is surely a mistake.

Clearly though, as the examples from Chapter 1 show, many people reason too holisticly for some situations—they just react without stopping to see whether they have good reasons for their reactions. In Chapter 1 we made life easy for you by structuring arguments into conclusions and premises. In this chapter, you need to begin the process of recognizing different types of arguments and formalizing them. You will then practice argument recognition throughout the remaining chapters of this book.

In formalizing arguments and learning other analytic techniques presented in this book, you will actually be learning more than specific skills. You will also be learning a strategy or discipline of mind with which to face complexity, a most valuable ability in a world of increasingly abundant diverse ideas and opinions. In a democracy, we assume that this clash of ideas and opinions is necessary for better ideas to emerge, but we also assume that better ideas can only emerge if the exchange of ideas is rational and focused.[1]

To evaluate an argument's reasoning we must first recognize its parts. Every argument has its "bottom line" or "main thrust," a basic point that its author wants to persuade us to accept. We call this basic point the *conclusion*. The elements of persuasion, or the reasons given for us to accept the conclusion, we call *premises*. In general, arguments have a tone like

"This is so (conclusion), because that is so (premises)."

or

"Because this is so and this is so (premises), therefore you should believe such and such (conclusion)."

In recognizing an argument's parts, as a strategy of analysis, the first thing you should do is avoid the temptation to react to individual statements. (Remember the problem with **1-7**, p. 24 in Chapter 1.) Try to understand before you judge. Ask yourself what the ultimate point of a passage is; try to determine what the author is trying to get us to accept. What is the main point of the argument? Consider the following two arguments and write on some notebook paper what you think the conclusion is for each one.

---

[1] Often the formalization or abbreviation process itself helps a great deal to achieve this goal, because some exchanges of ideas are too "full"; full of excess and tortuous verbiage, that is. See example **2-12**, p. 60 below.

EXAMPLE 2-1   We all know a democracy must guarantee freedom of speech to all citizens. However, some citizens of a democracy use their freedom of speech to attack the concept of a democracy. At times of great trial, be it war, economic upheaval, or confusion brought on by rapid change, such attacks can pose a serious threat to a democracy's survival. Thus, a true democracy will always be vulnerable to destruction from within.

EXAMPLE 2-2   The Bush administration's justification for our military role in the Middle East was that a small, defenseless country, Kuwait, was brutally invaded by Iraq, which possessed the fourth largest military force in the world. Should such a justification have been accepted by anyone? Syria invaded Lebanon, and we did nothing. Turkey invaded Cyprus, and we did nothing. Israel invaded and pirated land from Egypt, Jordan, and Syria, and we did nothing. The Chinese government brutalized a blossoming democratic movement, and we not only did nothing, we encouraged it by continuing business and trade as usual. The former Soviet Union intimidated its Baltic neighbors, and we did nothing. Besides, Kuwait was a backward dictatorship itself, stole oil from Iraq, and lied to the World Bank about their monetary support for Iraq during the Iran–Iraq war. Not to mention that until shortly before Iraq invaded Kuwait we were supporting Iraq militarily.

Notice that these passages have the tones we talked about above. The first one follows the pattern of "Because this is so and this is so (premises), therefore you should believe such and such (conclusion)." The author of this argument helps us identify the parts of the argument by using the **conclusion indicator** word 'thus.' The conclusion of the first argument is, "Thus, a true democracy will always be vulnerable to destruction from within." The use of language to communicate is not always successful and we need all the help we can get in understanding each other. Arguments often have "neon lights." Conclusion or **premises indicators**, or both, highlight parts of the argument.

Examples of **conclusion indicators** are:

therefore, so, hence, thus, accordingly, consequently, it follows that, we can infer that, which entails that, which implies that.

Examples of **premise indicators** are:

because, since, in as much as, given that, for the following reasons, follows from, may be inferred from, may be deduced from, may be derived from.

The second argument follows the pattern "This is so (conclusion), because that is so (premises)." This argument is typical of the much less formal situations that we find in everyday exchanges of ideas. There are no neon lights to tell us exactly what the author has in mind as a conclusion. In fact, the conclusion is not even explicitly stated, and we must interpret the author's major point. Identifying the conclusion in such passages tests your communication ability—particularly your understanding of the language used. It also will often draw on your experience or knowledge of what is being discussed. However, because logical ability and language ability are very closely related, the most important thing you can do to enhance your reasoning ability and your own defensive verbal posture is to learn to read and write better.

But even this will not ensure successful communication between people. Often, two people can use language correctly and still not understand each other due to the complexity of their topic, their different perspectives, cultures, values, or different experiential backgrounds. In such cases, communication is an ongoing process that takes work. Often follow-up discussion is needed, and sometimes we just need to ask the author of an argument, "What is your point?", "What conclusion do you want me to accept?"

What is the author's major thrust in the second argument? The author obviously doesn't think much of the Bush administration's justification for our involvement in the Persian Gulf War. The Bush administration's position is stated and then followed by the question, "Should such a justification have been accepted by anyone?" This is a rhetorical question, a question that implies its own answer. In this case the implied answer is "No." The author is saying that no one ought to have accepted such a justification, and following the rhetorical question are the author's reasons why we should not accept such a justification, along with an implied premise that our policy did not appear consistent. In such cases we must paraphrase the implied conclusion and premise; we must restate it in our own words in an attempt to capture the essence of what is implied as best we can. Here is one possible formalization:

EXAMPLE 2-2A

Conclusion (implied)	The Bush administration's justification for our military role in the Middle East was weak. (We shouldn't believe that the only reason for our involvement was that a small, defenseless country, Kuwait, was brutally invaded by Iraq, which possessed the fourth largest military force in the world.)
Premise (also implied)	This justification did not make sense in light of our past and present actions and was too simplistic in terms of Iraq–Kuwait relations.
Premises (remaining)	Syria invaded Lebanon, and we did nothing. Turkey invaded Cyprus, and we did nothing. Israel invaded and pirated land from Egypt, Jordan, and Syria, and we did nothing. The Chinese government brutalized a blossoming democratic movement, and we not only did nothing, we encouraged it by continuing business and trade as usual. The former Soviet Union intimidated its Baltic neighbors, and we did nothing. Kuwait was a backward dictatorship, stole oil from Iraq, and lied to the World Bank about their monetary support for Iraq during the Iran–Iraq war. Until shortly before Iraq invaded Kuwait we were supporting Iraq militarily.

We could disagree about whether this is an accurate presentation of the author's intentions. We could argue about it, and we could nitpick over some of the words used. But in most cases the context and the language used imply at least a focused range of interpretations that are close to saying the same thing. The interpretation of arguments is not always a black-and-white process and like most things in life sometimes there is considerable room for reasoned disagreement. The point is to get down a formalization in the form of "This is so (conclusion), because that is so (premises)." Then we can begin to discuss whether the formalization is accurate and evaluate the argument. As soon as the argument has more focus, then we can be more disciplined in analyzing its parts and the inferential linkage between its parts.[1] Are the premises true? Are they relevant to the conclusion? Do they give us good reasons to accept the conclusion? What is the relationship between the conclusion and the premises? Is a deductive or inductive relationship asserted?

Other Uses of Language

Before we go any further, we must recognize the obvious point that we don't argue all the time. Often, in my logic class, after a few days of analyzing suspect advertisements similar to those in Chapter 1, some of my students conclude that they can now find bad arguments everywhere they look. With excited anticipation of reward and confirmation that they have now achieved a special status above the masses, they sometimes come into my office after class and share with me what they consider obvious blunders of human reasoning. "Isn't this the stupidest thing you have ever heard?" one might say. Then I usually have to disappoint them with something like "Yes, based on what is considered to be scientific fact, those beliefs are silly. But there is nothing wrong with the logic, because there isn't any; I don't see a presentation of an argument. Perhaps there is one I am not seeing. Why don't you look at it again for homework, and see if you can interpret an argument in terms of premises and a conclusion."

I respond this way because often the passages found by students do not appear to be arguing about anything, but rather are alleged descriptions, statements of belief or opinion, or explanations, or bare value judgments without any support offered that the descriptions are true, the explanations the best possible, or the value judgments acceptable. Descriptions can be false, explanations can be incorrect, and values poor or unacceptable, but they are not arguments. For an argument to be bad we must have a group of statements that can be analyzed into premises and a conclusion, where an inferential claim is made that the conclusion, although controversial, should be accepted because of the evidence offered in the premises. Here are some examples of the use of language that are not arguments:

EXAMPLE 2-3   The Sun is about 4.6 billion years old. This is a lengthy time, and roughly half the stars in our galaxy are younger than the Sun. We

---

[1] Even if we disagree on the interpretation of an argument, at least the argument interpretation that we do have can be discussed. Even if we have misunderstood the author's intention, we can learn something by analyzing the reasoning of our interpretation.

should also keep in mind that at least half the stars in our galaxy are part of binary star systems. (*Description*)

EXAMPLE 2-4   The building collapsed completely during the earthquake, because it was built before there were construction standards for earthquake prone areas and its foundation and construction material were poor by today's standards. (*Explanation*)

EXAMPLE 2-5   Homosexuality is immoral. I can't help being totally disgusted when I see two men walking down the street with their arms around each other. I retch when I can't help imagine what they do behind closed doors. (*Value judgment*)

The first statement is a typical factual description of what astronomers believe to be true about our sun and galaxy. Although scientists believe there is an enormous amount of evidence to substantiate these statements as true, none of that evidence is presented here. So, there is no argument. However, these statements could become part of an argument, such as when the possibility of extraterrestrial intelligence is discussed or when these statements are challenged by very conservative fundamentalist Christians who do not believe the universe can be this old. Here is an example of how some of the statements in **2-3** can be used as part of an argument:

EXAMPLE 2-3A   The Sun is about 4.6 billion years old. This is a lengthy time and because the Earth was formed with the Sun, it follows that on Earth intelligent life took at least this long to evolve.

Notice that the factual statements are now linked by a conclusion indicator to a relatively bold and controversial generalization, that intelligent life on Earth required billions of years to evolve.

The use of the word *because* in **2-4** shows that language is flexible and words that we have designated as conclusion and premise indicators do not always function as indicators for parts of arguments. In an explanation, that part that is analogous to a conclusion is usually an accepted fact regarding an objective event that has taken place and is not controversial. In **2-4** it does not appear that anyone would be arguing about whether the building did indeed collapse; instead, what would be more controversial is the explanation (what follows the word *because*) for why it collapsed. In an argument, the situation is reversed: what follows the premises indicators should be less, or at least no more, controversial than the conclusion.[1] Thus, a rule of thumb for separating arguments from explanations is to first formulate the passage into the form "X, because Y." Then ask yourself if X is more controversial than Y. If the answer is yes, then it is most likely an argument. If the answer is no, then it is most likely best analyzed as an explanation.

---

[1] In general, when we attempt to persuade logically we should present evidence that we think will be accepted as true or noncontroversial. Arguments, however, can also be used to test controversial ideas by deducing or discovering implications to test. But this use of argument is different from direct persuasion. We will discuss this difference in more detail in Chapter 5 with the fallacy of Slippery Slope.

Often, relatively vague passages can be analyzed by testing different interpretations. For instance, consider the following:

EXAMPLE 2-6   There are no hummingbirds in this area. We haven't seen any hummingbirds all day.

Is this passage a description, an explanation, or an argument? It is probably not a description, because a link of some sort seems to exist between the two statements. But is it an explanation or argument? Should we interpret this passage to mean:

EXAMPLE 2-6A   (We can conclude that) There are no hummingbirds in this area. (Because we know that) We haven't seen any all day. (*Argument*)

or

EXAMPLE 2-6B   (Because) There are no hummingbirds in this area. That's why (explains the fact that) we haven't seen any all day. (*Explanation*)

Like a blurry picture we are given no context for **2-6**, so we must imagine a context in which this statement was made. There are at least two possibilities. For an argument (**2-6A**), perhaps some people have been hiking in a forest and they notice that no one has seen any hummingbirds. One of the members of the group infers (generalizes) from this accepted fact that there are no hummingbirds in this area. For an explanation (**2-6B**), we can imagine someone noticing that they have not seen any hummingbirds all day and asking other members of the group why this is so. Another member then explains that there are no hummingbirds in this area.

This example shows that although recognizing arguments will not be a simple black-and-white affair, we can have guidelines for analysis and discussion. In both cases, the statement, "There are no hummingbirds in this area" is more controversial than "We have not seen any all day." But in a context of persuasion, when we are identifying arguments, we are looking for a controversial statement that is supported by reasons for accepting that statement. Because in **2-6B**, no support is offered for the statement "There are no hummingbirds is this area," this interpretation is not an argument.

Context, background knowledge, experience, and cultural differences can all play roles in interpreting, recognizing, and understanding arguments. Recognizing arguments is part of the general task of intelligent and respectful communication, and it will often involve some work. Furthermore, in today's shrinking, interconnected, multicultural world we may not always be successful. However, we at least have a general guideline to help us with some of this complexity. Although technical arguments in logic, science, and mathematics do not always involve controversy in the same sense that we have been discussing, in recognizing arguments in the popular media the focus is most often on controversy. So, in recognizing arguments a helpful guide is to ask, "Is a controversial statement made (the conclusion), and is some alleged support offered for the acceptance of that statement (the premises)?"

Speaking of controversy, clearly, discussions of values and moral judgments are a vital but argumentative part of life. In **2-5** the volatile issue of sexual orientation is raised. But is this an argument? Although it appears to have something like a conclusion, a bottom line that the author wants us to accept—"Homosexuality is immoral"—the statements that follow do not appear to be supporting premises, but rather elaborations of the author's values in sexual matters. No reasons are offered, why we should have the same reaction to the homosexual life-style, or why we should not accept it as a permissible life-style, even if we share the author's heterosexual orientation. At this point, **2-5** is only a bare value judgment.

It does not follow from this example that values cannot be argued about, that they are merely expressions of subjective feeling, and that they cannot become part of reasoned discussion. Socrates (470–399 B.C.), one of the first Western philosophers, began his famous questioning with value issues. Prior to his gadfly antics in the city streets of Athens to make people think, Socrates was a soldier. He witnessed firsthand the violence and cruelty of 'which human beings are capable. Prior to Socrates, the philosopher Protagoras (ca. 485–410 B.C.) had claimed that the values and virtues that a society deem good are only so relative to that society, and it is presumptuous of a culture to believe that its way of seeing things is the only right way of living. The values we possess, according to Protagoras, are the result of our cultural conditioning and individual circumstance; what is true for you is true for you, and what is true for me is true for me. Socrates agreed with Protagoras that most people are presumptuous to believe that their values are the best, especially when we see that if they are forced to defend them, the reasons they give are very weak, usually based only on appeals to tradition, authority, or popularity. But because of his life as a soldier, Socrates reasoned that although there may be many different good ways to live, surely there are also many bad ways to live. If so, he reasoned, what general idea about what is "good" enables us to judge some particular way of life as good or bad? If there are no general principles that all people ought to accept, then "anything goes," and there is no such thing as a bad way to live.

Protagoras was what philosophers call a **relativist**. Relativists believe that objectivity is a myth; that there are no independent facts or values that everyone ought to accept; that truth and value are relative to the individual and cultural context. Protagoras believed that we cannot **reason** about values, we can only attempt to creatively **persuade** one another to see a way of looking at things. For Protagoras, life is a battle of perspectives and the goal of all discussion is more like art than science and logic—to mold reality from a particular perspective and to persuade others to adopt this perspective. Socrates was not a relativist; he established the foundation for **normative ethics**. The normative ethicist believes that some values (norms) can be judged as better than others; that values can be part of an objective rational discussion.

Many of my students, when hearing this issue for the first time, conclude that they are relativists, because they believe that Protagoras was right, that cultures do differ, and that we "ought" to learn to respect and be tolerant of other life-styles. But a little reflection shows that they are not relativists. They are really adopting a normative principle: We *ought* to value, learn from, and be tolerant of different ways of living. A consistent relativist is committed to saying that this principle is no better than one that says we ought not to value, learn from, and be tolerant of different ways of living.[1]

Another source of possible confusion is that a normative ethicist is not saying that all issues of value must be cases of objective discussion. Many believe that there are "levels" of evaluation and that some levels are matters of subjective taste as the relativist asserts. For instance, few normative ethicists would be interested in arguing about whether chocolate or vanilla ice cream is the best. What most normative ethicists want to argue is that in matters of **ultimate value** some commitments are more reasonable than others.

As an illustration that it is possible to reason about values, we can formulate the following where the evaluation of the homosexual lifestyle is part of an argument.

EXAMPLE 2-5A   The Bible says that homosexuality is a crime against nature and a crime against God. All crimes against nature are immoral and ought to be prohibited. So, homosexuality is immoral and ought to be prohibited.

Formulated this way, we now have an argument for a focused discussion. Are the premises true? Does the Bible really say this about homosexuality? Even if it does, why should we accept the Bible as authoritative on this issue? Is there such a thing as a crime against nature? What if science is able to demonstrate that sexual orientation is genetically determined, that people are determined at birth to be homosexual or heterosexual, that we have no choice in the matter? Even if the premises are true, do they offer good reasons (in the sense of a valid argument) for the conclusion that homosexuality is immoral and ought to be prohibited? To apply what we have learned thus far about arguments, the focus of discussion should not be on the reasoning of the argument. This argument is valid; if the premises are true, then the conclusion is also true. But the premises are clearly controversial, and most normative ethicists would say that the issues raised in the premises are clearly questions of ultimate value that need some sort of supportive evidence before we have good reasons to accept them.[2] These premises would need to be conclusions of other arguments. The important point is that there are "bare," unsupported value judgments and value judgments that are part of arguments. We are interested in recognizing the latter, so that focused discussion is then possible.

Here are some other examples of evaluations that are not yet part of arguments:

---

[1] A consistent relativist is also committed to believe that any method of inquiry, any method of arriving at beliefs is as good as any other. This implies that we don't learn about the world or even learn better ways to learn about the world. Most philosophers of science believe that scientific inquiry and the many methodologies devised by scientists throughout the centuries since the time of Socrates are examples of learning better ways of learning about the world. We will discuss this further in Chapter 3.

[2] Although the premises are controversial, the arguer apparently thinks the premises are noncontroversial or less controversial than the conclusion. Also, this series of statements is an argument and not an explanation, because the conclusion is not an accepted fact that requires an explanation. Note that a consistent relativist would be committed to saying that the beliefs expressed in this argument are as good as ones that would express more tolerance for alternative life-styles.

EXAMPLE 2-5B   I don't like sports. I don't like baseball, basketball, boxing, football, or golf. Every weekend the major TV networks in the United States construct their programming to torture me. No matter what the season there is some sort of "game of the week" that is supposed to be the focus of everyone's attention. And it is getting worse. Now, no matter what the country or time of day, there is some big game or contest that one cannot escape, be it a quiet rural neighborhood in central Kansas, a pub in Great Britain, or a hotel lobby in Japan. Ahhhh!!

EXAMPLE 2-5C   As Americans we believe in our (European) cultural heritage, not just because it is "ours," but because it is good. But our ability to preserve and transmit that common heritage depends on the continued existence of a majority population that believes in it.

The first one (**2-5B**) presents no reasons why we should not like sports, so it is not an argument. The second one (**2-5C**) is an interesting case that deserves elaboration. The first statement contains the value judgment, "America's European cultural heritage is good." Out of context, we might interpret the entire statement to contain either an argument (We ought to believe in our cultural heritage, because it is good.) or an explanation (Americans believe in our cultural heritage, because they know it is good.). However, the context from which this passage was quoted indicated that it is best interpreted as a value judgment being used as a premise for a larger argument with a very controversial conclusion.

In the original article, "Immigration: Threat to Our Cultural Heritage," Lawrence Auster[1] addressed the traditional liberal ideology of blind commitment to cultural diversity. He attempted to demonstrate (conclude) that this ideology, which has always been assumed to be an important contributing factor to the goals of democracy is actually "profoundly inimical to freedom" when translated into the contemporary context of unrestricted U.S. immigration of Latin American and Asian people. According to Auster, the natural forces of ethnic and cultural identity, power struggle for self-determination, and the hardship of assimilation into a new and alien culture have produced and are producing a "dangerous impetus toward ethnic chauvinism" and a "manifestation of . . . adversary culture." He claims we are finding a "massive assault on Western literature and thought" and a dangerous weakening of commitment to traditional American ideals that restricts freedom to express certain ideas. According to Auster, a "new, unfree America (is) being born of uncontrolled immigration." In other words, because more and more people immigrating to the United States are from Asian and Latin countries and do not share the political and philosophical ideals of European-Western culture, those ideals are not only constantly under attack but are in danger of being overwhelmed in a new atmosphere of intolerance for other cultures and an overzealous protectiveness of one's own. At

---

[1] Reprinted from *Newsday* in the Sunday edition of the *Honolulu Star Bulletin and Advertiser*, May 19, 1991, B1. See also Auster's "The Path to National Suicide: An Essay on Immigration and Multiculturalism," (American Immigration Control Foundation, 1990).

various universities in the United States, according to Auster, this cultural protectiveness became manifest in the phenomenon of "political correctness," where the ideal of free speech was under attack.

This issue is a deeply involved and sensitive one of great importance. It touches on one of the greatest issues of our time, not only for the United States but for the entire world: How to obtain a unity of purpose and commitment to global or national democratic values while different ways of living are brought together as never before in history; how to maintain a sense of unity as a fragile species with a common, contingent evolutionary history and an uncertain future, while celebrating cultural diversity and the need for cultural autonomy; how to reap the benefits while moderating the inevitable frictions produced by cultural differences; how not to encourage a destructive "us-vs.-them" attitude when cultural differences are brought together.

Personally, I disagree with Mr. Auster's conclusion that the liberal ideology of commitment to cultural diversity is hostile to freedom and destroying America. Although much too involved to describe in detail here, I can sketch the possible course an argument between myself and Mr. Auster might take. He would be able to get me to agree that completely uncontrolled immigration would hurt the economy—too many people, too soon, competing for too few jobs would create severe tension. However, I would question what I see as one of his premises. I would argue that Asian and Latin American immigration need not and ought not be a dismantling of European-Western culture and ideals, or a construction of a philosophical perspective that renders this tradition illegitimate for the twenty-first century. Rather, I would argue that we need a more humble perspective of the importance of the European-Western tradition: not only its political concepts, but particularly its primary products in terms of relating to nature, that of science and technology.

I would argue that this more humble perspective is vital. A greater danger than the frictions produced from cultural diversity, which are most likely only temporary, is the wholesale subversion of cultural differences by some of the peripheral values of Western culture—the cultural imperialism currently taking place as one culture after another adopts Western economic values and the engines driving these values, science and technology. Perhaps Western-European culture needs a good philosophical slap in the face to recognize its proper place as a most valuable contribution to the human experience, but not one that is exclusive and absolute. There are other valuable traditions besides science and technology, and wherever people have congregated on this planet and formed a society, insightful ways of relating to reality have been discovered.

I would argue that in the long run the advertisements we looked at in Chapter 1, and the daily unprotected onslaught and uncritical acceptance of such advertisements, are much more dangerous to the health of a society than would be a neighborhood full of families who speak different languages. I would worry more about all of these families buying the same video recorder at Sam Kung's, and the children of these families laughing at the "silly" beliefs of their grandparents from the old country. I would worry much more about the implications of government reports that describe the Japanese people as "creatures of an ageless, amoral, manipulative, and controlling culture" conspiring to dominate the world.[1]

I might not be able to persuade Mr. Auster to change his mind, but if Mr. Auster and I stayed on track, if we both used reasonable arguments for our positions, if we then identified the nature of our premises—statements of fact, values, explanations, and such—we would both learn a lot from each other. We would have an opportunity of learning precisely where we disagree and then be directed to a discussion of methodologies that might be used to resolve disagreements other than logical ones. We might also gain valuable insights into hidden assumptions we did not realize we were making and recognize that some of our premises need a great deal more thought. Anyone following our discussion would likewise benefit from such focus and perhaps be stimulated to suggest ideas or points of view that we had not considered.

My disagreement with Mr. Auster would no doubt involve explanations, values, and beliefs, as well as logical arguments. Life is tough, communication difficult, closure on major philosophical issues most often elusive. But the distinctions we have made thus far in language use at least give us areas of focus and a discipline of thinking that is surely better than pointless shouting, especially if our goal is to test our beliefs.

Meaning and Clarification

Because beliefs consist of statements, and statements consist of words, a little reflection reveals that the task of evaluating an argument and the terms contained in it are linked. In Chapter 1 I asked you to evaluate the claim, "Dunlop's SP-4 radial tire had the highest rating in *Car and Driver* tire test." We recognized that the terms *rating* and *test* were **vague**. Vague expressions lack clarity; their meanings are unfocused or fuzzy. To evaluate the claim of this advertisement these terms needed to be defined more precisely. How was the rating done? What were the specifics of the test? And so on. On the other hand, we found that the term *highest* was viciously **ambiguous**. An ambiguous term has two or more distinct meanings in a given context. The advertisers deliberately took advantage of an ambiguity: In this context, *highest* could mean either the *very* highest rating, meaning all other tires rated lower, or the highest category used, meaning that this was only a general category and several other tires could rate placement in this category, some or all rating higher than Dunlop within this category.

In this section we will see that in evaluating arguments and persuasive appeals, we need to pay attention to how arguments are packaged. We need to be "picky"

---

[1] This statement was from a foreward to a CIA report called "Japan: 2000." It was written by M. Richard Rose, president of the Rochester Institute of Technology. Although the inflammatory foreward was later removed from the final report, one wonders if the ideas have been removed from the minds of some U.S. political leaders. The story of this report was a Gannet News Service report reprinted in my daily newspaper (Jennifer Hyman, "Report to CIA: Japanese are amoral beings," *Honolulu Star-Bulletin*, May 25, 1991, A-1). At about this same time, Edith Cresson, French prime minister, was describing the Japanese as "ants" and "little yellow men" who "sit up all night thinking how to screw us."

and pay attention to the meaning of the words used and the style of presentation. Otherwise, we can easily be tricked into believing that evidence is presented when it is not, or that evidence is stronger than it really is. Consider this example:

EXAMPLE 2-7   Any intelligent human being living on the brink of the twenty-first century should support women's right to abortion. Abortion is not murder, it is simply the prevention of a future person.

The last sentence in this passage shows that we often offer definitions of key terms to provide clarity for our positions and eliminate vagueness and ambiguity. But it also shows that the meanings of terms are linked with beliefs, and definitions are as much in need of critical evaluation as arguments are. It is often appropriate in a discussion to explore a reasoning trail by asking "What do you mean by X?" where X is a key term within an argument. But we also have to ask if there are good reasons for the offered definition. In **2-7**, does the author have a right to define abortion as meaning "the prevention of a future person"? Is this the way the word *abortion* is commonly used? If not, are there good reasons for using the word in such a unique way?

Traditionally, most logic and English books distinguish between denotative and connotative definitions. A **denotative** definition of a term involves pointing to or naming particular objects or instances for which the term applies. This is how we start children toward language acquisition. If we want to teach a child the meaning of the terms *rabbit* and *chair*, we point to examples of these. An amazing process seems to take place during such language acquisition. We point to a few different examples of chairs, and fairly quickly a child "gets it." Even though we show only relatively few of the many different types of chairs, somehow the child grasps the essence of what all chairs are like.

Eventually, the child seems to understand what has been traditionally called a term's **connotative** meaning. Connotative meaning is the package of qualities and attributes intended by use of the term. For this reason connotative meaning is often called **intentional** meaning. When a child understands correctly the use of the word *chair*, he or she understands what is intended when a speaker uses the term, in this case an object for people to sit on that has different types of support. Notice that by merely pointing at different examples of chairs a child could easily be confused about the denotative meaning of *chair*. A child could easily be uncertain as to what we are pointing at. The examples we use could all be chairs made of wood, or with four legs, or of a particular color. The child must eventually learn that what we intend by the term is more flexible than these particular instances of chairs; that chairs can have different colors, different types of support, and can be made out of many different types of material.

Because we are interested in meaning to obtain clarity, and clarity presupposes agreement on meaning, it is important to see that there are different theories of meaning. Some philosophers, for instance, have denied that denotative definitions really exist or are ever very useful. These philosophers argue that only intentional meaning exists for terms and there never is any absolute, fixed, intended meaning for any term; that the package of the qualities and attributes intended by the use of a term is a "fuzzy" package, containing one set of qualities and attributes in one situation and a slightly different set in another.

For instance, the word *snow* will mean something different for a person living in Minnesota and a person living in Singapore. The package of qualities and attributes of Minnesota snow will probably be larger, containing more distinctions and shades of meaning. Thus, some philosophers argue that intentional meaning is always relative to a language-using group—a culture, subculture, or class of people. This is particularly apparent when we try to understand other cultures by learning the language of that culture.

Suppose I am trying to learn a language where the term *gavagi* is used for what I call *rabbit*. After a few days of exchanging pointing gestures with the members of this culture I conclude incorrectly that I understand what they mean by the term *gavagi*. Little do I realize that in this culture what the members intend by *gavagi* is "sacred, untouchable object that contains the spirit of one of our ancestors." Perhaps for me, *rabbit* means "pesky creature that used to eat grandma's vegetables, but tasted a little like chicken and was good for a barbecue when grandpa killed one for dinner." Clearly, in spite of our consistent pointing gestures the situation is unfortunately ideal for major miscommunication, especially if I ever go hunting with some of the members of this culture.

This little excursion into the relativity and flexibility of meaning is important. Most logic and English books claim that types of definitions can be classified and that rules provide "correct" definitions depending on the classification. For instance, **lexical definitions** are considered the essential meaning of a term or the way the word is used by most people. The goal of a dictionary is to provide the lexical definition of terms. Some examples of rules for providing lexical definitions would be: avoid circularity, avoid being too broad or narrow, avoid vagueness and ambiguity, avoid emotional terminology, be affirmative rather than negative, and indicate the context where appropriate. Such definitions are also considered classifiable as true or false. But because there are different theories of meaning, terms such as *correct*, *true*, and *false* may be too strong and subculturally biased.

Much of life's richness is derived from the deliberate use of fuzzy language. For instance, for someone who regularly watched late-night television in the United States during the late 1980s and early 1990s, the argument "The David Letterman show is better than that of Arsenio Hall, because Arsenio Hall is nothing but a long corridor designed by an Italian architect" might make a lot of sense (as well as being humorous). It implies a whole host of features about both men and their shows—how they dress, the depth of their personalities, the quality of their guests and entertainment, and the kind of people that watch each show. In logical analysis, our task of being "abbreviators" is not to inhibit, prohibit, or destroy such fun uses of language, or to limit all the creative ways that we use words to carve perspectives out of the silence of existence, any more than it is to ban all emotion from life. Our task is simply to provide clarity. Recognition of the existence of language-using groups and the flexibility of meaning will often be part of this task. So, if someone uses a term or a phrase in a unique way as part of an argument, our question is always the same: "Are there good reasons for using the term or phrase in this unique way?"

Another example found in logic and English books is that of **stipulative definitions**. A stipulative definition does not attempt to define a word the way most people use it, but specifies a relatively arbitrary operational meaning to a term, such as when an instructor says, "In my class an A grade will mean a course average of between 88 percent and 100 percent of total points." Such a definition is not totally arbitrary, of course. There may be good reasons why the instructor has chosen 88 percent rather than 90 percent as the low end for an A grade. Based on the instructor's standards for learning, past student performance, difficulty of the material in the class, the predictive value of achieving an A in this class and achieving an A in other classes, and the number of students who achieve A grades relative to the number of students who achieve other grades, the instructor may have decided that 88 percent is more appropriate than the stricter and more traditional 90 percent. But choosing 88 percent rather than 87 percent or 89 percent is probably arbitrary. Some cutoff point is needed, and the instructor has decided that the lower end for an A grade needs to be relaxed a little.

Notice that what we needed in the Dunlop "highest rating" example to properly evaluate the advertising claim was a stipulative definition for the phrase *highest rating*. When we discovered that an excellent or highest rating could be achieved by scoring between 300 and 350 points on the tire test, Dunlop's tire was placed in a different perspective relative to its major competitors, which actually scored higher within this category. By arbitrarily changing the stipulative definition of *highest* we could render the advertisement false. Suppose Dunlop's major competitors scored 325 points and above, but Dunlop scored only 310 points. If we arbitrarily changed the stipulative definition of *highest* to mean between 325 and 350 points, Dunlop's tire could no longer be given this rating.

In the early 1990s, Congress and the Food and Drug Administration (FDA) began to respond to consumer protection group concerns on food package labeling. By this time, food manufacturers had taken advantage of increasingly health-conscious consumers by making liberal use of words such as *light*, *low*, and *free* (as in reduced fat, low cholesterol, sugar-free). Without standards and stipulative definitions, these terms could mean just about anything. Thus, by 1993, the FDA issued rules based on a law passed by Congress, such that *free* must mean less than five calories, less than 0.5 grams of sugar, and less than 5 milligrams of cholesterol and 2 grams of saturated fat per serving. *Low* must mean less than 140 milligrams of sodium, less than 40 calories, 20 milligrams or less of cholesterol, and 3 grams or less of fat with 1 gram or less of saturated fat—each per 100 grams per of food. Without such stipulative definitions, a 1991 Lean Cuisine frozen dinner advertisement was able to claim that Lean Cuisine dinners "skimp on" calories, fat, and sodium. The advertisement claimed that these dinners contained only 1 gram of sodium per entree, but 1 gram is equal to 1,000 milligrams, 860 milligrams higher than what the FDA defined as *low*![1]

Sometimes, however, bureaucrats can be overzealous in their concern for precise definitions to protect the public. After the Exxon Valdez oil spill that destroyed the

[1] See Exercises, part II, number 4, Chapter 5, to see how the Lean Cuisine advertisement is representative of a common informal fallacy.

environment along large sections of Alaska's coastline, well-meaning regulation writers in the Environmental Protection Agency, in implementing the Oil Pollution Act in the early 1990s, defined *oil* as "anything that causes a film or sheen on water." This subjected any product that fit this definition to new rules on the shipments of hazardous goods, including milk, suntan lotion, and salad oil! The definition was eventually changed, but these examples underscore the point that providing definitions is more than just a philosophical ivory tower game of no consequence to business, industry, and the public.

Abortion discussions and decisions have involved a stipulative definition for the phrase "legal abortion." Because brain activity in the fetus can be detected after the first three months of pregnancy, many supporters of abortion have advocated that legal abortion should mean "within the first trimester," within the first three months of pregnancy. When these supporters say that women should have the constitutional right to legal abortion, they do not mean at any time during pregnancy. For these supporters of legal abortion, given that the mother is healthy and that the fetus is developing normally, there is a legal and moral difference between an abortion within the first trimester and one at eight and one-half months that terminates the fetus's development, just as there would be a legal and moral difference between an abortion within the first trimester and the discarding of a newborn baby in a trash can. Proponents have argued that even though three months is relatively arbitrary, there are good reasons for choosing a cutoff point around this time: That given the general right of a woman to control her own body and the general uncertainty of when life begins, and the physiological knowledge that brain waves can be detected in a fetus after the first trimester, three months is argued to be a reasonable legal and moral cutoff point.[1]

The important point about providing definitions is that even if we need to acknowledge the appropriateness of allowing for a great deal of flexibility in providing meanings for terms, we can still recognize the need for some flexible rules of thumb that highlight examples of unfairness in the use of words. For instance, the old television sitcom "All in the Family" showed a scene in which the star, Archie Bunker, a rather narrowly cultured man in his fifties who had only a sixth-grade education, was sitting at his kitchen table allegedly studying for a history test as part of his attempt to finally get his high school diploma. Rather than studying for the test, however, he was actually practicing a cheating technique that he planned to use. He was wearing a long-sleeved shirt, and under one sleeve was a roll of paper with key terms and dates on it. He was practicing to see if he could slyly slide the paper out without it being seen. When his daughter, Gloria, walked by and saw what he was doing, the conversation went like this:

EXAMPLE 2-8 Gloria: "Daddy, are you going to use those notes on your exam?"

Archie: "I certainly am, little girl."

---

[1] Those who oppose abortion, of course, object to giving any meaning to the phrase "legal abortion."

Gloria: "But Daddy, that's cheating!"

Archie: "It certainly is not, little girl. Cheating is when you are supposed to give something to someone and you don't. I'm supposed to give them the right answers, so I'm going to give them the right answers."

Gloria: "But, Daddy, you're not being honest!"

Archie: "I certainly am, little girl. I asked myself an honest question, 'Could you pass this test without these notes?' And I gave myself an honest answer, 'No you could not.'"

Clearly, Archie is not being fair in his use of the words *cheating* and *honesty*. His argument is a classic example of what logicians call **equivocation**, the inconsistent use of a word or phrase. Equivocations are often deliberately used to create humorous situations, as in old movie comedies when a butler tells the comedy star to "Walk this way," and the comedian proceeds to imitate the way the butler is walking rather than just follow him. But when we argue, and a word or phrase occurs in both a premise and the conclusion, the reasoning trail will not be valid if the words are used inconsistently. An illusion of evidence will be created, but with no evidence actually supplied by the premise. Clearly, Archie has not provided a good case for being honest in the way Gloria is using the term.[1]

Does the use of the word *intelligent* in example **2-7** add anything to the logical support of the abortion argument? Is it fair? When people argue, they want to be as persuasive as possible. They therefore choose words that "color" their positions in the most favorable light possible. But such coloring is not necessarily evidence. Consider this example:

EXAMPLE 2-9   Congress should support the president's request for the Peacekeeper missile. Our freedom, and the peace and freedom of the entire world for that matter, depends upon a strong military defense system.

What is a Peacekeeper missile? During the early 1980s, President Reagan wanted a reluctant Congress to fund his plan for building the MX nuclear missile. For its time, this would be an awesome weapon. It would be mobile, making it hard for an enemy to destroy before it was launched. It would be very accurate, enabling us to detonate a nuclear missile precisely over the location of an enemy's nuclear missiles. Also, each warhead of an individual MX missile would carry as many as fourteen nuclear bombs, such that once the missile was launched over enemy

---

[1] For practice, can you formalize in terms of premises and conclusion Archie's argument? Hint: Archie's defense can be reconstructed into two arguments with one conclusion being that he is not cheating and the other that he is being honest.

territory, the warhead would come apart and be able to independently target fourteen different military sites or cities. It would also cost tens of billions of dollars to construct, taking that much more money away from investment in economic development, education, and health care.

For critics of the MX missile, its construction also implied a radical and dangerous change in our nuclear defense policy. Our previous policy was called "MAD," an acronym for "Mutually Assured Destruction." As the name implies, by building up a redundant nuclear force that would enable us to destroy the major cities of a whole country many times over, our policy was to create a situation where it would be crazy for anyone to ever attack us. At one time, for instance, thirty-six nuclear bombs were targeted on Moscow, even though only one was needed to destroy the city and kill millions of people, and just one of our modern submarines could destroy more than two hundred cities. With this policy, we aimed our bombs at people, hoping that no sane leader would ever want all his or her people dead. This policy partly reflected the inaccuracy of pre-1980 missile guidance systems; a city is a relatively big place, and bombs don't have to hit it in the middle to destroy much of the city.

However, to hit reinforced concrete missile launching sites located underground, accuracy is very important. So, if we have an accurate missile that can knock out the enemy's missiles, why not aim our terrible weapons at other terrible weapons rather than at innocent people? The problem, according to Reagan critics, was the new strategy and psychology implied in this plan. From the other side's point of view, the MX missile was a "first-strike" weapon—it could destroy an enemy's missiles before they could be launched—implying that the country with such an offensive missile would then not only be interested in defending itself by ensuring nuclear war would never happen, but in winning a nuclear war as well. If you were on the other side and you knew that your enemy could now destroy the only means that you had of defending yourself, you would be more paranoid and the constraints that you had developed for not launching your own weapons would be lessened. In short, in an apparently dangerous situation you would be more "trigger happy," more likely to think "use them or lose them," to launch your weapons before they were destroyed by accurate incoming missiles. Those opposed to the Reagan plan argued that we would be spending billions of dollars to increase the danger of nuclear war, not lessen it.

Given this controversy, the Reagan administration made sure that all public discussions, speeches, and correspondence by administration officials used the term *Peacekeeper* rather than *MX missile*. What this example shows is that we can distinguish between the **cognitive** or descriptive meaning of a term and the **emotive** meaning or positive coloring added to a term. It is surely a more positive thought to think of peace than the detonation of a nuclear missile, but coloring the MX missile in this way did not offer evidence that its construction would bring about a more peaceful and secure world. For critics of the Reagan plan, the MX missile was not a weapon of defense and peace, but an expensive offensive weapon of war that would actually make war more likely, reduce economic investment and competitiveness, and produce fewer jobs and more homeless people. These critics may have been wrong, but coloring the debate with the word *Peacekeeper* was not evidence against the critics.[1]

Because politics is the art of persuasion, politicians and their speech writers spend a considerable amount of time coloring their positions in the most positive light possible. The Reagan administration referred to the Nicaraguan Contras as "freedom fighters," but to those Nicaraguans who defended the Sandanista government these Contras were often seen as terrorists who sometimes pillaged farms for food, raped women, killed innocent men, women, and children, and received illegal and immoral aid from the United States' CIA, which was interfering in the affairs of a sovereign country. When the Reagan plan to spend trillions of dollars on building a laser weapon system in space became controversial—not only because of its great cost ($6 billion per year) but also because of its questionable workability—the administration emphasized that its proposal was called the "Strategic Defense Initiative," and not "Star Wars" as used by critics. "Defense" sounds much more positive than "Wars." By the time of the Clinton administration, although the former Soviet Union was no longer our nuclear adversary, $3 billion was still being spent annually on the project, the name had been changed to Ballistic Missile Defense, and the program became part of a policy called Theater Missile Defense. "Theater"?![2]

The point is, does such coloring add anything to the clarity of the debates on these issues? Because positive coloring does not necessarily add any positive evidence, you should be on guard to separate the two.

The last example on the building of a weapon system in space also shows that sometimes it is in a politician's best interest to "neutralize" emotionally charged and controversial issues, to disguise negative features with a bland coloring or with what are sometimes called **bureaucratic euphemisms**. Consider these examples:

EXAMPLE 2-10   The United States demonstrated its military and moral superiority during the 1990 Persian Gulf war. Our surgical strikes and smart weapons overwhelmed Iraqi forces while keeping collateral damage to a minimum, and our own loses were mostly from friendly fire.

EXAMPLE 2-11   The president has fulfilled his campaign promise to the American people. With the exception of a few items of revenue enhancement, there have been no new tax increases. And this policy has worked. With the exception of a significant, but temporary, downturn in the economy we have avoided a serious recession.

---

[1] After billions of dollars had already been spent on research and development, the MX missile project was finally discontinued by President Bush and the U.S. Senate in September 1991. A year later the United States agreed to dismantle, also at great cost, those missiles already built.

[2] Incidently, our Department of Defense used to be called the "Department of War." In my state, our prison used to be called "Oahu Prison." After years of controversy on prison conditions and the embarrassing fact that three-fourths of the prisoners are of Hawaiian ancestry, it is now called the "Oahu Community Correctional Facility." In China, prisons are called "labor reform departments." Prisoners, some of them members of the pro-democracy movement, are forced to labor with no retirement benefits and no union negotiations, for an export economy for products sold to Japan, Germany, and the United States.

The front-line experience of war is a terrible thing. Firsthand experience of death and bodily mutilation forces us to acknowledge our own fragility and eventual death, and often at an even deeper level, confront the great puzzling issues of the meaning of life and death. It is not easy to get the black-and-white support needed to fight a war efficiently if people are emotionally confused and terrified. So our politicians and military commanders are trained to use bland phrases such as "collateral damage" to neutralize the fact that in any war vast numbers of innocent people will be killed who unfortunately lived in areas next to (lateral to) bomb sites. It also helps support the voters moral self-image to think that these people were killed as the result of a "surgical" process, implying that the military were doing the best they could not to kill them, even though they knew they would anyway. The politicians and military commanders also don't want to engender emotional confusion by describing the accidental killing and maiming of our own twenty-year-old boys by their own "smart" weapons, weapons with one-track robot minds programmed to smash only targets, any targets. So, they say our boys experienced "friendly fire."

It is also hard to maintain clarity of resolve for goals unless a simple "us-versus-them" attitude is maintained. So, in the case of the Persian Gulf War, they attempted to neutralize the fact that our forces killed approximately one hundred thousand Iraqi soldiers, mostly simple farm boys who didn't want to fight anyone, and that many were killed "surgically" in a "turkey shoot" on a Basra road attempting to return to their families.[1]

The world may be an evil place sometimes, and difficult paths may need to be followed that are often only the lesser of evil paths. A strong military and the U.S. action against Iraq may be examples of these lesser evils, but bland coloring with bureaucratic euphemisms does not add to the evidence for the choice the United States made, nor does it help explain why, after brutalizing the Iraqi people in the hopes of having them rise up against their evil leader, we then "backed off" and let that leader stay in power. Cynics claim that we needed stability in the region to ensure stable oil supplies, and a civil war would jeopardize that stability. This charge may have been wrong, but reference to surgical strikes and smart weapons should not be confused for evidence of moral superiority.

The favorable coloring of economic conditions and the actions used to deal with them is also high on the list of political strategies. Politicians do not get elected by promising taxes and economic hardships. Walter Mondale was clobbered by Reagan in the 1984 U.S. presidential election after candidly telling the voters that raising taxes would be necessary to reduce the government's enormous deficit that otherwise must be paid by future generations. George Bush learned this lesson well by dramatically stating repeatedly during the 1988 presidential elections, "Read my lips, no new taxes." When by 1990 taxes had been raised on numerous items that affected the entire population, the Bush administration was careful to refer to these increases

---

[1] Only 7 percent of the bombs used on Iraq were so-called smart bombs, even though militarily screened and censored media reports left the impression that all were. There are estimates that as many as fifty thousand Iraqi civilians were killed, mostly women and children. By December 1991, more than a year after the spectacular media coverage, a Pentagon report admitted that one-fourth of all United States casualties died from "friendly fire."

as "revenue enhancements." During this same period of time, when the country seemed to be in a recession—when economic indicators on employment and business activity based on traditional stipulative definitions of the term *recession* seemed to have been reached—administrative officials were careful not to use the "R-word" and instead referred to a "significant downturn" in the economy.

These examples are not meant to imply that we know that the economic and military policies of political leaders are always suspect. But their policies are always controversial, and clarity of debate is achievable only if we get to the real issues, only if we are sensitive to the words used and can peel away the surface dazzle of positive coloring or the obscuring screens of neutralized coloring and understand the positions of both sides of an issue before we judge them. For more examples of bureaucratic euphemisms, here are some famous cases: "termination with prejudice" (CIA reference to killing someone); "health alteration squad" (also used by the CIA, in this case to refer to an assassination team);[1] "enhanced radiation device" (U.S. military reference to a new type of nuclear bomb, a neutron bomb that would kill more people, but would not be as destructive to physical facilities).

While we are on this subject, note how lawyers sometimes create a favorable perspective for their clients. In August of 1991, the news media began covering a story of the Fort Lauderdale wife of a Broward County sheriff. She had been arrested for prostitution and having sex with more than fifty different men, many of them prominent figures in the surrounding communities. However, her lawyer argued that because she was a nymphomaniac and her husband was impotent, her action was not prostitution but "sexual surrogacy." Lawyers also help defend medical practitioners against suits with euphemisms for death. When an anesthesiologist made a mistake and a patient died, a report of the event referred to a "substantive negative outcome" for the patient. Other doublespeak phrases sometimes used are: "therapeutic misadventure," "diagnostic misadventure," and "negative patient-care outcome." And let's hope that if you need an operation your doctor will not need to write on your chart, "Failed to fulfill his wellness potential."[2]

But perhaps the academy award for an unjust euphemism should go to the Japanese government and its description of the use of young Korean girls as forced prostitutes to service Japanese troops during World War II as "comfort women." During the Japanese occupation of much of Southeast Asia as many as two hundred thousand young Korean girls were taken from their families. The girls were as young as eleven, and the Japanese purportedly preferred young, intelligent girls who did well in school. In harsh jungle conditions each was made to sexually service sometimes as many as sixty men per day at special "comfort stations." More than 90 percent of these young women died, and those who eventually were able to return to their homes were most often then rejected by their families and lived the rest of their lives as outcasts.

---

[1] In the 1970s during the Carter administration, after a great deal of controversy over the meaning of such terminology, the CIA was banned by law from engaging in any covert activity that would involve killing foreign leaders.

[2] For more examples of euphemistic doublespeak see William Lutz, *Doublespeak: From "Revenue Enhancement" to "Terminal Living": How Government, Business, Advertisers, and Others Use Language* (New York: Harper & Row, 1989).

In terms of the treatment of women throughout history, it is worth noting that no criminal law against male infidelity even existed until an 1810 French law, and this law only prohibited a married man from keeping a concubine in his house against his wife's wishes. Women have been killed, shamed, tortured, branded (literally tattooed by a husband to indicate a date of intercourse to ensure the paternity of his offspring), circumcised (female circumcision—a euphemism for removal of the clitoris or most of the external female genitalia to reduce female interest in the possibility of extramarital sex), and infibulated (the barbaric sewing up of a woman's labia majora nearly shut, so as to make intercourse impossible)— all in the purpose of fidelity to men.[1]

These examples should be sufficient to make you realize that the use of words is not a neutral activity in arguments, that a great deal of bias, slanting, and unfairness occur in the use of language and can interfere with objectively following a reasoning trail. It should also help you realize that being logical and emotional are not inconsistent human traits. That sometimes the rational thing to do is to be outraged over issues of great concern to us. Emotions do not always intrude upon the reasoning process. Feelings can act consistently with our reasoning skills and often help guide them.[2] Compare the statements: "The Japanese government used comfort women and stations during World War II to boost the morale of its soldiers" and "The Japanese government participated in institutional gang rape of children during World War II." Which statement is the most fair, the most accurate, the most rational? How should one have reacted to statements regarding "ethnic cleansing"— the murder of men, women, and children of a rival religion or ethnic identity— during the early 1990s war in Bosnia? It would not have been illogical to react with moral indignation and outrage.

Even an apparently straightforward descriptive statement can contain an entirely slanted world view. Consider this statement: "In the future, technology will be advanced to the point that astronauts will be able to take their wives and children with them on long space voyages." Do you see the not-so-hidden stereotyping? In the future won't some astronauts be women?

Before we leave this section, note one more common hurdle to the task of understanding prior to judging—the use of a jargon style. Here is a typical example:

EXAMPLE 2-12   The belief in an ontologically homogenous relation between statements about UFOs and extraterrestrial life is not evidentially probative. No one who takes the scientific evidence seriously should believe that we have been visited by intelligent extraterrestrials.

---

[1] For these and other examples of cultural double standards in extramarital sex see Jared Diamond, *The Third Chimpanzee: the Evolution and the Future of the Human Animal* (New York: Harper Collins, 1992), 85–98.

[2] For recent scientific evidence supporting the claim that emotion and reason often work together in rational human beings, see Antonio R. Damasio, *Descartes' Error: Emotion, Reason, and the Human Brain* (New York: Grossett-Putnam, 1994). Descartes is the author of the famous statement, "I think, therefore I am." Damasio presents neurological evidence that we should replace this statement with "I think and feel, therefore I am." People who have damage to areas of the brain that affect their ability to produce emotions also lose the ability to make rational decisions.

The use of such language is often a deliberate act of positive coloring in the sense that we are supposed to be left with the impression that anyone who can use such impressive language must know what they are talking about. We are so overwhelmed that we forget that we must understand the argument and its statements before we judge them. The first statement in **2-12** says only that the evidence linking statements about UFO sightings and real ETs is weak. So, all this argument amounts to is the claim that we should not believe that we have been visited by ETs because the evidence that we have is weak. Thus, rather than just accepting this argument because it is packaged in such an impressive style, the appropriate question to ask, and the appropriate reasoning trail to follow, would be to determine the nature of this "weak" evidence. The evidence might not be "evidentially probative," but the use of such an imposing phrase does not tell us much.

Later in this book we will learn how to use the techniques of formal symbolic logic to "straighten out" and clarify torturous statements. For instance, interpret this statement: "If it is not true that marijuana is not widely used or as harmful as alcohol, then we should follow the example of the state of Alaska and decriminalize its use." Although with a little thought you should be able to do this now, later we will see how to clarify this statement mechanically to mean, "If marijuana is widely used, but not as harmful as alcohol, then we should follow the example of the state of Alaska and decriminalize its use."[1] Here are some more examples that we will learn how to mechanically translate into a simpler style. Can you restate what these sentences are saying in a simpler style?

> **EXAMPLE 2-13**   "I went into the meeting not believing it would be illegal not to tell Congress the truth." Oliver North's answer to lawyer's question at his trial for his Iran–Contra role.

> **EXAMPLE 2-14**   It is not true that you can pass the final and not pass the course.

> **EXAMPLE 2-15**   Only if we don't make the car payment this month will we have enough money for both the medical bills and basic necessities.

> **EXAMPLE 2-16**   The view of the arbitration board that the fine against Johnson was justified but the suspension from the team was not is mistaken.

> **EXAMPLE 2-17**   It is not true that not being in the neighborhood on the night of the crime is a sufficient condition for not having knowledge of the crime.

What is Truth?

Up to this point in our discussion of meaning and reasoning we have been using the word *true* as if it were clear and uncontroversial what we were talking about. But

---

[1] Notice that in a full discussion of this issue, the term 'decriminalization' would need to be defined. In Alaska a stipulative definition was used, such that residents were allowed to legally have up to 4 ounces of marijuana for personal use.

obviously the fact that people disagree so often indicates widespread disagreement not only about what is true, but even over a definition of truth, over a conception of what it is, so we can know when we have it.

This section will be a little difficult, but in the interest of honesty and completeness we need to talk about truth. You may require a lot of guidance in this section from your instructor, but keep in mind that some things you read are difficult because life is difficult, and you should not expect to understand everything you read immediately. Many writings on philosophical topics need to be read and reread, discussed, and contemplated as part of an ongoing process. Just as we do not live in a black-and-white world, we do not always find closure, agreement, or complete understanding on every topic even after a lot of thought and discussion. In philosophy, the goal often is not the answer to a question, but an understanding of the significance of the question. Understanding the significance of a question can be just as valuable as having answers, because once the importance of a question is grasped a whole new perspective opens up. And from that point on, new ideas and experiences are possible. Philosophers disagree on what truth is, but by the end of this section you should understand (although controversial) how the term is used in this book, and see the relevance of this question for the major issues of our time.

First, a little history. The ancient Greek philosopher Plato (427–347 B.C.) was very concerned about how often people disagree and the negative implications this had for political harmony and the good life. With this in mind, he felt that the only statements that deserved the name *true* would be **self-evident** truths—beliefs that just by thinking about them convey an absolute assurance that they are true and that any alternatives are false. For Plato, the statement "There are nine planets in our solar system" would not be considered true in this sense. We could be wrong about this now—we could discover a tenth planet tomorrow—or something could change such that the belief was no longer true in the future. One of the nine planets could be destroyed by some cosmic catastrophe.[1] For Plato a truth that we could doubt, or was only partially true, or was only temporary did not make any sense. Truth was a definitive notion. It was a categorical notion; a statement was really true or it was not. For Plato, we could not say we had knowledge about anything unless we were *certain* about what we thought we knew. So, for Plato only statements such as "2 + 2 = 4" and "There are no round squares" would qualify as really true. We just know these statements are true by thinking about them; it is self-evident that 2 + 2 equaling 5 and a round square are impossible.

For Plato, another example of a self-evident truth would be "The shortest distance between two points is a straight line." Plato believed that we may learn about such self-evident statements from our experience, but once we do, we seem to have an insight into an "essence" that will be durable beyond anyone's experience and any human lifetime. Unlike the planet example, no future experience by any member of the human race, according to Plato, could show that 2 + 2 is not 4. If we all die, or even if the human race had never existed, Plato argued, the shortest distance

---

[1] If Plato were alive today he might cite as evidence the fact that during his own time there was widespread agreement that only five planets existed other than the Earth. Thus, he would argue widespread agreement over statements such as this does not guarantee truth. Something more is needed.

between two points would still be a straight line and 2 + 2 would still be 4. These would still be objective truths, whether or not there are any subjective human beings around.

Ever since Plato proposed this conception of truth, Western philosophers have had to grapple with whether there really are such self-evident truths common to all people and cultures, and whether, even if there are self-evident truths, there can be any about those things that are really important to us. Can there be any self-evident truths about what we should value in life, what we should believe about the meaning of our existence, whether God exists or not, and so on? Because of Plato's great influence, many Western philosophers have spent a lot of time attempting to derive important beliefs about values and meaning in life from alleged self-evident beliefs.[1] Some influential contempory philosophers see this attempt as futile and actually quite biased. A philosopher starts out as part of a particular culture, searching for self-evident beliefs that will justify all the important beliefs that people of this culture accept. After constructing an elaborate philosophical system that serves as a foundation for this culture, small wonder that the process is judged a success. But the next generation of philosophers, or those from another tradition, easily spot the transparent ethnocentrism of this alleged success, and then proceed to construct their own grand system. All of this would be relatively harmless if these philosophers had not been influential. Unfortunately, most of these philosophers have been reckoned "great" philosophers by key members of their culture, and, according to modern critics, their systems have been used as powerful rationalizing justifications for the actions of the kings, queens, and political shakers of history. It is easy to go to war, and establish inflexible institutions to enforce a conception of normalcy and justice, if you firmly believe that your way of life is based on a self-evident, essential foundation.

Another view of truth is offered by philosophers who call themselves **modern realists**. These philosophers have been influenced by the success of science in Western culture. The success of science makes it easy to believe in the existence of an objective physical world that we learn more and more about through the ages. Therefore, these philosophers offer what is called a **correspondence theory of truth**. According to this view, beliefs are true if we have a reasonable basis for believing that the terms used in our beliefs correspond at least approximately to real things in the objective world. So, when we say "There are nine planets in our solar system" or "The Earth is spherical" we point to the reasonable evidence that these beliefs are true. When realists are asked how we can define "reasonable evidence" they usually point to the methods of science and say that these methods guarantee that over time our beliefs will correspond more and more to the real world.

Unlike Plato, most modern realists claim to be **fallibilists**. They don't claim that our beliefs can ever be self-evident. The methods of science ultimately are founded upon **empirical evidence**. We must use our experience of the world to justify our beliefs, and because our experience is never complete, even our best established beliefs could be wrong. If we keep observing nine planets when we investigate our

---

[1] We can see the influence of this attempt in the U.S. Declaration of Independence: "We hold these truths to be self-evident. . . ."

solar system, then it is reasonable to believe that there really are nine planets. Furthermore, the realists argue, even though we could be wrong, our best established beliefs must be at least close to the real truth, they must reflect at least in some partial way an actual reality, otherwise it would be a miracle that our beliefs worked at all. Even if we discover that there are ten planets, the belief that there are nine planets would at least be approximately true. Realists consider their theory simply a commonsense theory of truth.

One of the major problems with this alleged commonsense view of truth, however, is that we have many important beliefs, such as our value judgments, that contain parts of sentences that do not seem to correspond to anything physical. In the sentence, "John is a good man," what does the term *good* correspond to? The difficulty of answering this question has led philosophers to either postulate some sort of mystical, nonphysical reality in which objects like good "hang out,"[1] or, for those philosophers who fancy themselves more down to earth, to reject values altogether as involving anything entitled to the name *objective*. Thus, the mystic philosophers have the problem of justifying to the rest of us some sort of hidden, nonphysical reality beyond the commonsense world we experience every day, and the down-to-earth philosophers must relegate some of our most important beliefs to statements about a secondary, subjective illusory reality.

Other criticisms of realism are: (1) How can we ever obtain a kind of "above-it-all," godlike view to see whether our beliefs indeed correspond to an objective reality? How could we ever somehow get outside of our minds and cultures and somehow look "down" from an above-it-all perspective at some divide and see whether our beliefs on one side correspond to reality on the other? Won't we always, so to speak, take our beliefs, minds, and cultures wherever we go? (2) As a symptom of this problem, the methods of science and the resulting beliefs seem culturally influenced. Different cultures at different times seem to have had different conceptions of what scientific meant, and this in turn produced different beliefs about what we should accept as true about the world. At one time it was thought sufficient to justify the claim that the Earth was spherical, because a circle was thought to be the most perfect mathematical object. It was thought that if reality is of divine origin, then it would be simple and elegant, and a circle is much simpler to work with than some other geometric shape. Today, such a claim would not be considered scientific; some sort of observational evidence would be needed, such as pictures from space. Shortly after Newton's theory of gravity became well established (the early eighteenth century), most scientists of this time credited only beliefs that contained observable entities as scientific. So, a belief that contained the word *atom* would not be considered scientific. Today, scientists theorize about a whole host of unobservable entities, including atoms, subatomic particles, and quarks.

For these reasons, some contemporary philosophers, who think of themselves as part of a "postmodern" philosophical movement, believe that both Plato's quest

---

[1] Although not a realist in the modern sense, Plato, for instance, believed in a nonphysical reality in which self-evident thoughts and concepts existed. Our thoughts about good would be true if they corresponded to concepts in this realm. Plato also ended up believing that the physical world was not even real; that it was only a kind of "shadow" of this other realm.

and that of the realist are not only futile but in fact have caused terrible harm. They argue that there is little difference in citing a divine or self-evident right to impose one's views and way of life on others and doing so by citing so-called objective evidence that one's views are the most reasonable. Narrow ways of life, self-serving power definitions of normalcy, inflexible institutions, and cultural imperialism are the result in either case. These philosophers argue that if you believe you are really right about something, either in the sense of self-evidence or objective evidence, then any kind of action becomes possible to impose those beliefs you believe are right. The solution, then, is to eliminate once and for all notions of truth in terms of "really right." In other words, the most honest and humane approach is to eliminate all notions of truth as objective!

In its place these philosophers offer a pragmatic, consensus, or **coherence theory of truth**. These philosophers recognize only beliefs that people have and that these beliefs can only be justified by other beliefs that people have. Situations never exist where a belief can be checked directly with a so-called objective world. There are only situations where one belief is checked against other beliefs to see if they are consistent with each other. If we believe that the Earth is spherical, and we offer as evidence space shuttle astronauts who look out the window of their spacecraft and take video pictures for the evening news that show a spherical Earth, we are not showing a real physical spherical Earth that corresponds to the belief that the Earth is spherical. Instead, we are comparing one belief with many others. We are comparing the belief that the Earth is spherical with the beliefs: (1) From this vantage point in space under the condition of being in the shuttle, the Earth "appears" spherical; (2) This appearance is reliable because we believe that our video cameras are reliable, that our beliefs about how reflected light works in outer space are reliable and nothing "funny," such as distorting optical effects, happens to light in space; (3) These beliefs are reliable because they are consistent with other shuttle pictures, pictures taken with different cameras by astronauts who went to the Moon, and pictures from other spacecraft, such as the Voyager spacecraft that went to the outer planets, and the Galileo spacecraft that flew by the Earth twice on its way to the planet Jupiter.

For these philosophers, our perspective is still a human perspective; we are not outside of our minds and beliefs when we take and interpret these pictures. And no matter how many particular beliefs of appearance are consistent with our general belief that the Earth is spherical, we are assuming (another belief) that nothing about the way light rays operate in space would distort these pictures.

There has not always been a consensus on this belief about light rays in space. Prior to the sixteenth century, some astronomers believed that light rays and perceptions based on them would be altered in space. They believed that reality was divided into two spheres, an earthly terrestrial sphere and an ethereal, god-like, or celestial sphere. On Earth, physical processes occurred according to certain physical laws, but these same laws did not apply to motions of objects in space, because the space beyond the Moon was considered to be a different realm of reality. It was thought absurd to believe that processes in a pure, god-like celestial realm would obey the laws of motion of an impure, earthly realm. The Moon, Sun, and planets were not even considered physical places. For this reason, some important people of this time had a hard time believing the results of Galileo's telescopic observations, which seemed to show that the Moon had mountains and that Jupiter had moons of its own. That the telescope worked on Earth, that it revealed distant details of physical objects, proved nothing to those who believed that the telescope would not work in a celestial realm because this celestial realm contained no physical objects in the first place. As evidence for this, those who objected to Galileo's telescopic interpretations would point to the fact that stars appeared smaller when viewed through a telescope. This provided clear evidence, they thought, that reality operated differently in the different realms and that the telescope distorted celestial truths. So, before Galileo's telescopic observations could be considered convincing, other beliefs had to change. People had to become convinced that there was no celestial realm, that the planets were physical places like that of Earth, and that the laws of nature operated in outer space as on Earth.

According to the pragmatic or coherence theory of truth, this shows that any particular belief cannot be considered in isolation from other beliefs. Every belief derives its meaning from being part of a "web of belief." What we observe through a telescope may **stimulate** a new belief, but only other beliefs can be used to **justify** that belief. If we are open-minded and observant, we are constantly receiving novel stimuli from the world as part of our experience of being human. This constant stimulation will constantly pressure the coherence of our webs of beliefs, as the novel experience of observing something through a telescope did during Galileo's time. New ideas will emerge that are inconsistent with the current web of belief, causing a constant process of tense readjustment. To keep the web of beliefs coherent will require a constant process of questioning which beliefs need be accepted and which ones rejected.

There is much more to these discussions of truth than this.[1] For our purposes, it is sufficient to note that the realists and the postmodern pragmatists seem to agree on one thing. We must dispense with the assumption in Plato's thinking that truth is a categorical notion. Instead, we must replace it with a notion that allows for relative degrees of evidence, support, or justification. Instead of viewing true or false beliefs as statements that are really just categorically true or false, we need to think of beliefs as either reliable and practical guides or unreliable and impractical guides for dealing with the physical world and each other. By a **reliable belief** we mean, then, a belief that seems to have certain features of evidence, support, or justification such that, based on past experience of evaluating beliefs, we find good reason for believing the belief will continue to work with our other beliefs and with new beliefs that result from novel experiences. For instance, as we will see in the next chapter, a reasonable person ought to believe that the statement, "Cigarette smoking is a principal cause of lung cancer" is a reliable belief, even though it could be found to be false someday.

You will probably need more examples before you understand this, and the next chapter attempts to give you concrete examples of reliable beliefs and techniques

---

[1] For further reading on these subjects, see Larrry Laudan, *Science and Relativism* (University of Chicago Press, 1990); Richard Rorty, *Objectivity, Relativism, and Truth* (Cambridge University Press, 1991); and Ronald C. Pine, *Science and the Human Prospect* (Wadsworth, 1989).

of support for them. For now, consider that what I am describing is an **attitude** toward truth as well as a theory of what it is. This attitude is consistent with the general theme of this book: our disagreements and confusions will have a better chance of resolution if we use unforced, open, focused discussions, and a free but disciplined following of reasoning trails. Because there is a little bit of Plato in all of us—we find it easier to have security, closure, and finality—it will be healthy for us to think hypothetically rather than categorically, and to question, to be open, to be humble about what we think is true. By hypothetical thinking, I mean a process that says, "We have a good reason to believe that X is true, because we have a good reason to believe that Y is true, and we have a good reason to believe that Y is true, because we have a good reason to believe that Z is true." And then, "OK, if you agree that the reasoning trail just outlined is valid, that we stayed on track and did not let our biases or any irrelevant coloring tricks distract us, let us now think together whether we need to question Z also, or do we have good reasons to stop here?"

This attitude is rare because it implies the need for a mental toughness that is difficult for most people (including philosophers) to accept. Our need for security makes it difficult to accept a constant hypothetical tentativeness where at no point along our reasoning trails can we stop and rest and say this is completely true. By denying that anything is completely true, we have to be constantly ready to think again, for if something were completely true no further thinking would be necessary.

The Platonist in the back of our minds usually wants more than this. Just because we agree tentatively on where to stop, doesn't mean we are correct about what we agree on. Lots of people agree that God exists, but they could be wrong. And lots of people agree that God does not exist, and they also could be wrong. But what is advocated here is not mere agreement. There is a difference between mere agreement that is forced, the result of a trick, or the result of unanalyzed biases, on the one hand, and agreement that is the result of painstakingly following different trails that have been open to free inquiry. Both kinds of agreement could be wrong. But the one that is the result of free and disciplined inquiry allows for more opinions and rigorous tests, and permits the novel stimuli of experience to "speak" to us. This approach may be based on a kind of faith, but I would argue it is a reasonable faith. We believe that the more a belief is made to stand up to genuine tests, the more likely it is to be reliable.

In 1616, the Catholic church decided that Galileo's ideas of a moving Earth and planets that were physical places should be prohibited from being taught and discussed. Because the church was one of the most powerful political institutions in the Western world, and because it funded a great deal of scientific research, this action was of great consequence. At the time, the church had some good reasons for believing that Galileo's ideas might be wrong. There were many unresolved issues, and even Galileo was unsure of the exact mechanism that would move our heavy Earth through space, keeping it in orbit around our Sun so far away, and keeping all of us from flying off the Earth because of its great speed of rotation. (No Newtonian conception of gravity yet existed.) But the agreement achieved by prohibiting Galileo's ideas was then the result of a forced agreement rather than a free and focused following of alternative reasoning trails. And like the father discussed in Chapter 1 who was afraid to test his daughter at Little League baseball, the Catholic church, by cutting off access to a reasoning trail, by not testing an alternative set of beliefs, was cutting off access to possible experiences of great benefit.

Bottom line time: In this book, we will be using *true belief* to mean a belief that is a reliable guide for dealing with each other and our experience (past, present, and future). By reliability we will mean a process of belief formation that is the result of both an ethical process—tolerance for different opinions, promotion of open inquiry, and undistorted communication—and a justification process: the belief in question has certain features of support that seem to have worked before in showing us beliefs likely to be reliable in the future.[1] With this in mind, we must now examine what it means for a belief to have certain features such that it is likely to continue to be reliable in the future, what it means to say that a belief is reliable given the evidence. How do we establish long-term reliability? This will be the subject of Chapter 3.

Before we switch to this subject, however, we can make a quick, but important, comment on the nature of logic from this perspective of the nature of truth. Logic is defined in most logic textbooks as a study of the principles used in correct thinking. But what is the status of these principles? Are they Platonic principles? Are they self-evident, necessarily true for all times and cultures? If Plato's conception of truth is unrealistic and unjustifiable, then such a conception of logic that gives an above-it-all, timeless, inhuman quality to its principles will be difficult to justify as well. Better will be a perspective that defines logic as a practical set of principles, or a shared set of thinking tools painstakingly derived from history and experience, that we have found beneficial for arriving at unforced agreement through the disciplined following of reasoning trails. Techniques of logical analysis are habits of adaptation that work in keeping us on track when we want to test our beliefs in an arena of open and respectful inquiry.

Key Terminology

Arguments
Premises
Conclusion
Premise indicators
Conclusion indicators
Descriptions
Explanations
Value judgments
Relativist
Socrates

Protagoras
Normative ethics
Vague usage
Ambiguous usage
Denotative meaning
Connotative or intentional meaning
Lexical definitions
Stipulative definitions
Cognitive meaning
Emotive meaning

---

[1] Philosophers call the study of this justification process *epistemology*. Literally, the study of, or justification of, knowledge.

Concept Summary

In appraising reasoning trails logicians slow our thinking down by formalizing such trails into **arguments**. Rather than just reacting quickly to an act of persuasion, logicians break the reasoning down into parts to see if the parts fit together well. First, we look for the main thrust of the argument, called the **conclusion**, and then identify the evidence offered, called the **premises**. Sometimes, indicator words help us in this process. Words such as *because* and *since* often indicate premises, and words such as *therefore* and *so* often indicate conclusions.

Not all uses of language involve arguments, so we must first distinguish arguments from other uses, such as descriptions, explanations, and unsupported value judgments. **Descriptions** can be true or false, but until an attempt is made to prove that what is described is true, no argument is involved. **Explanations** often use the same indicator words as arguments, but what follows apparent premise-indicating words such as *because* is a debatable set of reasons why an agreed-upon event happened. Explanations attempt to give reasons why something happened; they don't attempt to prove that something happened. In expressions such as "X, because Y," X is accepted in an explanation, whereas in an argument X is a controversial conclusion that is being argued for. **Value judgments** are also not arguments, but they usually express concerns of great importance to us, and so they are often parts of arguments. **Relativists** believe that it is impossible to rationally argue for the acceptance of one value judgment over another. **Normative ethicists** believe that it is possible to argue for value judgments; that some values are more rational to have than others.

In argument analysis it is also important to pay attention to the meaning of words, otherwise we can easily be tricked into believing evidence is presented when it is not, or that evidence is stronger than it actually is. **Vague** or **ambiguous** usage of words easily invite incorrect inferences. Much can be assumed unfairly in the uncritical acceptance of particular meanings of words. The acceptance of definitions is linked with beliefs, and hence can be controversial and is often in need of independent argumentation. Words can also be used to emotionally "color" controversial persuasive appeals, but such coloring must be distinguished from evidence. From this it should not be inferred that emotion lacks a proper role in deciding proper courses of action. Often, politicians and bureaucrats use a neutralized coloring to dull or desensitize matters of great concern to us. Such neutralization also should not be confused with evidence. **Denotative** and **connotative meaning**, **cognitive** and **emotive meaning**, and **lexical** and **stipulative definitions**, are some of the distinctions logicians use to regulate the fairness of language use in argumentation. The important point to remember is that bias, slanting, and unfairness in the use of words can interfere with the objective appraisal of a reasoning trail. We sometimes must be very picky and pay close attention to the words used in an argument, to get to the real issues that are often obscured behind the surface dazzle or neutralizing screens of tricky word usage.

At a deeper philosophical level, because we have distinguished between a reasoning trail and the truth content of that trail, we must also pay some attention to the meaning of the word *true*. The nature of truth is a great, perennial philosophical issue. The decisions philosophers have made about this concept have not just been part of a detached, philosophical, ivory-tower game, but have influenced the movers and shakers of history. In this book, I claim that a judicious and unpretentious conception of truth is assumed, one that sees our goal as accepting beliefs that are **reliable** or practical guides for life. By following disciplined reasoning trails and encouraging unforced, open discussion, it is possible to identify features of evidence, support, and justification for some beliefs such that these beliefs should be accepted as more likely reliable than others. This conception of truth implies an experimental attitude, a willingness to accept fallibility and self-correction, of valuing hypothetical thinking and tentative acceptance, rather than categorical thinking and a search for absolutes. From this perspective, logic is likewise viewed as a practical set of principles derived from past experience, principles that help us stay on track when we want to test for reliable beliefs.

EXERCISE I   True or False

1. Before a trail of reasoning can be logically appraised it must be formalized as an argument.

2. A logical argument is the same thing as a verbal quarrel.

3. In formalizing arguments there is always one absolute, black-and-white interpretation.

4. In analyzing reasoning trails into arguments, the word *because* is a premise indicator.

*5. The word *because* is always a premise indicator.

6. The word *so* is often a conclusion indicator, but it is not always a conclusion indicator or an indication that an argument is involved.

7. In an explanation, what is to be explained is less controversial than the explanation.

*8. Clarifying the meanings of terms is not linked with our beliefs about what is true.

9. The goal of logical analysis is to avoid all creative and emotive uses of language. Being logical is always inconsistent with being creative or emotional.

10. The goal of clarifying terminology is to provide understanding prior to judging.

11. In this book, a true belief will mean only those beliefs that are self-evidently true.

12. In this book, it is assumed that being logical is good, because logical analysis helps provide a focused way of testing our beliefs and arriving at reliable beliefs.

EXERCISE II

Determine which of the following passages contain arguments and which do not. For those that are arguments, structure each into premises and conclusions. For those that are not arguments, label each as a description, explanation, or value judgment. Explain (make a case for) your answers.

1. Last summer I visited Great Britain and France. This summer I plan to visit Japan and Taiwan. When this trip is completed I hope to make a decision about which country to specialize in for my international relations degree. I'm having fun and am really in no hurry to make up my mind.

*2. Answer from military spokesman to a news reporter's question during Persian Gulf War concerning why the Stealth fighter operates in complete radio silence during wartime: "The stealth fighter operates in complete radio silence in wartime, because it flies at night at high subsonic speed, can't be picked up on radar— and therefore is not monitored by AWACS early-warning radar planes."

3. If there is any implication that is clear from the U.S. Constitution it is that each of us has a right to control the processes that take place in our own bodies. This right to self-determination obviously covers everything that takes place in our bodies. A medical doctor can advise me on when to have an operation, but he or she cannot force me to have one. A woman has this same right to self-determination. Hence, abortion should be legal.

4. The government has a duty to protect the individual rights of all citizens. Because an unborn baby is as much a citizen as anyone else, the elimination of this citizen's life is tantamount to murder. Therefore abortion should be illegal.

5. People may think that I am strange, but I don't like ice cream and I don't like candy. I also don't like desserts. I also don't like sugar or cream in my coffee, and can't stand the sickly sweet taste of sodas.

*6. The greatest emotion that can be experienced by a human being is that of love, be it the bond between a parent and a child, two romantically involved people, or the brotherly feeling toward humankind in general. There is no higher value, no greater achievement, and no greater sense of fulfillment in life.

7. The latest test results on the educational achievement of U.S. students have again demonstrated poor performance compared to Japanese students of the same age. This is not surprising, because Japanese students know far more about the world, are more proficient in science and mathematics, and are more often fluent in a second language. The reasons for this are obvious: Japanese students watch far less TV per week than U.S. students, and Japanese teachers are paid and honored a lot more. In Japan, teaching is a venerable profession; in the United States it is better to be a basketball player and make a million dollars a year than to be a teacher and make only thirty to forty thousand.

*8. A human being becomes truly happy only when natural potential is developed. Because a feeling of other-directedness rather than a sense of selfishness is a natural potential that seems to be developed after basic necessities are met, and because the unconditional love for another is an act of other-directedness, it follows that we ought to value and encourage conditions that promote love.

9. Most cable television seems to offer nothing but sex, violence, and teen-age self-indulgent music. The TV in every American house in this day and age is nothing but an idiot box offering passive cheap thrills for narrow-minded bit brains.

10. Contrary to popular opinion, from a scientific point of view it is very unlikely that extraterrestrial intelligent life exists in our galaxy. This is so for the following reasons: The Sun is a star that is about 4.6 billion years old. The Earth was formed with the Sun, and intelligent life took at least this long to evolve on Earth. In assessing another star's chances for having planets with intelligent life, the star must be at least as old as the Sun. However, roughly half the stars in our galaxy are younger than the Sun. If intelligence takes 4 to 5 billion years to develop, it seems likely that these young stars cannot have intelligent life around them yet. We should also keep in mind that at least half the stars in our galaxy are part of binary star systems, and it is highly unlikely that lifebearing planets could survive the monstrous gravitational forces of binary systems. Accordingly, only a very small percentage of stars in our galaxy are suitable for planets with intelligent life.

*11. All planetary bodies in our solar system have received their fair share of meteorite hits throughout history, and scientists use the number of still-visible craters as a rough measure of the age of the surface. New observations of Venus from the Magellan space craft indicate that Venus is relatively free of blemishes created by meteorite impacts. By contrast, Mars and the Earth's moon wear old, pockmarked faces. The Magellan space craft has detected evidence of massive volcanic activity on Venus. Thus, recent volcanic activity on Venus must have resurfaced the entire planet.

12. John must have AIDS, because all people who have AIDS constantly have flu and cold symptoms and have an abnormal white blood cell count. John has had flu and cold symptoms for some time now and has an abnormal white blood cell count.

13. For more than forty years Congress has been controlled by the Democratic party, and look at the state of the country. It is time for a change. Vote Republican this November.

14. Frogs and other species of amphibians are disappearing in alarming numbers on Earth, because they are being eaten by aliens from another galaxy who have established secret outposts on Earth.

15. For more practice in structuring arguments, and to prepare for the analysis of informal fallacies, provide an interpretation in terms of premises and conclusion for some of the section II exercises in Chapters 4 and 5.

EXERCISE III

Analyze the passages below and identify key words or phrases that are ambiguous, vague, controversial, or unfair. Explain why these words or phrases need further discussion and how they could lead to mistaken inferences.

1. The administration's latest attempt to stimulate the economy and moderate the inflation rate was an incomplete success.

*2. In order to complete the glorious victory against the American-led infidels, the brave Iraqi forces strategically withdrew toward Baghdad.

3. Famous statement from an administration official during the Nixon-Watergate scandal, after the discovery and release of secret tape-recorded conversations with Nixon and his staff: "All former statements as to the level of administrative involvement and knowledge of the Watergate break-in are now inoperative."

4. In preparation for a nuclear exchange between the United States and the Soviet Union, the United States has adopted a policy of mutually assured discontinuation, whereas the Soviet Union has adopted a policy of counterforce and anticipatory retaliation.

*5. Famous statement from Betty Ford, wife of former president Gerald Ford, on why she was at a special hospital for drug abuse: "I over-medicated myself (with alcohol and valium)."

6. The misadventure of the termination of friendly force's live function was due to some incontinent ordinance.

*7. All intelligent people living on the brink of the twenty-first century, concerned with the moral growth of our nation, know that reproductive freedom for women is crucial for fulfilling the dream of our founding fathers. We should support on-demand abortion.

8. The antiabortionists say that we are murderers. But as pro-choice advocates, our goal is the defense of freedom and liberty.

9. The abortionists say that we want to restrict the freedom of others. But as pro-life advocates, our goal is the defense of the right to self-determination for those who cannot yet speak on their own behalf.

10. A 1993 statement from the managing director of the French Atomic Energy Commission urging a resumption in nuclear testing on Mururoa Atoll in Polynesia: "From a scientific and technical point of view, the resumption is necessary to maintain the means and the development of nuclear dissuasion."

EXERCISE IV

Essays. Write a descriptive essay addressing each of the following.

1. In July of 1991, Jeffrey L. Dahmer, a convicted child molester, was arrested for murder. At his apartment parts of eleven bodies were found. Police found skulls in a file cabinet and a closet, three headless torsos in a vat, three heads in a refrigerator, boxes filled with body parts, and a dresser stuffed with photos and drawings of mutilated bodies. Neighbors said that they experienced an overpowering stench coming from the apartment for months, and that from time to time they heard what sounded like a buzz saw and screams. Dahmer eventually confessed to the murders, plus cannibalism. From a moral point of view, explain how a relativist and a normative ethicist would interpret the actions of Dahmer.

2. This chapter contains several controversial passages and analyses of examples. In your judgment, are any of the words or phrases that I have used ambiguous, vague, controversial, or unfair? If so, identify them and explain why you think they are unfair.

3. Explain the difference between truth as a categorical notion and as a hypothetical notion. Include in your essay a discussion of the following conceptions of truth: Plato's notion of self-evidence, the realist's correspondence theory, the pragmatist's coherence theory, and the notion of beliefs as reliable guides.

ANSWERS TO STARRED EXERCISES:

I.

5. False. The word *because* can also be found specifying an explanation.

8. False. Meaning is linked with beliefs and the world views of language using groups. Hence, the uncritical acceptance of defini-

tions of terms is as presumptive as the uncritical acceptance of beliefs.

II.

2. Explanation. Although the words *because* and *therefore* are used in this passage, the fact that a Stealth fighter operates in complete radio silence seems accepted and an explanation is being offered for it. Because we don't know for sure the intent of the reporter's question, it is possible that this is an argument and the spokesman is attempting to justify the radio silence. The radio silence is agreed upon as occurring, but whether it is controversial or not is not clear. Most likely it is not controversial and is just being explained.

6. Value judgment. Although controversial, there is no attempt to justify love as the highest value.

8. Argument. Unlike number 6, there is an explicit attempt to justify and encourage love as supreme value. The phrase *it follows that* is a conclusion indicator.

   Conclusion: We ought to value and encourage conditions that promote love.
   Premise: A human being becomes truly happy only when natural potential is developed.
   Premise: Developing other-directedness is a natural potential.
   Premise: The unconditional love for another is an act of other-directedness.

11. Argument. This passage is most likely an argument for a best explanation. That Venus has very few craters is being explained by resurfacing due to volcanic activity, and an argument for why this is the best explanation is offered. The conclusion is controversial; there could be other explanations, but evidence is being offered that this is the best explanation. The word *thus* serves as a conclusion indicator. Here, "X (Venus has few craters), because Y (recent volcanic activity)" is the conclusion.

    Conclusion: Recent volcanic activity on Venus must have resurfaced the entire planet.
    Premise: All planetary bodies in our solar system have received their fair share of meteorite hits throughout history.
    Premise: Unlike some other planetary bodies in our solar system, Venus is relatively free of craters.
    Premise: The Magellan space craft has detected evidence of massive volcanic activity on Venus.

III.

2. The phrase *strategically withdrew* is being used to color the fact that the Iraqi forces were being forced to retreat. A conclusion that

Iraqi forces were victorious should not be inferred from such colored descriptions.

5. The phrase *over-medicated* is used to imply that the wife of a president of the United States was not doing something as serious (because her drugs were legal) as that of drug addicts who become addicted to illegal drugs. How would it sound if someone said, "I have been over-medicating myself with heroin, cocaine, and marijuana"?!

7. Although the phrases *intelligent people* and *moral growth* are presumptive, the phrase *reproductive freedom* is the most in need of argumentation. Those who are for legal abortion believe that a women should be free to choose to have an abortion or not, but those who oppose abortion believe that the fetus is a person and thus has a right (freedom) of self-determination. The use of this phrase colors the abortionists position in a positive way without giving any evidence for the conclusion.

Chapter
3

INDUCTIVE REASONING AND REASONABLE BELIEFS

We should relish the thought that the sciences as well as the arts will always provide a spectacle of fierce competition between alternative theories, movements, and schools. The end of human activity is not rest, but rather richer and better human activity.

—Richard Rorty

Deduction and Induction

In the discussion in Chapter 2 on immigration, if Mr. Auster and I were able to stay focused on the relevant issues, we would surely recognize that part of our disagreement is over issues of fact, general statements of fact, or, in the language of Chapter 2, over reliable beliefs. Often deductive arguments contain premises that are general statements of fact. Here is an example:

Conclusion: Smoking should be banned from all public buildings.

Premise: Smoking is a principal cause of lung cancer and other serious illnesses.

Supplemented with other premises, such as the generalization on a connection between secondhand smoke and these same serious illnesses, and the moral commitment that no one has a right to infringe on the health rights of another in the pursuit of what they deem their own happiness, the validity of this argument should not be in dispute. What is disputed by smokers' rights groups and cigarette manufacturers is the truth or reliability of the major premise and the assumption of the danger of secondhand smoke. How do we know these beliefs are true? When I ask my students this question, I sometimes hear, "Well, you know, '*they*' say these beliefs true." But who is being referred to and how does this group know these statements are true? The more informed student will respond that "they" are teams of specially trained medical researchers who have used scientific methodologies to study the link between cigarette smoking and illnesses such as lung cancer and emphysema for many years. True, but scientists often disagree. In fact, cigarette manufacturers have their own teams of scientists who claim that the link between cigarette smoke and serious illness has not been proven.

By the early 1990s, 36 percent of Japan's total population (almost 60 percent of the men) smoked cigarettes. At a time when U. S. government policy supported the belief that smoking cigarettes is very dangerous to health and the British government was putting "Smoking Kills" labels on packages of cigarettes, Japanese government policy asserted that the link between smoking and lung cancer had not been established and the Freedom Organization for the Right to Enjoy Smoking Tobacco called all the studies on cigarette smoking "gross exaggerations." Could this be another Flat-Earth story? Are the relativists right? Can any belief be supported well by evidence? Is it possible that anything is made true simply by believing it is true or as long as it is part of a cultural tradition? Is objectivity a myth? If the Japanese government changed its policy, accepting the same policy as that of the United States and Great Britain, would this mean that the link between smoking and lung cancer was formerly false in Japan, but now it is true? Or, that smoking was always harmful, and it just took many years for the culture and government of Japan to "discover" this objective truth?[1]

Skeptics will point out that many seemingly solid scientific ideas of the past were rejected by a later generation of scientists as false. Could this happen again? In Woody Allen's movie *Sleeper*, a former operator of a New York health food store has been put to sleep and cryogenically preserved for many decades and then brought back to life. He awakes to a relativist nightmare in which many of the beliefs and values that he took for granted in the 1970s are now seen as the myths of a more primitive time. One of the more startling beliefs of this new age for a former health food devotee is that smoking cigarettes is now thought to be good for people's lungs, like taking vitamins! How likely is it that this will happen? How do we know whether what science is telling us about cigarette smoking is objectively reliable?

Our task in this chapter is very important. Not only do we need to understand and be able to recognize the nature of different inferential claims, but we need to provide a philosophical justification for the claim that in spite of life's uncertainty, some beliefs are better (more reasonable to accept as reliable) than others. Specifically, we need to show that some inductive inferential links are better than others; that there are many beliefs that a reasonable person should accept—such as the link between cigarette smoking and serious illness, that the Earth is spherical and millions of miles from the Sun, that the Universe is enormous and very old, that life on Earth has evolved and that the human species has a contingent historical past. In short, we need to flush out and defend the claim that some ideas are more reasonable

---

[1] Even if it is true that anything is made true simply by believing it true, is this not an objective discoverable truth?

to accept than others, even though it is conceivable that we can still be wrong about future reliability when we accept a reasonable belief.[1]

Let's begin with some technical definitions and recognition exercises.

Deductive arguments: Arguments where the goal is to provide conclusive evidence for the conclusion; the nature of the inferential claim is such that it is impossible for the premises to be true and the conclusion false.

Inductive arguments: Arguments where the goal is to provide the best available evidence for the conclusion; the nature of the inferential claim is such that it is unlikely that the premises are true and the conclusion false.

Here are some examples:

Deductive argument: All U.S.-manufactured cars built after 1969 were equipped with seat belts at the factory.
John has a 1972 U.S.-manufactured Ford.
Thus, John's car was equipped with seat belts at the factory.

Inductive argument: After careful observation we have not seen any hummingbirds all day in this forest.
Therefore, there probably are not any hummingbirds in this forest.

Notice how each example fits the definition. In the deductive example the author is not claiming that John's car probably was equipped with seat belts at the factory. The tone of the argument is such that we should be sure that if the premises are true, the truth of the conclusion follows conclusively. John could possibly have a 1968 Ford, or the year when all cars had to be equipped with seat belts might have been 1973. (Remember, valid deductive arguments can have false premises.) But if it were true that every U.S.-manufactured car built after 1969 was indeed equipped with seat belts at the factory, and John has a car built after 1969, then there is no doubt that he has one of the cars equipped with seat belts at the factory. Deductive arguments attempt to provide certainty for their conclusions. As you should recall from Chapter 1, when the reasoning or inferential part of this attempt is successful, as in the present example, we call such arguments valid. When the inferential attempt is unsuccessful, we call such arguments invalid.[2]

In the inductive example, the nature of the situation and the use of the indicator word *probably* in the conclusion show that the author's intent cannot be to prove beyond a shadow of a doubt that there are no hummingbirds in the forest. Other inductive indicator words or phrases would be *likely, it is plausible that,* and *it is reasonable to conclude.* Inductive arguments can provide only "some" evidence for

---

[1] Remember that the rational force of a valid argument is seriously compromised if we can't also establish in some way that the premises are reasonable. And many premises can be justified only inductively.

[2] Remember that to completely fulfill the goal of a deductive argument, the argument must be sound. The reasoning or inference must be valid, and the premises must be true.

the conclusion; the conclusion always goes beyond the evidence provided in the premises to some extent. Clearly, it is possible for the premise to be true in this example, while the conclusion is false. We could walk all day in a particular forest and examine a large part of it, and agree that we have not seen any hummingbirds. But because our conclusion is a generalization about the whole forest, including many parts we did not observe, to make the conclusion false would require only one shy, elusive hummingbird, who perhaps moves quickly from our sight every time we approach. Furthermore, notice that even if we supplemented our argument with more true premises, such as a year of daily observations of the forest by a team of trained ornithologists specializing in hummingbirds and who know the signs of hummingbird habitation—food supply consumption, use of particular nest material, fecal droppings—still something could have been missed, and the conclusion could still be false.

To state, "No hummingbirds exist in this forest" is to make what logicians call a *universal generalization*. Further examples would be, "All crows are black," and "No politicians are poor." All that is required to *falsify* a universal generalization is one contrary instance to the generalization. Any finding of a single hummingbird, a white crow, or a poor politician would be a *falsification* of the above generalizations.

Technically, all inductive arguments are invalid. It is always possible for the conclusion to be false, and the premises to be extensive and all true. Many students will often draw an overly skeptical and cynical conclusion from this realization.[1] If most of our beliefs are based on inductive evidence, where our beliefs are conclusions of inductive arguments or premises in deductive arguments, and the conclusions of inductive arguments can always be false no matter how rigorous the effort put into substantiating our premises, then we really have no rational basis for believing anything as true or reasonable for acceptance! In other words, because the definition of a sound argument in Chapter 1 (deductive valid argument, plus true premises) is only an ideal case that is never realized in the real world—because we can never prove absolutely that the premises are true—some students become skeptical about accepting any belief as reliable.

The beliefs that most people carry around are based on very weak foundations and are consequently hasty universal generalizations. Therefore, such initial skepticism is often healthy. But in the present case most students go too far. My own suspicion is that this is a "pendulum" effect. Because many students have been taught by their parents, culture, or subculture to think in very black-and-white terms—to divide up the world between true beliefs (theirs) and false beliefs (other people's beliefs that they disagree with)—they are initially puzzled or confused by the claim that a belief can have evidence for it and be reliable, but not known to be true in an absolute sense. Thus, in recognizing for the first time that beliefs cannot be supported absolutely, they often swing all the way over to the opinion that nothing must be true. This epistemological swing is a way of psychologically protecting their beliefs from criticism and themselves from confusion. If nothing is true, then all beliefs are equal and the amount of effort we put into substantiating our own beliefs is irrelevant and we can believe whatever we want. If beliefs require evidence,

---

[1] So have some great philosophers!

but still cannot be known with certainty to be true no matter how much effort is put into gathering evidence, then this seems like a lot of work for nothing. We might as well save the time for watching music videos.

As you should recall from Chapter 2, this is the philosophical position Protagoras developed many centuries ago. Sophisticated versions of relativism are still defended today. For instance, the late philosopher Paul Feyerabend at the University of California at Berkeley argued that we live in a culture dominated by a scientific "mythology." He claims that so-called scientific objectivity is a myth, that modern scientific beliefs are no better than primitive beliefs, that all beliefs can be equally supported and hence can be considered true. According to Feyerabend, a careful analysis of rational evidence shows that all beliefs can be rationalized and that "anything goes." According to Feyerabend, this result is not just that of an ivory-tower debate of no practical consequence. Substantial concerns are at stake, because our world is rapidly becoming a global culture dominated by science and technology, and hence a cultural imperialism is obliterating cultural diversity.[1] Views such as Feyerabend's are used to support postmodernist's claims that sexual discrimination, racial injustice, and culturally biased curricula at colleges are produced by an oppressive Eurocentric perspective.

Feyerabend and the postmodernists raise perennial issues of great importance, which should be debated carefully in a society that values an open exchange of ideas. If nothing else, these claims show how apparent abstract issues are intimately linked to those things that concern us the most. However, it should be apparent that relativism is not the position defended in this book. I will argue that realizing that induction is not deduction does not mean that we cannot be rational in using induction. We can have good reasons for accepting some beliefs as reasonable to use for courses of action in an uncertain world, but we do need different *standards of appraisal* to judge deductive and inductive arguments. As we will see, in using inductive arguments the judgment of the worth of the inferential claim must always be relative to alternatives, rather than against some ideal of absolute certainty. In the above example, clearly the evidence of bird specialists who observe the forest daily for more than a year is *better* than that of an untrained group of hikers meandering through the forest on only one day. If you were on a committee that needed to make an important decision on logging the forest based on whether or not a rare, endangered hummingbird species lived in this forest, which evidence would you use?

Before we go any further examining the rational basis for inductive inference, let's get some practice recognizing and separating deductive and inductive inferential claims. Decide if the following arguments are either inductive or deductive arguments.

EXAMPLE 3-1 All Republican presidents have been in favor of a strong military.
President Bush was a Republican president.

---

[1] See Feyerabend's *Against Method*, his *Science and a Free Society*, and his *Farewell to Reason*.

It follows that President Bush was in favor of a strong military.

EXAMPLE 3-2   John is on the softball team and has short hair.
Dan is on the softball team and has short hair.
Kenji is on the softball team and has short hair.
It seems likely that all the members of the softball team have short hair.

EXAMPLE 3-3   All members of the softball team must have short hair.
Jay has short hair.
Therefore, Jay must be on the softball team.

EXAMPLE 3-4   If Hansen is the serial murderer, then his fingerprints would be on the gun.
His fingerprints were on the gun.
Therefore, it is clear that Hansen is the serial murderer.

EXAMPLE 3-5   Most presidents of the United States did not die in office. Therefore, it is doubtful that the twelfth president of the United States died in office.

Here are my answers. Example **3-1** is a valid deductive argument. The author is not attempting to persuade us that it is only likely that Bush was in favor of a strong military, but rather that the conclusion follows conclusively from the premises. Example **3-2** is an inductive generalization. From the particular cases of a few members having short hair and being on the softball team, the author generalizes that all the members of the softball team are likely to have short hair. Example **3-3** is an invalid deductive argument. Although the word "must" can be ambiguous, suggesting either high probability or conclusive necessity, it appears that the author of this argument is attempting to provide conclusive evidence for the conclusion, thinking that the premises are sufficient for knowing that Jay is on the softball team. If the author intended only something like "Jay is probably on the softball team" for the conclusion, then we would have to appraise this argument by inductive standards. But the tone of the argument appears to be deductive, even though the inferential claim fails from this perspective.

Number **3-4** (invalid deductive) is a good example of how easy it is to turn an apparent deductive argument into an inductive argument. As the argument is presented, the phrase "it is clear that" indicates that the author thinks that the premises provide conclusive evidence for the conclusion. They don't. It is possible that the premises are true, but Hansen is not the serial murderer. This could be a classic case of circumstantial evidence. There are many ways Hansen's fingerprints could have been on the gun. He could have picked up the gun shortly after the shooting or touched the gun in some way prior to the actual murder in which the real murderer used gloves, leaving only Hansen's fingerprints on the gun. Suppose in discussing this situation with the author of the argument, after reminding him of these possibilities, he backs off from his initial claim. He tells us that he meant it is reasonable to tentatively suspect Hansen, to bring him in as a prime suspect for questioning. If so, we would then appraise the author's argument based on inductive standards. How strong is the evidence, his fingerprints on the gun, for the conclusion that Hansen is the serial murderer? What other kinds of evidence are needed to strengthen our confidence in the conclusion?

Finally, example **3-5** shows that it is a misconception to believe that deductive arguments always move from general premises to particular conclusions and that inductive arguments always move from particular premises to general conclusions. Although this is often the case—**3-1**, **3-2**, and **3-3** fit this alleged rule—it is not always the case. Although **3-5** has a general premise (a statement about most presidents) and a particular conclusion (a statement about the twelfth president), the nature of its inferential claim is inductive. Because the premise states only that "most" presidents have not died in office, it is impossible to conclude with certainty that the twelfth president of the United States did not also die in office. The nature of the premise indicates that only probability is generated for the conclusion and that no attempt is made to provide certainty for the conclusion.

It is very important to decide if arguments are deductive or inductive, because different standards of appraisal apply for judgments on the worth of the reasoning. The distinction between invalid and valid deductive arguments marks a sharp dichotomy between inferences that are on track and those that are not. Like deductive arguments, inductive arguments attempt to take us (through inference) from what we think we know is true (premises) to what else is true (a conclusion). But in appraising inductive arguments we will see that judgments of good reasoning involve a comparative matter of degree, more of a sliding scale of bad to good, from very weak to very probable. As a simple rule of thumb, look for tones of arguments such as,

(deductive) "There is conclusive evidence that X is true, if the premises are accepted as true."

(inductive) "There is good evidence (if the premises are accepted) that X is probably true, a reliable belief."

Separating deduction from induction is not always easy, and as in general communication, disagreements occur. But standards of appraisal cannot be applied until some decision is made, until some focus has been decided upon.

Induction and Reliable Beliefs

As a further example of the tone of inductive reasoning and the philosophical nuances that you need to master for a full understanding of inductive appraisals, consider the following.

Suppose we bring into a room a barrel full of apples. Suppose someone tells us that the barrel contains one hundred apples. I reach into the barrel, and from the top, pull out one apple and place it on a desk. Upon inspection we can see that no one in his right mind would eat this apple. It is considerably overripe, being very soft and full of maggots. Although few people would purchase the barrel of apples for human consumption, suppose I gave you one thousand dollars to wager on the status of all the apples in the barrel. What I want to know is, at what point in gathering evidence (we are going to pull out more apples) would you be willing to bet your one thousand dollars on the conclusion that *all* the apples in the barrel are rotten. Here is my offer: if after only one apple is observed you bet that all the apples are rotten and you are correct, then you will win one hundred thousand dollars. If you bet after five are observed and you are correct, you win ninety-five thousand dollars, and so on. However, if even one apple is not rotten, I take the one thousand dollars back.

Clearly, this is an inductive situation. I am essentially asking you to tell me at what point in gathering evidence you believe the evidence is sufficient to warrant the generalization as reliable. In the first case, you have ninety-nine cases in which the conclusion could be wrong, even though the premise is true that one apple is rotten. Even after pulling out ninety-nine apples and (supposing) all were rotten, it is still possible that the last apple is not rotten,[1] that the conclusion "All the apples in the barrel are rotten" is false.

On the basis of only a single positive instance, would it be wise to conclude that all the apples are rotten? Wouldn't this be a very weak induction? Would you risk your one thousand dollars at this point?

Small amounts of evidence need not always be weak. A microbiologist, for instance, might be willing on the basis of this one apple to wager that all of the apples are likely to be rotten, if a few other pieces of information were provided. Given the general knowledge of the existence of bacteria and their ability to reproduce rapidly given the right conditions, if it were known at what temperature the apples were stored, and for how long, a conclusion that all the apples are rotten might not be unreasonable. If we also noticed that a significant amount of fluid was oozing from the rotten apple and the microbiologist was allowed to examine this fluid microscopically, scientists trained in this field might be willing to bet their one thousand dollars—reasoning that the fluid has (suppose) an extremely high bacterial count and must have been draining from the top of the barrel to the bottom (because of gravity) for some time.

Often in science a very small number of observations are made to do a lot of hypothetical work, especially when they are placed within a framework of a general consensus of belief of how other aspects of the world work. Around the world, physicists in huge laboratories the size of small cities accelerate subatomic particles and smash them into each other. With a very few key observations from these experiments, the theoreticians follow long excruciating mathematical trails until they arrive at statements about the first microseconds of the universe, approximately 15 billion years ago! Physicists believe that hidden within the depths of every atom is fossil-like evidence of how matter and energy were first formed to make our universe. By smashing subatomic pieces into each other they are attempting to mimic the first microseconds of our universe. Similarly, our microbiologist could apply other general scientific beliefs (also based on induction) to make a judgment

---

[1] Notice that in order to avoid a disagreement we would need a stipulative definition before we start for what we mean by "not rotten." Perhaps we would agree beforehand to have a neutral observer in the room, and if this neutral observer would be willing to eat an apple, we would designate it as not rotten.

on the significance of the observation of one rotten apple. These other beliefs are sometimes called **higher-order** inductions, auxiliary hypotheses, or background assumptions.

Much could go wrong, however. The general beliefs could be wrong. Perhaps the laws thought to govern bacterial growth don't work in this particular barrel. Perhaps gravity stopped working in this barrel. Maybe the rotten apple was placed on top of the barrel by an associate of mine just before the barrel was brought into the room. In judging the significance of one observation, our microbiologist would need to weigh the reasonableness of many other judgments. How likely would it be that gravity stopped working in just this barrel at just this time? How likely would it be that there is a conspiracy involved in this game; that the rotten apple was placed on top just before the barrel was brought into the room, or that one apple on the bottom was placed in a hermetically sealed package? As we noted in the previous chapter, a single observation or belief is not neutral but is part of a web of belief, and we can alter our interpretation of this single case depending on how firmly we believe it is reasonable to hold on to other beliefs.

At least we should be able to say that a bet by an informed microbiologist with only one observation is better than a bet by someone with no knowledge of bacteria and the microbiological process of spoilage. Also, a bet based on a single case would be better if it was known that spoilage in a barrel like this generally takes place from the bottom up, and better still if we had evidence that no conspiracy was involved—that we had good reasons to trust the person who brought the barrel into the room, that this person does not know me, that he retrieved the barrel from a warehouse that was secured from tampering, and so on.

However, without any beliefs like this to go on, without appeals to independent or higher-order inductions, concluding that all the apples are rotten from a single positive case is a very weak inductive inference. To make the inductive inference stronger more apples need to be sampled. If I pulled out four more from the top, for a total of five, and all of them were just like the first one, we would now have a better basis for concluding that all the apples are rotten. This type of evidence is called **induction by enumeration**; we gather or add up more positive premises or cases for our conclusion. In general, the more positive cases in favor of the hypothesis, the stronger the hypothesis. Would you bet your one thousand dollars at this point? You could still win ninety-five thousand dollars. Suppose all five apples are the same—very rotten, maggots, fluid draining onto the table?

Suppose we select five more apples. If all five are again simply pulled off the top, it is still possible that the apples on the bottom are not rotten. Thus, a stronger case could be made by choosing a **representative sample**, by selecting one from the top, one from the very bottom, one from each side of the barrel, and one from the middle of the barrel. If all five are rotten, this would strengthen the hypothesis considerably. Political pollsters use this technique. We have learned that it is much easier to determine the voting behavior of 50 million voters by examining a representative sample of twelve hundred voters (Democrats, Republicans, different races, income levels, ages, sexes, and so on) distributed throughout the country, rather than one hundred thousand predominantly liberal people living in New York. A small representative sample is much stronger inductively than a large unrepresentative one. Some positive cases are stronger and offer better tests than others. For this reason, philosophers of science, who study how inductive reasoning is used in science, distinguish between *mere positive cases* of support for a belief and *genuine confirming instances* for a belief.

Suppose a friend of mine believes in UFOs, and he claims to have some startling evidence—pictures—that we have been visited. Suppose I ask to see the pictures and when he produces them, I see that all the pictures show blurred images of fast-moving lights that could be spaceships, but also could be images of the planet Venus or of airplanes taken by an inexperienced photographer. When I complain that the pictures don't prove anything he responds, "Don't you see? The UFOs emit a special radiation that makes any pictures taken of them blurry!" Given his hypothesis, he has made blurred photographs of fast-moving lights positive cases for his belief in UFOs. But no scientist would consider this evidence genuine, confirming instances for a belief in UFOs. Similarly, in our hummingbird example above, suppose we simply looked on the ground within, say, a ten-square-foot area and noted that within each square foot we have found no hummingbirds. Each observation of no hummingbirds within a square-foot area of the forest floor would be a positive case for the belief that there are no hummingbirds in this forest. But because we would not be looking in the places an expert would know are likely locations for humming-birds, our observations would not be genuine confirming instances.

So, five representative apples are better, more genuinely confirming, than ten just off the top. Would you bet on the generalization at this point, that all the apples in the barrel are rotten?

I would bet at this point, if I could be reasonably assured of the barrel's independence from human interference and bias, that the barrel of apples was not tampered with, and that only natural processes were involved in whatever took place within the barrel. The inductive evidence is relatively strong and the reward very high— I can still make ninety thousand dollars. Keeping in mind that what we are interested in is not a deductive guarantee for our conclusion, but rather at what point is it *unlikely* that the conclusion is false, I would bet at this point, because pulling out more apples lowers my possible earnings and does not increase the likelihood of the conclusion being true that much. Without telling my students this, however, most usually remain unimpressed with the representative sample. They have had too much logic already and remain skeptical!

Suppose then we continued to sample the apples by pulling out forty more. Suppose that we now have fifty rotten, maggot-infested apples sitting on a table with fluid now running profusely off the table onto the floor. Suppose we pulled the apples from the barrel in a representative way, not just from the top down, but from all over the barrel. Would you bet now? Now we have both a lot of induction by enumeration and a greater representative sample. Suppose we also had a reasonable assurance that the barrel was not tampered with, the authoritative testimony of a microbiologist, and appeals to independent inductions as to the nature of spoilage in a barrel like this. You can still win fifty thousand dollars. Would it be wise not to bet? Most reasonable people (who have not had too much logic) would bet their "free" one thousand dollars at this point. But I suspect they would not bet their life savings or their cars. Even though it would be more likely now that all the apples are rotten, we know that it is still possible that some, even many, of the remaining apples are not rotten. Because of this, many of my students will still remain unimpressed and not bet.

Suppose we pulled out forty-nine more. Suppose that all ninety-nine sampled are rotten. Would you bet now? You still have a chance of doubling your money, of making another one thousand dollars. Do we know that the last remaining apple is rotten? Many of my students finally bet at this point and many people would probably even bet their life savings at this point if they could double their money, but there would still be considerable anxiety as the last apple was pulled from the barrel, because it is still remotely possible that this last apple is not rotten, that just by chance the spoilage had not yet reached this apple. It is still conceivable that the hypothesis, "all the apples are rotten," is false, even though we have an overwhelming number of positive cases supporting it. It is still conceivable, but no longer reasonable, assuming all the other inductive support noted above is still in place, including reasonable assurance that there has been no tampering with the barrel. From an inductive point of view, it would be irrational not to believe that the last apple is also rotten.[1]

There is much about the logic of induction and how it is used in everyday life and science that can be summarized with reference to this little example.

First of all, the kinds of inductions that we often actually use are much weaker than the standards most people impose on themselves in betting their one thousand dollars. Knowing that a universal generalization can be falsified or refuted by a single negative case, some of my students will not bet even after ninety-nine rotten apples have been observed. They would rather just keep their free one thousand dollars! However, we frequently have to act and "bet" on matters that are important to us, and with less evidence.

Suppose you were given the task of entertaining a very important person and you decided to take this person to a good restaurant. How would you go about deciding what a "good" restaurant is? You could use your own personal *experience*. Perhaps you went to a particular restaurant before, and the service and food were excellent. But if you only went once and tried only one offering from the menu, aren't you making a big inductive leap? Would this be much different than pulling one apple out of a barrel? How many times could you have gone to this restaurant? How many items could you have tried? Maybe the restaurant just had one particular good night when you were there. Suppose you didn't use your own experience, but called a friend instead. Suppose your friend endorsed a particular restaurant as excellent. Now you would be using the *testimony* of another person. But the same questions could be asked. How many times has your friend been to the restaurant? How many different items were tried? Suppose you called another friend who is restaurant critic for a local newspaper. Now presumably you would be appealing to an *authority*, someone who makes a living sampling restaurants. But authorities

---

[1] For further details on what scientists know about spoilage in a barrel of apples, see the entertaining, *Why You Can Never Get to the End of the Rainbow and Other Moments of Science*, Don Glass, ed. (Indiana Univ. Press, 1993).

often disagree and the same questions can be asked. How many times did this authority frequent the restaurant? How many items were tried? Was this authority there only one night? Was this test really an independent test? Did the operators of the restaurant treat this authority special knowing that this would be very important to their business? Suppose you read in the newspaper that a particular restaurant is the talk of the town and that it takes weeks to get a reservation. Now you would be appealing to **popularity**. Because so many people are going to this restaurant, you think it must be good. But how many times has each person gone to this restaurant? Are they really enjoying the food and service, or are they just assuming that they are because everyone is supposed to at such a popular restaurant?[1]

I am not suggesting with these examples that it is always unreasonable to use our own personal experience, the testimony of others, and appeals to authority and popularity. I am suggesting only that we make "bets" all the time apparently with a lot less evidence than the ideal standards of inductive reasoning. If you are sitting on a chair right now reading this book, how do you know that the chair will continue to support you? Your assumption that the chair will support you involves a universal generalization. You are generalizing from past experiences of successfully and safely sitting in chairs to the present moment. However, perhaps the chair will collapse soon, and you will suffer serious injury to your spinal column and be physically impaired for the rest of your life. Did you check the chair carefully before you sat down to make sure that it was not a fake chair like those used in the movies, that break easily when smashed over someone's head? Why not? You are inductively reasoning that, given your past experience, it is unlikely that someone has tricked you at this moment by replacing a normal chair with a fake chair. Perhaps you have sat in this chair before and everything has always been normal. But how many times have you sat in this particular chair compared to the number of times you could have sat in this chair? From one perspective, statistically speaking, no matter how many times you have sat in the chair, compared to the number of times you could have sat in the chair, the percentage is less than one apple out of a barrel of one hundred.

My point is not that you have been irrational by sitting in the chair. In fact, most people would think your action irrational if you started carefully checking every chair in the future for a hidden bomb (another possibility). The point is that we must and do use inductive reasoning rationally all the time, even though we lack certainty that our conclusions are true. If we waited until we had a situation comparable to ninety-nine out of one hundred apples, we would all be sitting in the corner of our rooms afraid to do anything. Lots of things could be true, but when the inductive evidence is overwhelming it is irrational not to believe or

---

[1] As a little aside, it might interest you to know that I am still somewhat famous among the employees of a particular prestigious Waikiki restaurant in Honolulu. Before deciding to be a college professor I worked as broiler chef at this restaurant. One night, this restaurant had a very bad night. Everything went wrong. We all were very nervous because a movie star had walked in unannounced with a lot of friends. This movie star was a very fussy prima donna, and we concentrated so hard making his dinner perfect that we not only made a mess of his dinner, but most of the others we served that night. For my part, I was fired for slightly undercooking his two-inch-thick lamb chop—I was trying to make sure that it was not overcooked—which I suppose is why I am writing this book rather than still working in restaurants! Anyone who ate there that night could have easily concluded that this restaurant was not all that it was advertised to be and would have been very disappointed if they brought special guests.

act.[1] *The point is not that we are unreasonable using induction or that we should always avoid generalizations (we can't), but how important is the conclusion we are considering and how can we make an inductive inference stronger when the conclusion is very important.*

A second point that should be understood from our barrel of apples example is that even though we may disagree about when it is reasonable to bet on the conclusion, we can at least agree that some levels of evidence and the inductive arguments based on them are objectively better than others. A representative sample is a more genuine test and amounts to a stronger inductive argument than a simple accumulation of positive cases. Lots of genuine positive cases are better than only a few. Positive cases become stronger if they are consistent with other established beliefs. Independent tests are better than tests that allow for the possibility of bias and tampering.

Many of my colleagues might disagree with me that it is reasonable to bet after ten apples, only five of which were the result of a representative sample. Instead, they may claim that given that we can still win fifty thousand dollars, betting after fifty have been observed in a representative manner is much wiser. But we would not disagree over what was the better case. Their inference would be stronger in the sense that the conclusion would be less likely to be false. However, they would also agree with me that some positive instances are better than others; that a test that involves a representative sample, even though the numbers were the same (five apples), is a better or more genuine test than one that simply selects five off the top.

In the previous chapter we saw how vague statements can lead to unfair acts of persuasion. A similar situation applies to the testing of beliefs. It is easy to find mere positive cases for a belief if that belief is vague. For instance, suppose a friend of mine believes in astrology (many astrological predictions are very vague). As evidence for the value of astrology, he cites the dramatic instance of a man whose horoscope told him it was a bad day to fly. On that day there was a terrible plane crash killing everyone on board, but the man lived because he followed the horoscope's advice. Although this seems a very precise prediction and a dramatic positive instance for the reliability of astrology, the weight of this positive instance changes considerably when we find out that horoscopes frequently tell people it is a bad day for them to fly without plane crashes occurring. If my friend responds to this by pointing out that "bad day" can mean a lot of things—delays in departure or arrival, lost baggage, poor service, a rough and uncomfortable flight, or a baby crying throughout the flight, then we see how very easy it is to find mere positive cases for a belief and how much harder it is to distinguish between genuine confirming instances for a belief and mere positive cases if our belief is vague. Because "bad day" is so vague, almost any occurrence would be a positive case for the astrological prediction.

If I believe that taking massive doses of vitamin C will prevent colds, and for the past one hundred days (from May to August), after taking massive doses of vitamin C I have had no colds, I have a lot of mere positive support for my belief.

---

[1] I once had a student who had a serious mental collapse. He called me from his home and said that he could not come to class anymore because he was convinced that the governor's wife liked the color red (she seemed to wear a lot of red dresses). And so the normal pattern of traffic signals had been changed in our state—red meant "go" and green meant "stop." He said we could not be sure if everyone had been informed of this change, so it was too dangerous to come to the college!

But have I genuinely tested my belief by selectively not including the winter months? Even if a belief is not vague, mere positive cases can easily be generated by selective testing.

So, although we cannot have a nice, clean black-and-white line of separation between good and bad inductive arguments, as we can with the distinction between valid and invalid deductive arguments, we can have a sliding scale of reasonableness where some inductive arguments are objectively more convincing, more reasonable, than others. As we noted above, the issue is not whether we can ever be certain using induction—we can't, induction is not deduction, it is always conceivable that our conclusion could be false—rather, the issue is when can we reasonably accept a conclusion given the alternatives. *It is a hopeless, unrealistic, utopian task to have as the goal of life and reasoning the separation of absolutely true beliefs from all conceivable beliefs. Nor are we interested in separating the reasonable from the conceivable; rather, we are interested in actions (we have to bet on a lot of things) based on separating the reasonable from given alternatives.* Hence, given important inductive situations, especially those in the domain of science that deal with generalizations concerning human nature, the goal is not certainty but tentative acceptance, constant testing, and better inductions.

A third point that emerges from our apple barrel example is that false ideas work. This realization should both humble and encourage us. It should humble us, because it should caution us against being too confident and egocentric about our own beliefs. It should encourage us because the nature of induction is consistent with the notion that life will always be a game of self-correction, growth, and learning. An important illustration of this point is the use of induction in science.

Scientists often test their theories using a combination of deduction and induction. Once a theory is formulated, a prediction is deduced about what should be observed under particular conditions if the theory is true. If we propose a simple generalization that all the apples in the barrel are rotten, then we would predict that one selected from the middle should be rotten. If one selected from the middle is not rotten, we can deduce that our generalization is false.[1] The testing procedure is ultimately based on induction, however, because if an apple is pulled from the middle of the barrel and observed to be rotten, this *confirms* the hypothesis inductively, but it does not deductively prove it. Compare the following forms of reasoning:

EXAMPLE 3-6   If all the apples are rotten, then one from the middle is rotten.
One from the middle is not rotten.
Therefore, it is not true that all the apples are rotten.
(*valid deductive argument*)

EXAMPLE 3-7   If all the apples are rotten, then one from the middle is rotten.

---

[1] A simple falsification of this sort assumes that the generalization is simple. In science a complex theory will consist of many generalizations—usually the main hypothesis along with auxiliaries and background assumptions. Hence, because we can deduce only that at least one of the premises is false of a valid argument with a false conclusion, given a false prediction from a complex theory, we can infer only that part of the theory (one of the generalizations) is false.

One from the middle is rotten.
Therefore, all the apples are rotten.
(*invalid deductive argument*)

EXAMPLE 3-8   One apple from the middle is rotten.
So, all the apples are rotten.
(*weak inductive argument*)

The logic of this situation shows that we need more than simple confirmation to accept a belief as reliable. It is possible to deduce true conclusions (the apple will be rotten in the middle of the barrel) from a premise that may be false (all the apples are rotten). That an idea or belief "works" by simply making a correct prediction or having positive confirming instances in its favor is no guarantee of its truth or long-term reliability. Simple confirmation for a belief offers little assurance that the belief will continue to work and that a little critical investigation might not reveal a better-supported idea or belief, one that will work even better and have more long-term reliability. I might be a vice president for a tobacco company, have smoked two packs of cigarettes all my adult life, and not been sick a day in my life. The belief in no proven link between smoking and serious illness works for me. So far. As Bertrand Russell, a British philosopher and logician, once remarked, "The man who has fed the chicken every day throughout its life at last wrings its neck instead." Our goal should be to make better use of inductive reasoning than a chicken.

All major scientific theories are based upon inductive reasoning. Scientists know that the fact that a theory works and has been reliable is no guarantee that it will continue to be so. Because it is possible to deduce true predictions from a false theory, no matter how long a theory has been successful in making predictions, it cannot be known to be true absolutely. It could be found false tomorrow. In fact, in our relationship with nature the logical situation is much worse than our apple example. In nature, the situation is analogous to a barrel full of an infinite number of apples.

However, some things are very important to us and we have to place our bets as best we can. Because our health is very important to us and we began this section asking how science can reasonably advise us on a link between cigarette smoking and serious illness, let's draw out the scientific story on cigarette smoking a little. In the process you should be able to see how science attempts to make inductive inferences more reliable, and the delicate balance between cautious methodology and boldness of inference necessary for self-correction in science.

Induction: A Case Study

In the 1970s, the warning label on a package of cigarettes changed from "Caution, smoking *may* be hazardous to your health" to "Caution, smoking *is* hazardous to your health." This change showed that within the scientific community a consensus had been reached that, among other things, cigarette smoking is a principal factor in *causing* lung cancer.[1]

To clarify terminology first, what does "principal factor" mean? It does not mean that every person who smokes cigarettes will get lung cancer. To say, as is believed today, that 75 percent of all lung cancer cases are caused by smoking is not the same as saying that 75 percent of the people who smoke will have, and eventually die of, lung cancer. Some smokers may die of other causes, such as other forms of cancer, accidents, or old age. However, it is now believed that cigarette smokers are about 140 times more likely than nonsmokers to get lung cancer. This implies that it is possible to have lung cancer and not smoke, and not have lung cancer and smoke. The probability of dying from lung cancer is very low if you do not smoke, and the probability of dying from lung cancer is also somewhat low if you do smoke. But the probability of dying of lung cancer if you do not smoke is so low, that even though the probability of dying of lung cancer as a smoker is also low, it is 140 times higher. So, "principal factor" means "major" factor; that most lung cancer cases are caused by cigarette smoking. How has this been established? How has it been determined that this is a reliable belief?

Many decades ago public health officials began to notice a significant increase in the number of lung cancer cases at the same time cigarette smoking was increasing. In the 1930s there were about three lung cancer deaths per one hundred thousand persons per year. By 1955, this had increased to twenty-five deaths per one hundred thousand persons per year. At first this was primarily a male disease; many more men smoked than women. As more and more women began to smoke the death rate for women began to increase. At this point, cigarette smoking was strongly correlated with lung cancer. But **correlations** can be coincidental. Increased use of suntan lotion by teenagers in the 1960s and 1970s is strongly correlated with increased cases of teenage venereal diseases, but an increase in the number of teenagers is more likely the cause of both. In the case of cigarette smoking, factors such as air pollution, daily stress, and the pace of life had also increased. Thus, because science is interested in making inductive inferences as strong as possible, strong correlations are only a good place to start an investigation, not conclude one. At this point, the evidence linking smoking and lung cancer was only circumstantial. There were lots of positive cases, but there was no genuine test at that time.

Scientists have learned, often painfully, that nature can often trick us. They have learned that sometimes when on the surface two occurrences seemed to be linked in the sense of one causing the other, there is actually a third event causing the illusion of a link between the first two. The fact that cigarette smoking and lung cancer have been increasing together could be an illusory link, and the real cause of both could be the increased stress (both environmental and mental) of modern life. So, we learn to adopt methodologies of inquiry that will increase our chances of separating beliefs that reflect only nature's tricks from more reliable beliefs about

[1] By 1992 in Great Britain the front of all cigarette packages displayed the message, "Tobacco Seriously Damages Health," and the back had one of the following: "Smoking Causes Cancer," "Smoking Causes Heart Disease," "Smoking Causes Fatal Diseases," and "Smoking Kills."

nature's real processes. We learn to adopt methodologies that will submit our beliefs to genuine tests.

One such methodology is called a ***controlled study*** or experiment. A controlled study is deliberately designed to reveal coincidental correlations by allowing possible hidden factors to reveal themselves and destroy the initial correlation. A controlled study controls the factors or ***variables*** that might be the hidden causes by keeping them the same and under observation in at least two groups of cases, and then allowing only one variable to be different. To genuinely test the link between cigarette smoking and lung cancer, at least two groups of people would need to be examined in such a way that the people selected for both groups would not be significantly different overall in where they live, what they eat, and the stresses they experience. They would not all live in the same place, eat the same food, or have a similar life-style. Rather, overall there would be no significant difference in the two groups. For instance, some people might live in the country and some in the city, but the proportion of each should be the same in each group. Ideally, the only difference overall would be that one group smokes cigarettes and the other does not. In this way, if a third factor is the real cause of lung cancer, because this cause will be in both groups, there will not be a higher incidence of lung cancer in the group that smokes and the correlation between the cigarette smoking and lung cancer will be destroyed.

One of the first systematic controlled studies on cigarette smoking was carried out by medical researchers E. C. Hammond and D. Horn.[1] More than 180,000 men in the United States between the ages of fifty and sixty-nine were interviewed, targeting their smoking habits, general health, and living environment. Each man was randomly assigned to an observation group in such a way that there was no apparent difference in each group other than in one group all the men smoked cigarettes and in the other group no one smoked cigarettes. At the beginning of the study none of the men had any serious illness. After a period of time, the doctors compared the death rate due to lung cancer between the two groups.

To simplify the results, the statistics can be presented from the point of view of the proportion of deaths per one hundred thousand men over a ten-year period. In the group of men who had never smoked, the study indicated that thirty-four cases of lung cancer resulted per one hundred thousand men over ten years. In the group of men who had smoked cigarettes, the study indicated that 4,719 cases of lung cancer resulted per one hundred thousand men over a decade (Figure 3-1). This was a significant difference (140 times more). Furthermore, the study showed that the more a man smoked—from less than one-half pack per day to more than two packs per day—the more likely he would be one of the 4,719 after ten years. Could the high percentage of lung cancer deaths among the men who smoked be a coincidence?

Compared to an ideal study, the Hammond and Horn study could be criticized on several points, which the tobacco companies were quick to do. Although a basic environmental variable was controlled in this study—no significant difference existed

---

[1] Hammond, E. C. and D. Horn, "Smoking and Death Rates," *Journal of the American Medical Association*, 166: 1159–72, March 8, 1958; 166: 1294–1308, March 15, 1958.

FIGURE 3-1 HAMMOND AND HORN CIGARETTE SMOKING STUDY

Ages: 50–69 *at the beginning of the study.*

Health: *No apparent serious illness at the beginning of the study.*

Result: *Ten years later, 140 times more instances of lung cancer for men who smoked cigarettes.*

4,719 LUNG
CANCER CASES

100,000 MEN
(CIGARETTE SMOKERS)

34 LUNG
CANCER CASES

100,000 MEN
(NONSMOKERS)

in each group in the number of men who lived in the city and those who lived in the country—several other important factors were not controlled. For instance, it was possible that all of the men who died of lung cancer, including those who had never smoked, ate a diet significantly different from those who did not die of the disease. Or, they may have had unhealthy occupations, whether in the country or the city, that placed unusual stress on the lungs. Perhaps all the men with cancer had stressful occupations involving a great deal of daily anxiety. Finally, and this has been a point the tobacco companies have used consistently, the variable of heredity was not controlled. That is, it is possible that all the men who had cancer were prone to some form of cancer genetically, and this would be revealed by finding other incidents of cancer in their family histories. Parents, grandparents, and relatives of the men who succumbed to lung cancer might have shown a high percentage of cancer, if this was investigated.

Such criticisms serve a very positive function. Critical examinations that reveal weaknesses serve as a basis to make the next study better. Since the Hammond and Horn study, hundreds of similar studies have been conducted, each incorporating tighter and tighter controls based on possible oversights of the previous studies. By the early 1980s, studies included controls on all of the variables just mentioned. The results have not changed much. By 1985, more than 140,000 people a year died of lung cancer, an impressive majority smoked cigarettes, and for the first time lung cancer for women exceeded that of breast cancer. In the early 1990s, the results of a forty-year study of British doctors showed that only three of every two hundred men who smoked twenty-five cigarettes a day from the age of thirty-five onward lived to age ninety, whereas thirty of every two hundred nonsmokers survived to ninety. And for those who think that serious illness from smoking is only produced in one's later years, this same study showed that in the forty-five to fifty-four age group, 1.14 percent of the smokers died each year (thirty times higher) compared to .0389 percent of the nonsmokers.[1] Other diverse corroborating factors have also been identified in other studies, ranging from the effects of secondhand smoke to analysis of cigarette smoke, which revealed more than two hundred toxic substances and radioactivity.

This history reveals other factors that make inductive inferences stronger. One controlled study, no matter how thorough, is not enough. Such studies need to be criticized and **repeated**. They also need to be repeated by **independent investigators**. If Hammond and Horn repeated their study themselves, they might be too interested in seeing their results replicated, and this could seriously bias the design of any new studies, including how interviews are conducted and results interpreted. Also, they received a large grant for their first study, and their reputations and futures as professional scientists would be at stake. Finally, as in our barrel-of-apples example where knowledge of bacteria was used to strengthen an inductive inference, when independent investigators have worked in an entirely different area using techniques of inductive inference to establish beliefs that **corroborate** a particular belief, that belief is strengthened. We have learned independently that toxic and radioactive substances are linked to cancer, and learning that these substances are in cigarette smoke strengthens the case that nature is not tricking us, but is showing us something about what happens when cigarette smoke is repeatedly put into human lungs.[2]

Does this prove that cigarette smoking is the principal cause of lung cancer? No. There is no such thing as a scientific proof, if "proof" means something that is known with absolute logical certainty. A controlled study can never be completely controlled. The number of ways people can differ, the number of possible variables, is infinite. In all the studies on cigarette smoking, it is possible that some obscure third factor was the real culprit. All of the people who had lung cancer could possibly have had some subtle factor in common that went unrecognized in every study. Perhaps the real cause was that all of the lung cancer victims were given bubble gum on their fifth birthday, and it just so happened that the gum contained a chemical that weakened their immune systems and left them susceptible to a particular virus later in their lives, which then led to lung cancer fifty or so years later! Worse, perhaps lung cancer is very slow in developing and some other factor starts it at an early age, and the developing cancer somehow causes most people to smoke! These and millions of other "off the wall" possibilities have never been

---

[1] This study was conducted by Richard Doll of the Imperial Cancer Research Center at Oxford University. For a summary of Doll's work, see his "The Lessons of Life," *Cancer Research: the Official Organ of the AME*, Vol. 52, N7, April 1, 1992, 2024.

[2] As another example of corroboration, consider that controlled studies have shown a link between cigarette smoking in women and cervical cancer. Also, when normal human skin cells are grown in a laboratory and then infected with human papillomarvirus (a virus associated with cervical cancer) and then doused with low doses of N-nitrosomethylurea (a cancer-causing chemical derived from the nicotine in tobacco smoke), when injected beneath the skin of mice these cells produce malignant tumors in most cases resembling those of cervical cancer. *Science News*, March 18, 1989, page 166, and April 6, 1991, 215.

tested.[1] The link between smoking and lung cancer cannot be known in the sense of being known beyond any logical or conceivable doubt. The question is, however, can we not say that we know that cigarette smoking is a principle cause of lung cancer beyond a reasonable doubt? Can we not say that after decades of research, we possess a reliable belief that will likely stand up to further test and upon which we can base a course of action? That it is very unlikely that in the year 2020 we will discover that nature has been tricking us and that smoking is actually good for our lungs, like taking vitamins? People can continue to smoke for all kinds of reasons, but they should not fool themselves on the risk they are taking. Given the alternatives, it is clear what a reasonable person should believe: Long-term cigarette smoking is causally linked with serious illness, and it is irrational to believe that no evidence exists for this claim, or that we will someday find that smoking is good for our lungs.[2]

Relativists use the element of uncertainty inherent in inductive reasoning to claim that we can believe whatever we want to believe, that all beliefs are true and can be made reasonable with enough effort. As we have seen, however, though we cannot be absolutely certain, decades of careful research show that linking cigarette smoking with serious illness is a much more reliable belief than tobacco company claims that the link is only a coincidence. Consider another example. Suppose we were on a jury charged to judge the case of a man suing the Adolph Coors Company because he claimed to have found a dead mouse in a can of Coors beer while drinking it. Independent scientific evidence is introduced by a coroner showing that the mouse died about a week before the man called a local TV station saying that he felt something against his mouth while drinking the beer, and three months after the can was sealed at the Coors brewery. Also, the mouse had external wounds consistent with being stuffed through the can's pop-top opening, and tests showed that the mouse died from a single blow to his neck and did not drown. Finally, testimony reveals that six months earlier the man making the charge had been in an accident with a Coors beer truck and he is bitter that his insurance claim is still unsettled.

Based on this testimony, it seems most likely that the man killed the mouse with an ordinary mouse trap and stuffed it through the opening in the beer can. Based upon this evidence, it would be irrational to believe that the man was telling the truth. Even though it is still conceivable that the date of death for the mouse is

---

[1] A more likely and interesting possibility is the role of radioactive radon gas prevalent in some homes. Radon gas also is believed to cause lung cancer. It is emitted naturally from the Earth and collects in small amounts in some homes that are well-insulated from cold winters. Recently, however, scientists have conducted studies controlling for radon gas. The tentative conclusion at this stage in the research is that radon and smoking appear to "synergistically" increase the risk of lung cancer. See John S. Neuberger, "Residential Radon Exposure and Lung Cancer: An Overview of Published Studies," *Cancer Detection and Prevention*, 1991, 15, no. 6: 435; and "Residential Radon Exposure and Lung Cancer: Evidence of an Urban Factor in Iowa," *Health Physics*, 66, (March 1, 1994): 263.

[2] Often people who smoke believe that "This can't happen to me." Perhaps objective facts need to be supplemented by subjective experiences. For a grim description of what it is actually like to experience lung cancer—the pain, surgery, and eventual death—see John A. Meyers, *Lung Cancer Chronicles* (New Brunswick: Rutgers University Press, 1990).

wrong, that the mouse received the wounds from the machinery at the brewery, that the man's bitterness toward the Coors Company is just a coincidence, that all the testimony is part of an involved conspiracy to support the Coors Company, there are no reasonable grounds for believing these possibilities. With enough effort we can imagine how the man might be telling the truth and that all the evidence against him is either circumstantial or part of a conspiracy, but *imagining a conceivable scenario* is not evidence. If we were on the jury, even though we could be wrong, we would need to select the most reasonable belief from the given alternatives. Based on the evidence, some beliefs can be judged more reliable than others. Even though we can be wrong, there are many situations in which it is irrational not to "bet," irrational not to choose the best alternative given the evidence.

Logic and Creativity

An essential theme of this book is that logic is a tool that helps us stay on track when we want to test our beliefs for reliability. Do not infer from this that logic is some sort of way of life that differs from or interferes with the creative, artistic life; that logic is always to be used in isolation of, or in place of, creativity and imagination. In the hat example of Chapter 1, to begin the logical trail that led to the conclusion that the blind man must have on a white hat, out of all possible things to think about and out of all possible places to start, it was necessary to notice that the first man could not have seen two red hats, otherwise he would have known he had on a white hat and gone free. Once we see this premise, it seems trivial and it is hard to see how we could have seen anything else, how we could have thought about anything else and followed a much different trail. But the starting point of a trail always seems easy after it is recognized.

How do we have insights into a perspective that allows us to start a trail of thinking? Have you not wondered at some time or another where the human race has gotten all its ideas? I remember as a small child watching an aunt make wheat bread. I remember bothering her with question after question about what she was doing: "How is it made? What are the ingredients? Wheat? What is wheat? A plant? From the seed of a plant?" How could human beings ever figure out such a thing? How did we connect up the seed of a common weed to make the bread for my peanut butter and jelly sandwich?

Jean-Paul Sartre, in his existentialist novel, *Nausea*, tells the story of a man who has decided, "Science! It is up to us." This man has decided to read every book in his city library so that he will know everything there is to know. He starts with the first book on the first shelf, and by the time Sartre's hero Antoine Roquentin meets up with this "self-taught man" he is into the L's and reading *Peat Mosses and Where to Find Them*, by Larbaletrier. According to Sartre,

> He has read everything; he has stored up in his head most of what anyone knows about parthenogenesis, and half the arguments against vivisection. There is a universe behind and before him. And the day is approaching when, closing the last book on the last shelf, . . . he will say to himself, "Now what?"

He will know many facts, but will he have made any connections? Will he gain any perspectives from all this with which to follow some new interesting trails of thought?

As a final example for these introductory chapters, as one to tie together our discussions of deductive and inductive logic, truth and the attempt to find reliable beliefs, and as an example to prepare you for why logicians and mathematicians believe you should experience firsthand the beauty, the agony, and the ecstasy of following formal symbolic logic trails (as you will be doing in Chapters 9 and 10), I would like to share with you an example of the use of the human mind at its best. This illustration contains some mathematics (geometry), but with a little bit of effort and a few pictures you should be able to understand and appreciate some exemplary reasoning.

Sometime during the third century B.C., in the city of Alexandria, Egypt, another admirer of books by the name of Eratosthenes was reading about a curious fact. Eratosthenes was the head librarian of the famous library of Alexandria. The knowledge stored in this library was far ahead of its time, and he had access to many great thoughts and facts. The fact that preoccupied him this day was rather ordinary on the surface. He read that in the city of Syene, the site of modern-day Aswan, on the day of the summer solstice, June 21, the longest day of the year, at exactly noon, the Sun shone directly over a particular water well and cast no shadow. The average person would probably not even notice this and would surely consider anyone strange who made a big deal over shadows while everyone normal was just trying to go about the day's business. But for someone it was at least curious. Someone had to notice the event and connect it with the summer solstice. For the brilliant Eratosthenes, nature was taunting us with one of its secrets. Magic with which to begin trails of inquiry is everywhere, but usually we are too busy and too "practical" to notice. Perhaps you have to have the mind of a child and see everything as fresh, new, and exciting. In this case, there was a very exciting connection between the shadowless well and a very big thought.

It is doubtful that it happened this way, but for dramatic effect imagine that Eratosthenes first came across this little item of knowledge a few minutes before noon, June 21. We can imagine him looking up from the book with a passionate, startled look and staring blankly into space. He then rushes outside to find a tall, straight pole and observes at exactly noon a shadow cast by the pole. He then jumps up and down in ecstasy at this great discovery: There was no shadow in Syene, but at exactly the same time there was a shadow in Alexandria. What is the connection?

Combining a few premises based on reasonable inductions with logical and mathematical deduction, Eratosthenes discovered, on the basis of this curious fact, not only that the Earth is spherical, but that its circumference is very large—approximately twenty-five thousand miles. It was an ordinary fact, but a very big fact when placed at the starting point of a deductive reasoning trail.

He had a great deal of help in making this discovery. Not only did he have help in obtaining all the premises—someone needed to measure the distance from

Alexandria to Syene,[1] and someone had to observe and note the connection of a shadowless well at Syene with the date and time—but the intellectual environment in Alexandria that made possible the connection of these facts owed its existence to the ancient Greek philosophers who believed that a few facts could go a long way with a little deduction.

The city of Alexandria was founded by the Greek conqueror Alexander the Great. With him came the intellectual tradition of faith in reason, the passion to explore the physical and biological world of Alexander's tutor Aristotle, the belief in natural law of the pre-Socratics, and the triumph of the mathematical teachings of Pythagoras, Plato, and Euclid—particularly geometry. With the deductive science of geometry and a belief, fostered by the Greek philosophical rationalists, that with mathematics we possess a mystical power to transport our minds and "see" (like our blind man from Chapter 1) what our eyes may never see, Eratosthenes was able to create a reliable belief about the nature of our Earth that was far ahead of his time.

First of all, what could account for there being a shadow in one place, but not at another? For one possible logical trail, if the Earth was flat, the Sun could be relatively close to the Earth. When the Sun was directly over the well in Syene, it would be at an angle in relation to the pole in Alexandria, thus casting a significant shadow (Figure 3-2). A more likely possibility was that the Sun was very far away, so far that by the time its light reached the Earth, the light rays would be virtually parallel to each other.[2] If this were so, then another way to account for the shadow in one place, but not the other, would be to assume that the surface of the Earth was curved (Figure 3-3). And if it was curved, it was reasonable to assume that its curved surface met to form a sphere. This was not a new idea. Pythagoras had argued as early as the fifth century B.C. that the sphere was the most perfect shape for the Earth, and Aristotle and others had noted that during an eclipse of the Moon, the dark shape made on its face was that of a disk. (They reasoned that the eclipse was caused by the Earth's shadow moving across the Moon when the Earth was positioned between the Moon and the Sun.)

As the master keeper of this tradition, Eratosthenes knew that circular objects can be measured and analyzed using a method of proportion. (For the purpose of illustration we will assume he knew about degrees. Actually he used fractions.) He also knew from the geometry of Euclid, the alternate angles formed by a line intersecting two parallel lines would be equal. Thus, he reasoned, by measuring the angle of the shadow in Alexandria, we would automatically know the angle made by drawing a line from Alexandria to the center of the Earth and another line from Syene to the center of the Earth (Figure 3-4). This angle would then deductively tell us not only the proportion of space taken up by the angle compared

---

[1] Supposedly, Eratosthenes had to hire people to walk off the distance. The distance is approximately five hundred miles, so Eratosthenes's premise had to be the result of an average or approximation, a type of inductive generalization, from the results obtained from individual measurements of the distance.

[2] This premise was also based on inductive and deductive reasoning. But by this time astronomers had already worked out a reasonable basis to believe that the Sun was very far away, even though the estimate of how far the Sun was from the Earth was far less than what we believe to be true today.

FIGURE 3-2

One way to explain why there was no shadow in Syene, yet at the same time a shadow in Alexandria, is that the Earth is flat and the sun is relatively close to the Earth.

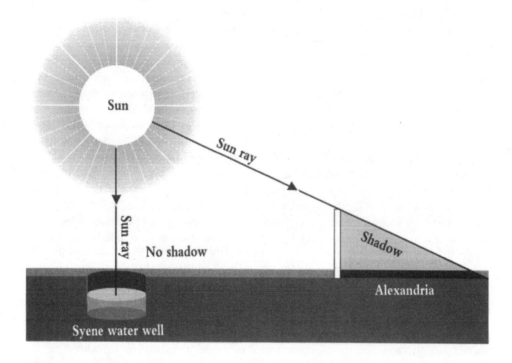

to the entire 360-degree circumference, but also the proportion of space taken up by the distance from Alexandria to Syene compared to the circumference of the entire Earth. Using a simple sundial, he measured the angle created by the shadow to be 7 degrees, approximately 1/50 of a circle. Thus, if the distance from Alexandria to Syene was 1/50 of the circumference of the entire Earth, then the distance around the entire Earth must be fifty times more than the distance between the two cities. Because he had ascertained that the distance from Alexandria to Syene was about five hundred miles, the Earth must be approximately twenty-five thousand miles in circumference (500 × 50 = 25,000).

No one at the time could confirm this conclusion with direct experience. Although the observation of a lunar eclipse supplied observational evidence supporting belief in a spherical Earth, and it was reasonable to assume at this early date that the Sun was very far away, no one possessed the means to circumnavigate the globe in the third century B.C. Nor were there pictures from space showing a beautiful blue sphere. The direct factual evidence for a large, spherical Earth was hardly overwhelming. At this time, whether one believed in the validity of Eratosthenes' picture of the Earth depended more on one's faith in the power of following logical and mathematical trails than the facts. Like our blind man in Chapter 1,

FIGURE 3-3

Another explanation why there was no shadow in Syene, yet at the same time a shadow in Alexandria, is that the sun is very far away from the Earth (so far that the rays of light are virtually parallel by the time they reach the Earth) and that the Earth has a curved surface, which implies that the Earth is a sphere.

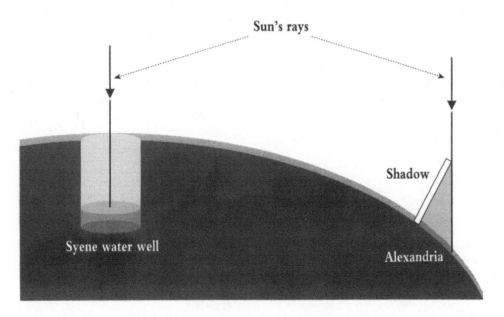

Eratosthenes could not see the entire Earth. But if his premises were correct, he could see it in another way.[1]

A tradition that placed great faith in the following of reasoning trails, curious facts noted by someone else, reasonable background assumptions based on inductive reasoning accepted at the time, accident and serendipity,[2] all of these supplied the ingredients for Eratosthenes' discovery. But they had to be connected by finding the right place to start. It is easy to see the connection after someone has made it. It is easy to see all the pieces after someone has laid them out. The number of potential pieces and trails, however, are always infinite, and one can possess all the facts in the world, like Sartre's self-taught man, and unless one can connect them, all that results is "Now what?"

Eratosthenes had many biases, but these biases helped him make connections. He accepted the rationalism of the ancient Greeks and the idea that mathematics

---

[1] Eratosthenes also had to assume that Syene and Alexandria were on the same meridian of longitude, on the same line that would connect these cities and connect the north and south poles. This assumption is not only necessary for an accurate measurement of the Earth's circumference, it is wrong. Hence, by today's standards of measurement, his result, although remarkable, is not accurate. Modern measurements give a 24,800 mile circumference.

[2] Luck has played a role in many scientific discoveries. Was Eratosthenes looking for facts about water wells and shadows? What if he had not read the particular book that described the shadowless well in Syene?

FIGURE 3-4

By using basic geometry, a few facts, and a lot of insight, Eratosthenes was able to show that if the five hundred miles from Syene to Alexandria was 1/50 of the entire circumference of the Earth, then the Earth must be approximately twenty-five thousand miles in circumference (500 × 50 = 25,000). By measuring the angle a of the cast shadow in Alexandria to be 7 degrees, he knew the angle b to be 7 degrees or about 1/50 (7/360) of a circle. Although the angles in this illustration are exaggerated, note that a crucial premise in Eratosthenes' conclusion is the geometric principle that a line L intersecting two parallel lines creates alternate angles that are equal (a = b).

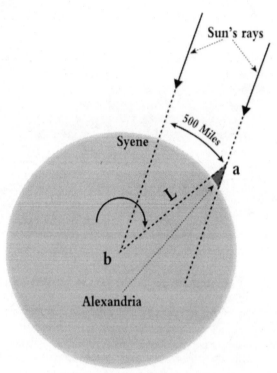

and deduction granted us a special power to transport our understanding to invisible places. Today, physicists and astronomers continue this tradition. They tell us reasonable stories of incredible places trillions and trillions of miles away, of stars bigger than half our entire solar system, of billions of galaxies, some thousands of times larger than our Milky Way with its 100 billion stars, of neutron stars where the matter is so dense that a spoonful would weigh many tons, of black holes, places that seem to be "holes" in our three-dimensional reality, and of quasars that seem to have the brightness and energy of millions of galaxies put together.

For most of the people of Eratosthenes's time, it was hard to believe that the Earth was so large. The distances they were used to extended no more than a thousand miles. Eratosthenes's conclusion conflicted with what these people were

FIGURE 3-5

Andromeda galaxy: like the galactic merry-go-round of our home galaxy, billions of stars swirl around the great galaxy of Andromeda. At more than 500,000 times farther away from Earth than the nearest star to us, massive stars in this galaxy, many trillions of miles apart, appear as packed-together white dust. Photo courtesy of Palomar Observatory/Caltech.

familiar with, and no one could take a picture from space confirming that he was right. But the reasoning trail he followed gave him a reliable basis to believe that the Earth was really this big. Today, astronomers follow similar reasoning trails and tell us a story about the size of our universe that is also hard for most people to believe. Relatively close to our fragile home, they tell us that our Sun is about 93 million miles away. If you could drive in a magic car at sixty miles per hour you would spend 175 years getting to the Sun. Yet the nearest star to us is three hundred thousand times further away than this. Astronomically speaking, the thousand or so stars that we can see at night with the naked eye are all relatively close to us. Most of them are part of a little section within the Milky Way galaxy. Stars on the other side of our galaxy are twenty thousand times farther away than the closest star to us. Our galaxy in turn is a swirling island home in a vast sea of similar island homes. The closest major galaxy to our Milky Way, the Andromeda galaxy, is more than 550,000 times further away than the closest star. So, if it would take 175 years

FIGURE 3-6

Just as Eratosthenes discovered the Earth was big, modern astronomers have discovered that the universe is very big. They believe billions of galaxies exist. This is a picture of a cluster of galaxies called the Coma cluster. It is relatively close to Earth, yet these galaxies are over a hundred times farther from Earth than the Andromeda galaxy. Photo courtesy of Palomar Observatory/Caltech.

to get to the Sun traveling at sixty miles per hour, it would take $300{,}000 \times 550{,}000 \times 175$ years to get to Andromeda! And yet Andromeda is a close neighbor in an immense astronomical neighborhood that extends billions of times further away than the distance to the closest star (see Figures 3-5 and 3-6). Like Eratosthenes, with some careful observation and by following logical and mathematical trails, we have continued to push our understanding to unimaginable places, places we will undoubtedly never visit.

Scientists who work in both astronomy and physics also tell us a reasonable historical story of how our existence on this fragile blue speck we call Earth is the result of billions of lucky physical accidents that have taken place over an incredibly long time—about 15 billion years. To name but one, it is now believed that our Moon was formed by an accidental collision between the Earth and a large asteroid or a large chunk of debris left over from the formation of our solar system. Without this collision, our atmosphere would be very different; there would be no tides, and most likely no life on Earth.[1] Similarly, evolutionary biologists study life's history on Earth, carrying out the implications of Darwin's theory of natural selection, and describe for us the myriad, contingent trails involved in species development. To name but one important part of just one never-to-be-repeated trail, they tell us that if another asteroid or comet had not struck the Earth about 60 million years ago, leading to the extinction of the dinosaurs, the human species would never have evolved.

Putting these beliefs all together, modern science has constructed for us a humbling, but reasonable world view of where we are and where we have come from. Our Earth is a very small speck of fragile dust in a very old and large universe, and our importance is far from obvious. This world view could be wrong, but the quality and quantity of the evidence, and the rigor of the inductive methodologies used, are the same as with the cigarette-smoking example described above. Thus, like the cigarette example, the beliefs that make up this world view are some of the most reliable beliefs we have, at least in broad outline. Most contemporary philosophers of science believe that no matter how much logical deduction and mathematical analysis is used in science, at some point the world must be checked by observation for the confirmation of a belief. But many great scientific and technological discoveries are instituted by people sitting at desks, following the elegant trails of mathematical equations. Many times these trails lead to uninteresting conclusions.[2] Often they lead to a conclusion that is inconsistent with experience, and we must start over and examine new trails. But they are powerful tools that sometimes let us see very far.

Key Terminology

Inductive arguments
Deductive arguments
Universal Generalization
Falsification
Relativism
Standards of appraisal
Induction by enumeration
Independent or higher-order inductions
Representative sample
Positive confirming instance
Genuine confirming instance
Evidence from personal experience

Appeals to testimony
Appeals to authority
Appeals to popularity
Correlation
Causation
Controlled study or experiment
Variables
Repeating experiments
Corroboration
Eratosthenes
World view of modern science

Concept Summary

Logicians distinguish between deductive and inductive reasoning. Although both forms of reasoning attempt to take us by inference from what we think we know

---

[1] See *What If The Moon Didn't Exist? Voyages to Earths That Might Have Been*, Neil F. Collins (Harper Collins, 1993).
[2] Remember this when you study Chapters 9–10. You will need to think like Eratosthenes, but many of your trails will lead nowhere.

to be true to what else is true, unlike deductive reasoning, where in principle conclusive support can be given for conclusions, **inductive reasoning** can never provide conclusive evidence for inductive conclusions. **Technically all inductive arguments are invalid.** Rather than conclude from this that it is irrational to use induction—we don't really have this choice because we use induction all the time—we must recognize that different methods of appraisal apply to judging the worth of inductive and deductive inferences, and that the goal of induction should be to learn how to make this form of reasoning stronger.

Judgments of the worth of inductive inferential claims must always be relative to alternatives; some inductive arguments are better than others, even though there is no clear demarcation between good and bad inductive arguments as there is between valid and invalid deductive arguments. Without any other supporting inductions, **induction by enumeration**—finding more positive cases for a belief—is better than having only one positive case for a belief. Using techniques of **representative sample** are better than simple induction by enumeration, and a distinction can be made between finding **genuine confirming instances** for a belief and **mere positive cases** for a belief. Beliefs based on inductive inference are also stronger if they are supported by independent, **higher-order inductions**.

It is important to have the proper attitude when appraising inductive claims. It is important to recognize that evidence can exist for a belief even though it is not known absolutely whether the belief is true. Because most people tend to think of belief acceptance as categorical—they either totally accept a belief or they don't—we must learn that it can be rational to tentatively accept a belief, even though we may find the belief false or unreliable someday. Rather than seek beliefs that give us rest, decisiveness, and closure, the goal of inductive reasoning is to constantly test and correct, to constantly be open to learning how to make our inductive inferential claims better, to accept beliefs that, based on their support, are likely to be reliable guides to the future.

The distinguished human activity that incorporates the best techniques of inductive reasoning is science. After centuries of being tricked by nature, scientists have devised self-corrective inductive techniques to test beliefs and theories that will make these beliefs and theories likely to be reliable guides to the future. Using creativity, insight, observation, deduction, and these techniques of induction, science has not only created powerful technological tools and reliable beliefs related to our health, but also a startling world view about our evolution and place in the universe. By establishing reasonable inductive theories and assumptions and then following long deductive reasoning trails, science allows us to "see" a reality that otherwise would forever remain hidden from us.

**EXERCISE
I**

True or False

*1. With enough evidence, inductive arguments can be made as strong as valid deductive arguments.

2. The amount of evidence is so overwhelming that we now know with absolute certainty that cigarette smoking causes lung cancer.

3. The observation of a correlation between an increase in smoking cigarettes and increased rates of lung cancer (3/100,000 lung cancer rate in the 1930s and 25/100,000 by 1955) was a good place to start an investigation on the possible causal relationship between cigarette smoking and lung cancer, not conclude one.

4. Induction by enumeration is stronger inductive evidence than a representative sample.

5. In a controlled study, it is possible to control every single variable.

6. A conclusion based on a controlled study that is repeated several times by independent investigators is better than one that is done only once.

*7. False beliefs can work in making true predictions, and it is almost always possible to find mere positive cases in support of a belief.

8. Inductive arguments always involve inferences from particular premises to general conclusions.

9. To be reasonable and logical means never to be imaginative, because logic and imagination are incompatible.

*10. Inductive arguments must always be judged relative to alternatives, and there is no definitive demarcation between good and bad inductive arguments as there is with valid and invalid deductive arguments.

11. Universal generalizations can be falsified by a single negative instance.

12. Because of the nature of inductive reasoning, we should always avoid believing in universal generalizations, and there is no rational basis for accepting any belief as reliable.

EXERCISE II

Identify which of the following are deductions and which are inductive generalizations.

1. Harris has a Honda, and it is reliable. Nguyen has a Honda, and it is reliable. Kanishiro has a Honda, and it is reliable. Therefore, all Hondas must be reliable.

*2. All Fords are expensive. So, Tanya's Ford must be expensive.

3. All of the apples in the barrel are rotten. So, one in the middle must be rotten.

4. A principal cause of lung cancer is cigarette smoking. So, John's smoking is dangerous.

5. April smokes and has poor health. Bob smokes and is sick a lot. Yuriko smokes and has trouble breathing in the evening. So, smoking contributes to poor health.

*6. Julio smoked for forty-five years and then had lung cancer when he was sixty-five. So, his smoking must have caused his lung cancer.

7. Presidents of the United States are always wealthy. So, Bill Clinton must be wealthy.

8. In the past, all governments that overspent on defense and neglected the welfare of their people eventually collapsed. Hence, any present or future government that overspends on defense and neglects the welfare of its people will collapse.

9. The Supreme Court is about to make euthanasia legal. Hence, it is inevitable that suicide clinics will begin to pop up next to abortion clinics.

10. Many women who have had abortions later regret it. So, we should prevent all women from having abortions for their own good.

EXERCISE III   Interpret each of the following as a deductive or inductive argument. Structure the argument into premises and conclusion and then give a brief argument for your interpretation based on the presence of indicator words or the apparent nature or the inferential claim.

*1. Because the United States and the Soviet Union have not been able to agree for the past forty years on how to significantly reduce their nuclear arsenals, it is unlikely that they will be able to do so in the near future.

2. All Republicans support the president's Supreme Court nominee, and John Eless supports the president's Supreme Court nominee. Hence, there is no doubt that John Eless must be a Republican.

3. My family has bought Chrysler cars since the first ones were made. These cars have always been reliable. Thus, U.S.-manufactured cars are better than Japanese-manufactured cars.

4. Most presidents of the United States have been rich. Therefore, the winner of the presidential election in the year 2000 will probably be rich.

*5. Iraq's acceptance of the United Nations agreement is a necessary condition for a cease fire. A cease fire is necessary to stop the bombing of Iraqi cities. So, unless Iraq accepts the United Nations agreement, the bombing will continue.

6. "John is having an affair with Jill," thought Deirdre. "When I called John this evening, his phone was busy. When I then called Jill her phone was also busy."

7. It is most likely that the murderer knew the victim. First, there were no signs of forced entry into the house. Second, according to the neighbors, the victim's dog always barked at strangers, and

on the night of the murder no one heard the dog bark. Third, two half-empty glasses of wine were found in the living room where the victim's body was found.

8. If we pay our medical bills and car payment this month, we will not have enough money for basic necessities. We have to have enough money for basic necessities. So, if we pay our medical bills, we can't make the car payment this month.

9. At Kansas State University only females weighing more than 110 pounds are members of the softball team, and all members of the softball team must have at least a C grade point average. It must follow that any female at Kansas State either weighing less than 110 pounds or with a grade point average below a C is not on the softball team.

*10. People argue that sex education should be taught in public schools. Why is this necessary? The worn-out reason is that a lot of parents do not talk about it at home, so it must be taught in school. Yet, since this trend started in the 1970s, VD and pregnancy among teenagers and even preteens has sharply gone up. Why? I thought sex education was supposed to reduce these problems, not increase them. The answer is that a sex education course is a "how-to-do-it" course, nothing else. Sex education is stupid.

EXERCISE IV

Essays. Appraise the following arguments from an inductive point of view by describing how you might investigate each claim such that genuine positive support could be separated from mere positive cases of support. Hint: First think of possible cases that would most seriously weaken each claim, and then devise a hypothetical experimental test or investigation that would force nature to reveal the negative case if it exists. (Remember that in the cigarette smoking example, if we found that of those who had lung cancer all lived in a polluted environment, this would seriously weaken the claim that smoking cigarettes caused the lung cancer.)

1. Homosexuality is obviously morally wrong and disapproved of by God. Why else would so many homosexuals have AIDS?

*2. A recent survey of women with college degrees showed that many of the husbands of these women have had heart attacks. So it seems that a higher price than expected must be paid for a college education for married women.

3. A recent survey at a major university showed that students who took large doses of vitamin C for one year found that they had fewer episodes of cold and flu compared to the year before when they took no extra vitamin C. Therefore, taking large doses of vitamin C must be effective in preventing cold and flu episodes.

4. An article in *Reader's Digest*, entitled "Why Marijuana is Your Enemy," cited the work of government-sponsored scientist Robert Heath. Heath claimed to have done a controlled study using rhesus monkeys that demonstrated clear signs of a causal connection between marijuana use and brain damage. Six rhesus monkeys in one group were given daily doses of marijuana, and six rhesus monkeys in another group were not. After a specified period of time all the monkeys were killed and their brains were analyzed. All six monkeys in the marijuana group showed enlarged synapses—a key part of brain function—implying brain impairment. (Note: A rhesus monkey weighs only about six pounds, and Heath administered the marijuana to the monkeys by injecting directly into the monkeys' lungs the equivalent of twenty-five human marijuana cigarettes per day of pure THC, the psychoactive ingredient in marijuana.)

5. Analyze exercise III, number 10 (the sex-education claim).

ANSWERS TO STARRED EXERCISES

I.

1. False. Technically, all inductive arguments are invalid. With inductive arguments, no matter how much evidence we have in terms of true premises or positive support, it is always possible that the conclusion could be false.

7. True. From a generalization such as "All the apples in the barrel are rotten" we can correctly predict that some in the middle are rotten, even if the generalization is false. This and other examples (the UFO example, for instance) show that we need techniques of analysis that demonstrate more than mere positive confirmation for a belief.

10. True. Life is tough and uncertain, but we have to act, and the best we can do in most situations is choose the most reasonable alternative. The goal is to learn how to learn, how to make inductive claims stronger.

II.

2. Deductive. Deducing from a general statement: "All Fords are expensive," a particular instance: "Tanya's Ford must be expensive."

6. Inductive. Concluding a causal connection: Julio's smoking caused his lung cancer, from the mere time sequence: he smoked, then got lung cancer, is a type of generalization.

III.

1. **Conclusion:**   It is unlikely that the United States and the Soviet Union will be able to agree on how to reduce their nuclear arsenals.

   Premise: They have been unable to do so for the past forty years.

Inductive argument. Using past experience, this argument is generalizing about the future. The indicator word *unlikely* helps us understand that an inductive claim is being made.

5. **Conclusion:** Unless Iraq accepts the United Nations agreement, the bombing will continue.

   Premises: Iraq's acceptance of the United Nation's agreement is a necessary condition for a cease fire. A cease fire is necessary to stop the bombing of Iraqi cities.

   Valid deductive argument. If the two necessary conditions mentioned in the premises are true, then the conclusion is true. The indicator word *so* helps identify the conclusion. The author is deductively carrying out the implications of the necessary conditions mentioned in the premises. If Iraq wants the bombing to stop, there must be a cease fire. If Iraq wants a cease fire, then they must accept the agreement. So, if they want the bombing to stop, they must accept the agreement (Either they accept the agreement or the bombing does not stop.)

10. Although there are other possible interpretations for this passage, one argument contained within it is the following.

    Conclusion: Sex education courses taught at public schools have caused an increase in VD and pregnancy among teenagers.

    Premises: Since sex education courses started at public schools, there has been an increase in VD and pregnancy among teenagers.

    Inductive argument. The author is generalizing that there must be a causal connection between sex education classes and teenage sex problems from the fact that sex education courses and an increase in teenage sex problems happened at the same time. The author may think that this is conclusive evidence, but because the argument generalizes from the information contained in the premises, the argument is an inductive argument. In Chapter 5 we will see that this argument is a Questionable Cause fallacy, a type of very weak induction.

IV.

2. First of all, we don't know if the percentage of husbands who had heart attacks is an unusual number. Because this passage is concluding that women with college degrees somehow cause their husbands to have heart attacks, we would first need to survey a control group population of married women without college degrees and compare the rate of heart attacks for their husbands. Then, even if the percentage of heart attacks for husbands of women with college degrees is high compared to the

control group, we would need to design a survey to see if a third cause was not responsible for this result. In other words, the fact that women with college degrees is correlated with husbands with high heart-attack rates does not mean that the former caused the latter. For instance, it is possible that women with college degrees are attracted to men with certain characteristics, such as the desire and ability to have challenging but stressful careers. (It is also possible that men who have personalities or characteristics implicated in other studies as possible causes of heart attacks are attracted to educated women.) In short, various controlled experiments would need to be conducted in which different characteristics of the husbands of college-educated women are tested as the possible cause. For instance, two groups of college-educated women and their husbands could be compared. In one group would be women and husbands with stressful occupations, and in the other would be women with husbands with less stressful occupations. If a higher rate of heart attacks were found in the first group, this would confirm the hypothesis that stress is more likely a factor in heart attacks than just being married to a college-educated woman.

Chapter 4

INFORMAL FALLACIES I

All effective propaganda must be confined to a few bare necessities and then must be expressed in a few stereotyped formulas.

—Adolf Hitler

Until the habit of thinking is well formed, facing the situation to discover the facts requires an effort. For the mind tends to dislike what is unpleasant and so to sheer off from an adequate notice of that which is especially annoying.

—John Dewey, *How We Think*

Introduction

In everyday speech you may have heard someone refer to a commonly accepted belief as a fallacy. What is usually meant is that the belief is false, although popularly accepted. In logic, a fallacy refers only to an argument that is popularly accepted. Here is our definition: A *logical fallacy* is an argument that is usually psychologically persuasive but logically weak. By this definition we mean that fallacious arguments work in getting many people to accept conclusions, that they make bad arguments appear good even though a little commonsense reflection will reveal that people ought not to accept the conclusions of these arguments as strongly supported. Although logicians distinguish between formal and informal fallacies, our focus in this chapter and the next one will be on traditional *informal fallacies.*[1] For our purposes, we can think of these fallacies as "informal" because they are most often found in the everyday exchanges of ideas, such as newspaper editorials, letters to

---

[1] Most logic books distinguish between formal and informal fallacies. Formal fallacies are deductive arguments whose invalidity can be detected immediately by mere inspection of the argument form. Informal fallacies are said to depend more on the content. We will be blurring the distinction between formal and informal somewhat, because we will show that the best way to learn informal fallacies is to identify the "essence" or form of each one.

the editor, political speeches, advertisements, disagreements between people, and so on.

In analyzing informal fallacies we will have our first opportunity to see the importance of logical abstraction and categorization. As Hitler noted in the opening quote to this chapter, people are easily persuaded by using a few simple psychological formulas. Situations may change, and the people may be different, and so the content may change, but the basic tricks are the same. To help defend ourselves against these tricks, logicians identify these formulas, break them down into parts, analyze the logical mistakes, and give them names. Like knowing what a chair is, once you recognize one you know all chairs. Although recognizing fallacies will not always be a simple black-and-white process of identification, categorization saves time and intellectual effort, and provides us with a guide for criticizing arguments and knowing how to make them better.

So, we will be doing more than just giving names of fallacies and providing a few examples. We will now practice in earnest the slow, deliberate discipline of logical analysis by looking at each fallacy in terms of the following format.

Argument Structure: Conclusion (identify)
                            Premise(s) (identify)

Label & description: A one-sentence description of why the fallacy name fits. One sentence describing what is taking place in the premises or conclusion consistent with the label.

Argument Analysis: An analysis that provides an argument for why the conclusion is not well supported in terms of identifying one of the following:
1. **Reasoning:** Even if the premises are accepted as true, the inference to the conclusion is poor or weak.
2. **Questionable premise(s)**: The premise(s) are presumptive or unfair in some sense; the truth of the premise(s) can be easily questioned.
3. **Suppressed Evidence**: The premises are true and the reasoning valid or good, but specific relevant evidence is omitted, such that if it were provided, it would make a major difference in accepting the conclusion.

Logicians have identified hundreds of informal fallacies. We will map out the logical essences of twenty of the most common ones in terms of the above format. Notice that labeling the fallacy is a very small part of the above process of analysis. Most important is identifying arguments and applying what you have learned by providing an argument for why another argument is weak—applying the concepts of validity, invalidity, and soundness for deductive appraisal, and criteria for reliable beliefs for inductive appraisal. Notice, for instance, in the argument analysis section above the distinction between (1) and (2) requires that you know the very important difference between criticizing an argument's form or reasoning, and criticizing the argument's premises. Some of the informal fallacies we will be analyzing will be weak in reasoning, and we need not waste time and intellectual effort worrying whether the premises are true or fair. On the other hand, some fallacies will have valid reasoning, but questionable or unfair premises, so we should focus only on criticizing the premises. This distinction is not only crucial for a complete understanding of logic, but as a purely practical matter it is a powerful tool for staying on track in criticizing an argument. The most prevalent mistake students make in criticizing arguments is to confuse the two—to criticize the truth or fairness of the premise or premises when they should be criticizing the reasoning, or to criticize the reasoning when they should be criticizing the premise or premises. With the exception of suppressed evidence (we will discuss this fallacy in Chapter 5), because each fallacy will be weak in either the reasoning or premises, once you know the essence of a label, you only have to think about it once, so to speak. You just apply the fallacy analysis consistently.

So far, this presentation is probably rather abstract, so let's tie it down with an example. Consider the following argument:

EXAMPLE 4-1   According to the national Uniform Crime Report, the number of women arrested rose by 66.1 percent between 1970 and 1980, compared to only 6 percent in the number of men's arrests. Before the rise of feminism in the 1970s the percentage of women arrested had consistently been lower than that of men. After the rise of feminism the crime rate for women clearly went up. Hence, there can be little doubt that the rise of feminism in the 1970s caused an increase in female crime.

In following the format outlined above, the first thing we should do is identify the conclusion. With this argument the conclusion indicator "hence" makes this easy, and the sentences preceding the conclusion are the premises. Structured this way, the argument is a classic example of a Questionable Cause fallacy. Note its essential features. The conclusion states a causal connection between two events, that the rise of feminism caused an increase in female crime. The premises merely describe an association in time of the same two events, that feminism started in the 1970s and at about this same time female crime apparently increased. All Questionable Cause fallacies will have these features: a causal connection in the conclusion and a temporal association in the premises.

Now think about what is wrong with this inference. Suppose you sneezed right now, and shortly afterward a door in the room you are in suddenly slammed shut. Just because these events are associated in time—you sneezed and then the door slammed shut—it would not be wise to jump to the conclusion that your sneeze caused the door to slam shut. Lots of things happen around the same time other things happen. You might have been tapping your fingers against your desk, scratching your nose, blinking, coughing, typing, or sipping coffee. Or, the wind outside your home increased suddenly sending a gust through an open window in the room where the door slammed shut. So, the bare fact that something happens before something else is poor evidence that the first caused the second. This type of evidence by itself would make for a very weak inductive inference.

Although we will cover Questionable Cause fallacies in more detail in Chapter 5, here is what a complete analysis will look like in the above format:

EXAMPLE 4-1A

Conclusion:	The rise of feminism in the 1970s caused an increase in female crime.
Premises:	According to the national Uniform Crime Report, the number of women arrested rose by 66.1 percent between 1970 and 1980, compared to only 6 percent in the number of men's arrests.
	Before the rise of feminism in the 1970s the percentage increase of women arrested had consistently been lower than that of men.
	After the rise of feminism the crime rate clearly went up.
Label & Description:	Questionable Cause. There is a causal connection in the conclusion, and a temporal association in the premises.
Argument Analysis:	The reasoning of this argument is weak. Although the premises are relevant and may be true, they provide insufficient evidence for accepting the conclusion. Just because two events happen together in time does not mean that they are necessarily connected causally. The events happening together could be a coincidence and other factors could have been involved, such as changes in the economy disadvantageous to women or changes in police methodology in recording statistics. Until more evidence is cited to show that this one change (feminism) is more likely to be the cause than other changes happening at the same time, this argument is a very weak inductive argument.[1]

All of the fallacies covered in the next two chapters will have an ideal essence that we can use as a guide for consistent analysis. The essence of the Questionable Cause is: Causal statement in the conclusion; temporal statement in the premises; weak inductive reasoning, but premises at least relevant to the conclusion. To be

---

[1] Although not relevant to analyzing this argument as a Questionable Cause, we could also question the presumption that a 66.1% increase in female crime represents a dramatic increase in female crime compared to that of men. A large percentage increase (for women) does not necessarily translate into a large number of female arrests if the total number arrested in the previous decade was very low, and a small percentage increase (for men) can still translate into a large number of men arrested if the total number arrested in the previous decade was already high.

called a Questionable Cause, a fallacy must have these features, and once these features have been identified the weakness is attacked always as a weak inductive inference.[1] Here is way to summarize and picture the essence of Questionable Cause:

EXAMPLE 4-1B

Conclusion:	A caused B.
Premise:	A happened, then B happened.
Label & Description:	Questionable Cause. There is a causal connection asserted in the conclusion and only a temporal association in the premise.
Argument Analysis:	Reasoning. Although the premise is relevant to the conclusion, it is insufficient to support the conclusion. Weak inductive inference.

Although both the interpretation of argument structure and the charge of fallacy require argument, notice that in this formal analysis the goal of the first three steps is descriptive. The conclusion and premise or premises are identified, and then a description is given of the formal characteristics of the premises and the conclusion. In the fourth step an explicit argument must be presented for why the formal characteristics identified are always logically weak, and the particular content must be connected with the formal characteristics. Soon you will be doing this yourself. You will not just passively label and describe, but must learn to argue, to stay on track, and to logically persuade. You will not learn fallacies the way a child learns not to do something, just because his or her parents say so. You must argue for a particular interpretation and make a case that a particular argument should not be accepted. Finding a fallacy is much more than just name-calling; it is the beginning of a dialogue. So, *you can't accuse someone of committing a fallacy unless you provide an argument.*

However, if you learn the essence of each fallacy, you will have a lot of help. You will always have a ready-made focus or theme for your argument. For instance, in the case of a Questionable Cause, the theme is always the same: You argue (no matter what the content) that even though the premises are relevant and may be true, the inference from the premises to the conclusion is insufficient because the reasoning is a weak induction. You still have to do some work with the content. You must combine what is being discussed with the formal weakness. In the case above, we had to think of other things that might have happened at the same time the rise of feminism occurred, to show that it is a weak induction to jump to the conclusion that feminism was the cause. However, at least you will always have the focus of the formal weakness to guide you in what content to think about.

---

[1] Although we could attack the premise and question or quibble whether feminism really began in the 1970s, we don't need to, and it would be off track to do so for a Questionable Cause analysis.

The Value of Abstraction

In following the process of analysis outlined above I will be forcing you to abstract. Although I often hear students complain that philosophy is "too abstract," abstracting is one of the most valuable intellectual processes that you will learn in your academic career, no matter what your field. Learning a bunch of facts is useless unless the facts can be connected with a pattern. Without patterns, forms, structures, or concepts you will always be lost in a jungle of individual trees and never be able to see the forest and where you are. People who are successful in their careers, whether in management and business, science and technology, or even art, music, and literature, are able to see beyond the details and confusing particulars of daily experience to the "essence" of things. Although it is true that philosophers often seek such essences for the pure, intrinsic joy of just knowing, the practical value of such essence-knowing should not be forgotten.

Western culture has always recognized a close relationship between the seeking of abstractions by philosophers and the timesaving discoveries of mathematicians. Like philosophy, mathematics is highly abstract, and for this reason people often conclude that mathematics just isn't their "thing," that it is too hard, too alien a discipline. Many students seem to have the opinion that algebra was invented to torture them in math classes. One purpose of this book is to show you that struggling in mathematics and formal logic reflects an attitude problem rather than an intelligence or aptitude problem. It necessitates "therapy" rather than some sort of neurological fix. Most students who have this attitude simply do not realize how easy is the analytic "game" that lies behind logic and mathematics. Mathematicians have a corny little saying that underscores this: "By an inch it (mathematics) is a cinch, by a yard it is hard." Later in this book, when we bring to a climax the process of logical abstraction with symbolic logic, I will show you that it is much easier to follow or create a reasoning trail using rules than it is to discover which trail to follow. "Remember Eratosthenes" will be our motto. If you remember this example from Chapter 3, you will understand that the mathematics was the easiest part; the process of connecting the facts and seeing where to start were much harder.

For now, to help your motivation a little, let's look at a few examples of the time-saving aspects of mathematical abstraction.[1] Multiplication can be seen as simple shorthand addition. Rather than adding up eight sets of eight things one by one, we learn that all we have to do is remember the rule $8 \times 8 = 64$. Likewise, algebra is actually shorthand arithmetic. We learn that shorthand notations such as 2^x mean multiply the number 2 by itself x times, so that 2^8 means multiply eight twos together for the result 256. The little x is called an exponent, and this shorthand algebraic expression is called exponential notation. This notation is a lot faster than writing out and calculating $2 \times 2 \times 2 \times 2 \times 2 \times 2 \times 2 \times 2 = 256$. We also learn that if we multiply any number with an exponent by the same number with a different exponent that $B^x \times B^y = B^{(x+y)}$. This little bit of formal mathematical

---

[1] Also think how much time Eratosthenes saved. Compare how much time it would take to walk around and measure the circumference of the Earth compared to multiplying 50×500!

knowledge comes in handy when you want to calculate the number of atoms in the entire universe. Yes, that's right, the entire universe! Because we know the approximate number of atoms in a gram of hydrogen (10^{24}), and the number of grams of hydrogen in an average star (10^{33}), and the number of stars in an average galaxy (10^{11}), and the number of galaxies in the universe (10^{11}), we simply add it all up as follows:

number of atoms/gram of hydrogen — 10^{24}

number of grams of hydrogen/star — 10^{33}

number of stars/galaxy — 10^{11}

number of galaxies in the universe — 10^{11}

number of atoms in the universe — 10^{79} $(24 + 33 + 11 + 11 = 79)$

If we were to write this number out it would be a 1 followed by 79 zeros. Even if it were physically possible, think how long it would take to count this number of atoms. Although our number is only an approximation,[1] counting this number of atoms one by one would surely lead to an even less accurate approximation we would make from losing track trying to count this enormous number and handing the project over to different generations (it would take many generations!). A little formalization can save an enormous amount of time, and, like Eratosthenes, we can know some amazing things by simply following a few formal rules.

Fallacies of Relevance

To begin our formalization of fallacies, we will start with those fallacies that violate one of the most important aspects of good reasoning—what I have metaphorically called staying on track and correctly following a reasoning trail. In Chapter 1 we used the example of the little girl baseball player and her father's reasoning to make the point that a lot of bad reasoning involves shifting attention away from what is **relevant** for testing our beliefs; that a lot of bad reasoning is simply a psychological excuse for not testing our beliefs, to not think about what we really should think about. Many informal fallacies are called fallacies of relevance because they shift attention away from the heart of an issue and distract us from the type of evidence we should be seeking in order to establish a conclusion. Technically expressed, all *fallacies of relevance* have premises that are logically irrelevant to the conclusion.[2]

Note that Questionable Cause is **not** a fallacy of relevance. Its premises at least have the virtue of being relevant to the conclusion. Because the cause of an event does happen prior to the event, premises that discuss a temporal order of events

---

[1] This result is simplified because it discounts less abundant elements and features of the universe that are not stars, such as interstellar dust, quasars, and exotic dark matter. Neutron stars, though, don't count, because they are not made up of atoms! The gravitational forces that formed these stars were so great that the atoms were crushed. However, if our estimate of the number of features or the number of atoms within any of the other features of the universe changes, we can quickly recalculate the total.

[2] The premises are psychologically relevant though, and this is why they are so often successful in persuading people to accept conclusions they ought not to accept.

are relevant to a discussion that those events are causally related. Questionable Cause is an example of a fallacy of weak induction. Although fallacies of relevance and fallacies of weak induction both have weak reasoning, the distinction between relevant and irrelevant premises is crucial for staying on track in criticizing arguments and focusing on what kind of evidence supports what kind of conclusion. Compare the Questionable Cause fallacy with the following.

Appeal to Popularity

One of the strongest psychological forces in human nature is the desire to "belong," to be accepted, to have friends, and to be part of a culture where behavior is somewhat the same so one is comfortable in how to act and think. For most people it is uncomfortable to feel "out of place," different, or weird. As with most human desires, there are probably very good evolutionary reasons for the naturalness of the need to belong. It is a myth that emotional attachments are intrinsically illogical; that logic and emotion are always opposed. Logic and emotion are meant to work together. We know from evolutionary anthropology that our pre-human ancestors lived a very harsh existence. Compared to the competition, our ancestors were frail creatures. Many animals had more powerful physical characteristics; they could run faster, see better, and had better offensive and defensive bodily weapons (claws, fangs, horns, and such). To survive, our ancestors needed to belong to each other and to think; they needed to cooperate and care about each other, and to devise, reason, and calculate. And they needed to apply these tools of survival at the right time and place most of the time, or we would not be here contemplating the relative merits of reason and emotion.

Problems emerge when we use the wrong tool for the job at hand. There are clearly times when it is less sensible to reason analytically than to just feel. It would be most inappropriate to tell a grieving mother and father that they are not being logical about their son's death in a war. Recall the discussion of bureaucratic euphemisms in Chapter 2. There are times when it is more reasonable to be emotional. This would be especially true if the parents were told that their son was killed by "collateral damage from friendly fire." On the other hand, we can be seriously distracted from the proper evaluation of something important to us if we react too quickly because of a powerful feeling. Consider the following argument.

EXAMPLE 4-2   We should support President Bush's Supreme Court nominee, Clarence Thomas. The American people are obviously very happy with the way the president is running things. Since his demonstrated leadership during the Persian Gulf War, his policies are obviously the policy of the vast majority of Americans. The president's approval rating is now one of the highest in history, even higher than former President Reagan. Most Americans are tired of hearing the far-left nitpicking from pro-abortionists, the hyper-liberal analysis from the editors of major newspapers, and the self-appointed righteousness of so-called civil rights leaders. Like President Bush, Middle America, the real America, wants this nomination.

In the spring of 1991, then–U.S. President Bush nominated Clarence Thomas to the Supreme Court. Thomas, who is African-American, would be replacing Thurgood Marshall, who is also African-American. The nomination was controversial. Unlike Marshall, Thomas held conservative views on such civil rights issues as affirmative action and equal employment opportunity. Furthermore, abortion rights advocates were suspicious of his views on abortion, and, perhaps most important, many legal experts questioned the depth of his experience with and scholarly production on constitutional issues. Clearly, the focus of the above argument was that we should support Thomas. So, in structuring this argument the conclusion would be: We should support President Bush's Supreme Court nominee, Clarence Thomas. The remaining sentences were offered as supporting premises.

What's wrong with this argument? Was it on track? Was it discussing the real issues? Note that the main focus of the premises is that the vast majority of U.S. citizens were happy with the way President Bush was handling things. He was popular after the Persian Gulf War and most of his decisions at that time were popular. The thrust of this argument was that because the Thomas appointment was a popular decision, anyone thinking about it should also have accepted it. It is not intrinsically wrong for people to be happy with the actions of their president, but the real issue here was why should we have been happy with this decision. Because this argument did not discuss why a majority were supporting the Thomas appointment, this argument was not on track, it was not discussing the main set of issues: the qualifications of Clarence Thomas and whether or not a candidate for the Supreme Court should or should not have certain views on civil rights and abortion. Although the argument discussed why Bush was popular in general, it did not discuss the logically relevant reasons why a majority were supporting Bush on this decision. What were the reasons related to Thomas's qualifications that a majority was supporting him? Did Thomas have the special legal and analytic skills, the depth of thinking, and a distinguished background in the law normally required for the highest court in the land? The American Bar Association had assessed Thomas as only "qualified," a lower grade than usual for a Supreme Court nominee. Five previous nominees, including one rejected by the Senate, were all ranked as "well-qualified." Was Thomas biased toward conservative views, or was he an innovative, independent thinker who believed that the old methods of busing and affirmative action had not improved the lot of minorities?

When majority support is appealed to in the premise or premises of an argument, but no reference is given to the logically relevant reasons for the majority support, an *Appeal to Popularity* fallacy is committed. This argument is best classified as an Appeal to Popularity (rather than to Authority, as in the next fallacy covered), because its basic tone is: Support Thomas because most people do. Here is how criticism of this argument would be formalized.

EXAMPLE 4-2A

Conclusion: We should support President Bush's Supreme Court nominee, Clarence Thomas.

Premise:	Because a vast majority support Bush's decision to nominate Clarence Thomas.[1]
Label & Description:	Appeal to Popularity. The premise cites popularity ("vast majority") as a reason for the conclusion.
Argument Analysis:	This argument is weak in the reasoning. Even if it is true that a vast majority supports the Thomas decision, the inference to the conclusion is weak. We are not given the logically relevant reasons why a majority is supporting this decision. What are Thomas's qualifications, and how do these qualifications fit the job of a Supreme Court justice? The majority may have very good reasons for supporting Thomas, as may President Bush, but these reasons need to be discussed. But they are not discussed, so it is premature to accept this conclusion.

And here is the formal essence of all Appeals to Popularity.

EXAMPLE 4-2B

Conclusion:	Do X, believe X, or (in the case of an advertisement) buy X.
Premise:	Because a majority does X, or believes X, or buys X.
Label & Description:	Appeal to Popularity. The premise cites popularity (list the key phrase) as a reason for the conclusion.
Argument Analysis:	Reasoning. Even if it is true that a majority supports something, the logically relevant reasons for that support need to be discussed. Develop by specifying the reasons that should be discussed.

One of the most important mistakes students make in criticizing Appeals to Popularity is to attack the premise rather than the reasoning; that is, to question the truth of the premise rather than the inference given the truth of the premise. We could indeed question the truth of the above premise. We could ask for evidence, such as a political poll, showing that a majority of U.S. citizens support the Thomas decision. But this focus would not be on track in terms of criticizing an Appeal to Popularity. All Appeals to Popularity should be criticized by focusing on the inference: even if the premise is true, the inference is weak and the conclusion not supported because the premise is irrelevant to the conclusion; the main issue or

---

[1] In analyzing a fallacy we need list only the key premise or premises that contain the fallacy appeal. Most often, arguments will mix relevant and irrelevant appeals, and these arguments are clearly better than a bare appeal to popularity. However, even when this happens, isolating the irrelevant appeals is a reminder that we should not be persuaded by these appeals alone.

issues are not being discussed. In other words, we need not criticize the truth of the premise, because the focus should be on the relevance of the premise.[1]

With a little reflection on what you have seen in the popular media (TV, magazines, newspapers), it should be apparent that many advertisements use Appeals to Popularity. "Nissan is number one in the state of California," "Kool cigarettes are number one in Hawaii," "Everyone is voting for Mayor Fasi," "Visa is welcome everywhere," "It's Miller time" (showing what is supposed to be a popular activity—having a Miller beer after work). Note how these are fallacies of relevance. If you want to buy a car, what is most relevant to know, that many people are buying the car or that a lot of people are buying the car because it is a good car? Kool cigarettes may be the most-often purchased cigarette brand in the state of Hawaii, but should you smoke? And if you do, why Kool? The polls may show that a majority of people are voting for Mayor Fasi, but what are his qualifications, what is his political record, and are his policies consistent with what you want? Visa may be welcome everywhere, and being accepted by merchants is relevant in this case, but is it the best credit card to have? Does it have the lowest interest rate and yearly service charge? The whole world may drink Miller beer when people get off work, but should you?

The version of Appeal to Popularity discussed thus far is nicknamed the **bandwagon appeal**, a historical reference to a wagon that held the band of musicians for a parade that everyone followed. The psychological appeal is to our natural desire to be a comfortable part of a large group. But many people pride themselves on being smart enough to see through such blatant crowd-pleasing and herd-mentality appeals. They pride themselves on having risen above such mob appeals, of being their own person. However, just like those who need to feel like part of a large group, these people need to have their identity reinforced, so propagandists and advertisers are ready for this pride with an elitist or snob appeal. Rather than appeal to a majority in the premise, the **snob appeal** solicits a sense of being popular in a distinctive way, of being different, of being more handsome, more beautiful, more intelligent, or more appreciative of the finer things in life. It is no accident that most advertisements show handsome men, beautiful women, or attractive couples. They hint that you, too, can be better than the average person, just like the models, if you use the advertised product.

In the 1960s, a magazine advertisement for Camel cigarettes showed a scene at the beach with a number of people in line to buy a snack. The caption asked simply, "Can you tell which person in this picture is the Camel smoker?" The first person in line was an overweight woman wearing a bikini bathing suit, her rolls of fat overflowing over the edges of the suit. Next was an overweight man with no tan,

---

[1] Some logicians will categorize Appeals to Popularity (and Appeals to Authority, the next fallacy) as fallacies of weak induction. As such, the conclusion is treated as a kind of generalization where the reasoning is something like this: "Well, if a majority of people are in favor of X, there must be (inductive generalization) something about X that is good." But treated this way, the premise is then relevant to the conclusion, because all fallacies of weak induction have premises that are relevant to the conclusion. One main task of this book is to help you learn the discipline of staying on track, and in everyday acts of persuasion there are so many ways that attention is shifted away from the relevant issues for testing beliefs, so I have chosen to view these fallacies as fallacies of relevance. If we are going to conclude that a policy, product, or course of action is good, then we want to know what that something is that makes it good.

wearing a ridiculous-looking Mickey Mouse™ hat. Then a skinny woman with ugly glasses askew, trying to hold onto a tube-shaped flotation device with a duck head with her irritable kids hanging all over it trying to pull it away from her. Finally, at the end of the line was "Mr. Beautiful," impeccable tan, rippling muscles, smart-looking dark glasses, hair stylishly combed, and very patient and cool. Right, the Camel smoker.

Often advertisers will switch back and forth between bandwagon and snob appeal versions of Appeal to Popularity with no change whatsoever in the product. Miller beer has done this throughout the years. In the 1960s and '70s, Miller beer was advertised as "The champagne of bottled beer," a beer obviously not for the masses. But by the 1980s, as noted above, Miller was supposed to be the beer of choice for the average person getting off work: "It's Miller time." As media techniques and special effects have progressed, verbal appeals have been replaced with multimillion-dollar dramatic, MTV-like video montages of beautiful people being friendly and happy together at the beach, a party, or going out on the town ("Pepsi . . . It's just right," "The night belongs to Michelob"). Political campaigns have adopted the same techniques, with discussion of the issues and a candidate's qualifications replaced with sound bites and carefully selected short media clips, showing the candidate doing something popular. There is nothing wrong with being entertained, but there is a time and a place for entertainment and times when it is more appropriate to do a little work and think critically and discuss matters relevant to important decisions.

Appeal to Authority

As the pace of life quickens, as opinions on right and wrong, true and false seem to multiply exponentially, as the amount of information available to us begins to feel like an enormous wave that will produce chaos and insecurity rather than organization and clarity, it is natural to seek shortcuts and secure foundations. There just doesn't seem to be enough time to sort through it all. One shortcut that we often use is to turn to people whom we admire for advice, people whom we think of as being more experienced on a topic, or in general more intelligent, or happier and more organized in their lives. Or, like a child that imitates his parents, we look to famous people for ideas on how to think and behave.

Like the natural psychology of wanting to belong, which fuels Appeals to Popularity, our desire to seek out shortcuts through expert opinion and advice from others surely has a practical foundation. There isn't enough time for every person to assimilate all the relevant information for every decision, to research every topic related to every alternative course of action. However, there are times when this need for fast advice can be misplaced, the acceptance of advice being too fast and the shortcut in the reasoning too short. An *Appeal to Authority* fallacy is committed when an improper appeal is made to alleged expert advice. Consider the following argument from a husband attempting to persuade his wife to buy a Mr. Coffee automatic coffee machine.

EXAMPLE 4-3   Wife: "Which coffee machine should we buy? There
                        are so many to choose from in this store."

               Husband: "Let's buy a Mr. Coffee machine."

               Wife: "Why? There are so many other kinds."

               Husband: "Because Joe DiMaggio (referring to a TV
                        commercial) says it makes the best coffee he
                        has ever tasted."

The conclusion of the husband's argument is that they should buy a Mr. Coffee machine, and the main reason he gives is that it is endorsed by Joe DiMaggio, a famous former baseball player and husband of Marilyn Monroe. But Joe DiMaggio is an expert on baseball, not coffee. Furthermore, note that the husband is not discussing in any depth the most relevant aspects for their decision: the quality of the product, length of warranty if any, price, and the comparison of these features with other coffee makers. The TV advertisement has worked on the husband. The positive psychological feelings he feels toward Joe DiMaggio have been transferred to Mr. Coffee. Joe always seems like such a nice guy and always seems to be in control, he must know what he is talking about. But the argument is not on track, the husband has not focused on the relevant items for buying something. Although some reference is made to the coffee maker being the best, no comparative reasons are given as to why it is the best. And because Joe DiMaggio is not a proper authority on coffee, his reasons why it makes the best coffee need more scrutiny. Here is a complete analysis:

EXAMPLE 4-3A

Conclusion:	Buy Mr. Coffee.
Premise:	Because Joe DiMaggio says it makes the best coffee he has ever tasted.
Label & Description:	Appeal to Authority. There is an appeal to authority (Joe DiMaggio) in the premise.
Argument Analysis:	This argument is weak in its reasoning. Even if the premise is true—that Joe DiMaggio is not just acting and is really endorsing Mr. Coffee—Joe DiMaggio is not a proper or relevant authority on coffee. What should be discussed is product quality, price, and a comparison of these features with the competition.

Although this is not an earthshaking example of great consequence, note how often these simple appeals are found in the popular media. Many advertisements feature famous people—popular singers, sports and TV personalities, and movie stars—because appeals to authority work. By the time he was twenty-eight, basketball player Michael Jordan was endorsing fifteen products, including Wheaties,

McDonald's, Nike, and Gatorade.[1] For each endorsement he was paid millions of dollars, allowing Jordan to add an extra $35 million a year to his $60,000 per game-minute salary. People are routinely distracted from what they should think about and accept instead the simple shortcut of identifying with a famous personality.

Political advertisements also use Appeals to Authority. In the United States, both the Democratic and Republican parties use entertainment personalities to endorse their respective presidential candidates. During the 1970s and early '80s, the Republicans would use Frank Sinatra, Bob Hope, and Charlton Heston, and the Democrats would use Gregory Peck and Henry Fonda. By the 1992 presidential campaign, the Republicans were still using Bob Hope and Charlton Heston, but the youthful-appearing Democrats used Glenn Close, Richard Gere, Barbra Streisand, and Wynton Marsalis to endorse Bill Clinton.

The fact that competing products and political candidates are endorsed by different famous personalities shows that most relevant and decisive are issues at a deeper level than mere endorsement. When something is important, and surely voting for a political leader is, it is time to turn on our critical information-gathering ability, to discipline ourselves and focus on what is really important. Why is the candidate being endorsed by a famous personality? What are the candidate's qualifications, his or her stand on the issues, and why are these qualifications and positions better than his or her opponent's?

But what if the authority used is a **relevant** or **proper authority**? What if the authority figure is an acknowledged expert on that which is being endorsed? Consider these arguments.

EXAMPLE 4-3B   Carl Sagan has stated that biological evolution is a fact, not a theory. As an eminent scientist, he must know what he is talking about.

EXAMPLE 4-3C   Vitamin C has considerable potential usefulness in cancer therapy. The Nobel Prize–winning chemist and molecular biologist, Linus Pauling, has endorsed the benefits of vitamin C for general health and cancer therapy for years.

Scientist Carl Sagan became a somewhat famous TV personality in the 1980s: He appeared on the Johnny Carson show, wrote several well-received books and a series of articles on science for *Parade* magazine, a Sunday newspaper supplement in most U.S. cities, and starred in and led the production of the popular science series *Cosmos*. He has contributed significantly to educating the general public on the world view of modern science through his belief that the average person can understand technical scientific subjects if the details of these subjects are explained in nontechnical language. Darwin's theory of human biological evolution played an important role in this education, especially its implications for understanding

---

[1] The Gatorade spot was the most elaborate and psychologically vicious. It showed video clips of many of Jordan's most famous basketball moves interspersed with him drinking Gatorade and young kids trying to play like him. This was accompanied by a jingle ("I want to be like Mike"), and ended with the caption, "Be like Mike. Drink Gatorade." Nutritional critics referred to Gatorade as "overpriced, colored sugar-water."

human nature, our chances of survival, and the value of species diversity, preservation, and respect for nonhuman life. If the Darwinian theory is true, then the human species is the result of billions of chance or contingent events that would not be repeated again anywhere in this vast universe if we were to destroy ourselves with the weapons we build with our "intelligence."[1] Nowhere is this point better summarized than by Loren Eiseley in his classic book *The Immense Journey.*

> Lights come and go in the night sky. Men, troubled at last by the things they build, may toss in their sleep and dream bad dreams, or lie awake while the meteors whisper greenly overhead. But nowhere in all space or on a thousand worlds will there be men to share our loneliness. There may be wisdom; there may be power; somewhere across space great instruments, handled by strange, manipulative organs, may stare vainly at our floating cloud wrack, their owners yearning as we yearn. Nevertheless, in the nature of life and in the principles of evolution we have had our answer. Of men elsewhere, and beyond, there will be none forever.[2]

Scientists such as Sagan and Eiseley are adamant in their support for Darwin's theory not only because they believe it places the human species in a context that acts as an enlightening antidote to the anthropocentric and us-vs-them attitudes that produce so much destruction and violence on our fragile Earth, but because the inductive scientific evidence for the theory is overwhelming. Not only do numerous examples exist of genuine confirming instances for the theory recorded in the fossilized rock pictures of animals that lived millions of years ago, but independent areas of investigation, such as embryology and DNA research, point to the same conclusion: the few million species of plants and animals alive today are the lucky descendants of hundreds of millions of extinct species.

Consider, though, the rational gap that exists between merely citing Carl Sagan as an authority who believes that Darwin's theory is true and accepting the theory as true by being aware of all the evidence, or at least a substantial part, that Sagan is aware of. Although appealing to a relevant or proper authority is clearly better than appealing to a famous person who is not an expert on what is being endorsed, given an important issue or decision, the same inference problem exists as with that of appeals to an ***irrelevant*** or ***improper authority***—the reasons the authority has for endorsing something are most relevant and should be discussed. To stay on track we should discuss and test beliefs by learning about the evidence authorities use as a basis for their endorsements. Carl Sagan, famous scientific TV personality, should not persuade us. Rather Carl Sagan *and* the scientific evidence should persuade us.[3] Note also that people who merely accept Sagan's authority as decisive

---

[1] In my opinion the awareness of the full ramifications of our ***historical contingency*** is one of the most important items one should learn as part of a college education. Although only one rather dramatic event, consider that if the dinosaurs were not destroyed by a comet striking the Earth 60 million years ago, the human species would never have evolved. For more on our historical contingency (cosmological, biological, and cultural) I modestly suggest *Science and the Human Prospect* (Wadsworth, 1989), by Ronald C. Pine.

[2] Loren Eiseley, *The Immense Journey*, Vintage Books ed., (New York: Random House, 1959), page 162.

[3] Another reason that we should be sensitive to even expert appeals to authority is that experts should be forced to explain what they know in terms the lay public can understand. Claims that something is too complicated for the average person to understand are most often excuses not to try to communicate, and such claims should never be used as a convincing premise.

deprive themselves of an important education; they are deprived of learning the amazing story of our historical contingency, the scientific detective story conducted daily all over the world in deciphering fossil pictures, and the intricacies of DNA, the blueprint for each of us and all life on Earth.

Most scientists accept Darwin's theory of evolution, but this mere fact alone should not persuade us.[1] Although few scientists doubt the theory of evolution, just as in the case of famous personalities endorsing different products or presidential candidates, in many other areas relevant experts disagree. Although it is true (4-3C) that Linus Pauling was a Nobel Prize–winning chemist and molecular biologist, his views represent a distinct minority of expert opinion on the overall medical value of vitamin C. Most medical doctors believe that the evidence shows that although vitamin C is an essential vitamin, taking massive doses of it will only produce rather expensive urine, and not the general health benefits that Pauling claimed, such as fewer episodes of cold and influenza and a potential treatment for cancer. But the fact that Pauling's views represent a minority opinion or that a majority of experts disagree with those views is not most relevant. Again, discussion should focus on the relevant reasons—in this case scientific studies—that Pauling and the medical community are using to base their respective claims, including whether scientifically fair controlled studies were conducted and what they showed.

There are undoubtedly times when it is wise to accept appeals to relevant authorities. If a small child is close to death after a car accident and doctors say they need to operate immediately to save the child's life, this is probably not a good time to be too critical and wait for second opinions and more research. Also there are times when the issues are not that important to us and we just do not have the time to do our own research. However, we should at least know what we are doing when we accept appeals from relevant authorities, that the quality of our reasoning could be better, that given more time it would be better to find out the expert's reasons for endorsement.[2] If you were dying of cancer you would probably take the time to find out if vitamin C really worked; you would want to know if there were any scientific studies that showed that people with cancer got better when they took massive doses of vitamin C.

Whether relevant and proper or irrelevant and improper, here is the formal essence of all Appeals to Authority:

EXAMPLE 4-3D

Conclusion: Do X, believe X, or (in the case of an advertisement) buy X.

---

[1] Note that a blend of popularity and authority is used when a majority of authorities believe the same thing, and this fact is used as a premise for a conclusion. This shows that often we need not quibble over fallacy labels. Often there will be appeals, the classification of which seem to overlap different fallacy labels. A snob-appeal version of Popularity may seem very close to Provincialism (see below), appeals to a majority of authorities could be classified as an Appeal to Popularity, or an appeal to Loyalty (see below) may seem to be an appeal to Popularity. However, if the classification seems arbitrary, the important focus will be that the critique of the weakness will be the same. Whether popularity or a majority of authorities, the focus should be on the omitted relevant reasons.

[2] Note that inductive reasoning would play a role in judging the reliability of expert opinion. The past record of the authority's endorsements would be relevant and important in appraising the authority as a genuine expert.

Premise:	Because Y says so (Y = a relevant or irrelevant authority, and Y endorses X.)
Label & Description:	Appeal to Authority. There is an appeal to authority in the premise (note Y).
Argument Analysis:	Reasoning. Even if it is true that an authority supports something, the logically relevant reasons—the reasons and arguments of the authority—for that support need to be discussed. Develop by specifying the reasons and arguments that should be discussed.

Note the similar features to that of Appeals to Popularity. The important feature (reference to an authority) for categorization is always in the premise, and the truth of the premise is not questioned in criticizing the argument. We could question whether Joe DiMaggio really uses Mr. Coffee and whether it really produces the best coffee he has ever tasted, and we might discover that he is only being paid to act, that he does not use the product at all. However, this would be off track for a critique of Appeal to Authority. For this label, the reasoning should be attacked, not the premise. The major focus for all Appeals to Authority is that *even if* the premise is true, the reasoning is weak—a poor inferential link exists between the premise and the conclusion because relevant matters for accepting the conclusion are not discussed.[1]

Traditional Wisdom

Unlike most animals that have instinctual, hard-wired behavior patterns for survival, the human species is capable of learning, of gathering information about the world and passing on that information from one generation to the next. Isaac Newton claimed he was able to discover the principle of gravitational attraction and work out its mathematical treatment because he stood "on the shoulders of giants." Newton was referring to the scientists and philosophers who had come before him and whose work was crucial for allowing him to make the final connections. A large part of our success on this planet has relied on the discoveries of our ancestors.

However, we are also capable of learning that some of the ideas accepted as true or reliable by our ancestors were mistakes or do not always work when applied to new and larger areas of experience. Thus, an uncritical acceptance of past "wisdom" is similar in its inferential weakness to that of appeals to Popularity and Authority. A *Traditional Wisdom* fallacy is committed when an action or belief is inferred to be good simply because it was considered good in the past. No support is given

---

[1] Sometimes the actor being paid has acted previously as an authoritative personality relevant to the product being endorsed. Actor Robert Young was best known for his roles in the TV shows *Father Knows Best* and *Marcus Welby, M.D.* In both shows, he played the role of a very stable and wise person that people could turn to in times of confusion and agitation. Later, in a Sanka coffee commercial, he seemed to play the same role endorsing the caffeine-free benefits of this product. In the commercial, although there was no direct reference to him being a doctor, he wore the same clothes and acted the same as he did in *Marcus Welby, M.D.*, endorsing Sanka as a cure for upset people who were about ready to strangle their dogs or kids.

as to why the action or belief was considered good in the past and whether it still reliably applies in the present. Here is an example given to me by a former student who at the time was studying to become a police officer.

EXAMPLE 4-4   Policeman: "Captain, why do we have to issue twenty traffic citations per month?"

Captain: "Because when I was where you are I had to issue twenty citations per month, and my superior before that had to, and his superior before that also had to, that's why!"

According to my student, as a new recruit in training he had to drive around in a very conspicuously marked police car. Veterans, however, were able to buy their own cars, which were unmarked most of the time, and they could quickly place a portable police light on top of their cars when they needed to go into action. Hence, my student found it very hard to make the established monthly quota, because most people slow down and drive more carefully when they see a marked police car behind them. He also wondered why there was a quota system in the first place. And why twenty? Should he give a ticket to someone who was driving only a few miles over the speed limit just to be able to make his quota? It seemed to him that in one month he might give a ticket to someone who was only driving five miles over the speed limit, because he needed to make his quota, but that in another month he might let someone go who was driving ten miles over the speed limit because he had already made his quota.

Note that the captain's argument did not answer my student's question. The original reason for the policy was not given (it was probably not even known or remembered), and its applicability to the present circumstances was not discussed. Just as in the case of the Little League baseball example in Chapter 1, the policy's relevance for the present is not tested. A very good reason might have existed for the original policy, and once we discovered what it was we could be in a better position to see whether it still applies. Thus, like the other fallacies of relevance discussed so far, this argument is not on track—the reasons for the original policy are not discussed nor the present applicability of the policy.

Here is a structured analysis of the captain's argument, followed by the formal essence of all Traditional Wisdom appeals.

EXAMPLE 4-4A

Conclusion:	Policemen should issue twenty citations per month.
Premise:	Because policemen have always been responsible for issuing twenty citations per month.
Label & Description:	Traditional Wisdom. There is an appeal to traditional wisdom in the premise ("policemen have always").
Argument Analysis:	The reasoning is weak. Even if it is true that twenty citations have been a standard for all policemen to follow for some time, the original reasons for this

policy are not revealed and the issue of whether present circumstances warrant a continuation of this policy is not discussed. What were the initial reasons for a quota system? Why the number twenty? Should there still be a quota system?

EXAMPLE 4-4B

Conclusion:	Do X, believe X, or (in the case of an advertisement) buy X.
Premise:	Because X has always been done, believed, or bought.
Label & Description:	Traditional Wisdom. There is a traditional wisdom appeal (X "has always") in the premise.
Argument Analysis:	Reasoning. Even if the premise is true, the inferential link is weak. Develop by pointing out that the original reason for X and its current applicability should be discussed.

Again, note that the premise is not attacked on the basis of its truth, but on its relevance. In analyzing Traditional Wisdom appeals we are not interested in whether it is true that something has been traditionally accepted, but rather in learning why something has been traditionally accepted and whether it still applies. Nor is our intention to reject all appeals to the alleged wisdom of the past, but rather to critically appraise traditional beliefs and policies and preserve those that continue to be reliable.

Provincialism

In addition to our natural desire to belong to a general group, we usually have specific group allegiances and identify with people we think are most like ourselves. We not only identify with a general culture and then various subcultures, but also with our own time and place. Children at a certain age will naturally fear strangers and a general us-vs-them attitude probably played a major positive role in the struggle for survival of our prehuman ancestors. However, traditionally a person is said to have a provincial attitude if he or she overidentifies to the point of cultural nearsightedness, thinking that a particular way of living is the only possible way to live.

From a modern perspective that takes into account the many groups, subgroups, allegiances, associations, and cliques that exist in a diverse social environment, a **Provincialism fallacy** can be thought of as a more specific version of an Appeal to Popularity. A provincial appeal is misused when it takes us off track by accepting what an "in-group" believes or does without discussing the reasons for the beliefs or actions.

In the early 1970s, a pivotal election for governor of Hawaii occurred, the result of which was destined to change the state's political structure for decades. Because Hawaii was a state run primarily by Democrats, the primary election race was decisive for who would become governor. At that time, the primary race was very close between George Ariyoshi and Thomas Gill for the Democratic nomination. Just days before the voting an advertisement surfaced on a local Japanese-language radio station, part of which reminded the Japanese-speaking listeners that if George Ariyoshi was elected, he would be the first governor of Japanese ancestry in the entire history of the United States. Ethnic Japanese make up a large voting block in the state of Hawaii, so one can easily imagine that this powerful appeal had a significant affect on the outcome of the election, which George Ariyoshi won. Although in many ways ethnic Japanese now constitute a majority in terms of political power and influence, like other immigrants who came to work in the sugar plantations for the rich Republican caucasians, for generations their families had faced a long, hard struggle to achieve respect and economic security. For many it is likely that having one of their own become governor was a fitting consummation of their struggle and accomplishment.

Surely there is nothing intrinsically wrong with such sentiments, and it is admirable for a people to be proud of their ancestry. But note how such sentiments in this case were misplaced, not just for those who were not Japanese, but for those who were Japanese as well. It would not have been a fitting consummation of the struggle of ethnic Japanese if George Ariyoshi turned out to be a poor governor and leader or if his policies led to disastrous results for the state. It would not have been a fitting consummation to their struggle if George Ariyoshi ended up embarrassing other ethnic Japanese. What was really relevant for ethnic Japanese was George Ariyoshi's qualifications, not merely his ethnicity. Here is a structured analysis of this advertisement as a Provincialism fallacy, followed by its formal essence.

EXAMPLE 4-5

Conclusion:	Vote for George Ariyoshi.
Premise:	Ariyoshi will be the first governor of Japanese ancestry in the entire history of the United States. (Implied: you should vote for Ariyoshi if you are Japanese; support one of your own.)
Label & Description:	Provincialism. There is an in-group appeal (ethnic Japanese) in the premise.
Argument Analysis:	Reasoning. Even if the premise is true that George Ariyoshi will be the first governor of Japanese ancestry in the entire history of the United States, the reasoning is weak because there is an appeal to the identity of Japanese ancestry at the expense of an overall evaluation of Ariyoshi's qualifications for governor. His position on the issues, his past accomplishments, and evidence of leadership ability should be discussed.

EXAMPLE 4-5A

Conclusion:	Do X, believe X, or (in the case of an advertisement) buy X.

Premise:	Because this is an alleged appropriate thing for Y (an in-group) to do, believe, or buy.
Label & Description:	Provincialism. There is an in-group appeal (describe Y) in the premise.
Argument Analysis:	Reasoning. As in the case of the previous fallacies, develop a case that the inference is weak because the relevant issues are not discussed, that the relevant reasons why an in-group should support something should be discussed.

As Lincoln's famous adage that you can't fool all the people all the time reminds us, sometimes provincial appeals can backfire politically. Daniel Akaka, later a United States senator from the state of Hawaii, once ran for lieutenant governor with an advertisement that not so subtly reminded the voters that he would be the first Hawaiian to achieve such a high office. Non-Hawaiians, and perhaps even many Hawaiians as this was a general advertisement on TV, were insulted, and Akaka lost badly.

During the early 1990s, the rap megastar Ice Cube could be seen in advertisements endorsing St. Ides malt liquor, a high-alcohol beer. In magazines targeted for inner-city black residents, he would be holding a can of St. Ides while flashing a gang sign. Message: If you want to be "in," if you want to be one of the "home boys" drink St. Ides; other domestic beers are for wimpy white boys. He also appeared in television commercials that contained rap lyrics implying that the beer increased male sexual success.

Provincial appeals effortlessly capitalize on a kind of "new racism"—an endorsement of radical cultural pluralism that in showing respect for cultural identities actually disguises feelings of superiority or inferiority as respect for the differences of others. "You do your cultural thing, and I'll do my cultural thing, but my cultural thing must be preserved at all costs because it is better than yours."

However, you should not generalize from these examples that all provincial appeals are racial. Many advertisers target specific groups of people in marketing their products. For years Marlboro cigarette ads have been aimed at men who think of themselves as rugged and independent, showing scenes of adventurous, muscular men riding horseback in spectacular wilderness settings. More recent ads show men riding Black Stallion 4x4 Jeeps and ask you to join the Marlboro Adventure team. One of the most famous examples of in-group targeting was the Virginia Slims cigarette advertisements found in magazines during the 1970s and '80s. During a time of the feminist movement and a general acknowledgment that women have had to overcome gender discrimination to develop their potential and to accomplish their goals, these advertisements typically showed two scenes: an old-fashioned black-and-white photograph of a woman working hard for a mean and oppressive husband and a color photograph of a slim, attractive modern woman, holding a Virginia Slims cigarette, looking independent and self-assured. The main caption of the advertisement read, "You've come along way baby," and a smaller caption read, "Slimmer than the fat cigarettes men smoke." If you identified with this new woman, then you were supposed to smoke Virginia Slims.

These two advertisements are examples of cigarette companies' attempt to target specific groups of people. The traditional Marlboro ad appeals to men who identify with being strong and independent; the relatively new Capri ad appeals to young women.

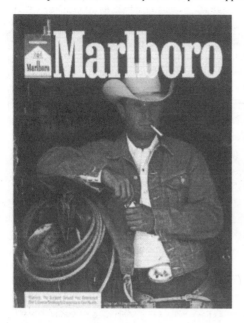

Many advertisements imply that our time is the best and that if you are "in" with modern ways you will buy their product. Other examples of provincialism would be when we are urged to be "pro-choice" on abortion because this is the position of enlightened, intelligent, modern people, and urging support for President Clinton's policies in the first few years of his presidency as a vote for the future accepted by a new generation. Or, when Miller beer, noted already for its switching back and forth over the years between snob and bandwagon appeals, advertised its new Ice Beer at the same time as the Clinton appeals—"New Rules, New Beer, New Light Ice Miller Beer."[1] Note that such wisdom-of-our-time appeals are the psychological opposite of the wisdom-of-the-past appeals in Traditional Wisdom. Like different authorities endorsing different positions or products, such psychological opposition shows that what really ought to be decisive and persuasive is something deeper than a mere claim of the wisdom of the past or intelligence of the present.

---

[1] Notice that when the appeal is to an in-group of supposedly intelligent people—people who are supposedly above the herd mentality of average people—it matters little whether we classify the fallacy as a snob-appeal version of Popularity or Provincialism. Technically, the fallacies can be distinguished because Provincialism applies to a wider range of identifications than "smart people." In a concrete case, it may not matter which label is used. The important point, regardless of the label, is that relevant issues are not being discussed.

Appeal to Loyalty

The movers and shakers of history are powerless without a sense of solidarity, purpose, and cause engendered in those they lead. Nationalism and loyalty to one's country are the result of focusing the psychology of group identity on a flag and its history. Every country has stories of great sacrifice by loyal heroes in its history books. Yet, because heroes are killed by other heroes, and war and armed combat carry such great consequences, should we not be sure of our reasons for loyalty? Like the fallacies previously discussed, an appeal to a blind commitment of loyalty misplaces a natural psychological reaction when it shifts attention from what we should focus on in making a decision or discussing an important issue. An *Appeal to Loyalty* fallacy is committed when an argument is given that distracts us from the issues by appealing to loyalty in the premises.

Here is an example from former U.S. President Richard Nixon, commenting on the significance of American prisoners of war in Vietnam returning to the United States after years in captivity:

EXAMPLE 4-6   "When a POW (prisoner of war) can return after six and one-half years with the phrase 'God bless America' on his lips, it (the war in Vietnam) has all been worthwhile."

The Vietnam War was highly controversial, and many people opposed the involvement as not in the best interests of the United States. The issue was not whether to be loyal, but determining the loyal thing to do—support the war effort or try to stop it? Given that the war cost $676 billion, that 15 million tons of munitions and 100 million tons of chemical herbicides were used, that eight thousand aircraft were lost, that thousands of innocent civilians were killed by friendly fire and collateral damage, that 53,813 U.S. soldiers were killed, and more than three hundred thousand wounded, and that the war was estimated to cost $12,000 per U.S. family (in 1970 dollars) in terms of taxation and inflation[1]—were the foreign policy reasons for this effort persuasive? Nixon may have had very good reasons for continuing to support military involvement, and we may admire the commitment of others and sympathize with their suffering for a cause, but the reasons are not given here, and our sympathy is not evidence that the cause was just. Loyalty to a cause does not justify itself.

Below is a formal analysis of Nixon's argument followed by the formal essence of all Appeals to Loyalty.

EXAMPLE 4-6A

Conclusion: The U.S. military involvement in Vietnam was justified.

---
[1] According to a study by Tom Riddell of Bucknell University.

Premise:	POWs supported their country even though they faced extreme hardships (implied: supporting the military involvement in Vietnam was the loyal thing to do.)
Label & Description:	Appeal to Loyalty. There is an appeal to loyalty in the premise.
Argument Analysis:	The reasoning is weak. Even if it is true that a number of people remained loyal to a controversial cause, this is not relevant to the justification of that cause. Attention is shifted away from a discussion of the issues as to why we should also be loyal to this cause. Was the military involvement in the best interests of the United States? The reasons in terms of foreign policy objectives are not described that would justify our loyalty to this cause.

EXAMPLE 4–6B

Conclusion:	Do X, believe X, or (in the case of an advertisement) buy X.
Premise:	Because supporting X is the loyal thing to do.
Label & Description:	Loyalty. There is an appeal to loyalty in the premise.
Argument Analysis:	Reasoning. Even if supporting X is an act of loyalty to a cause, such support is irrelevant in terms of shifting attention away from the reasons for supporting that cause. That X is supported as a loyal action or belief does not explain why it should be. Criticize and develop by describing the issues that should be discussed.

Defense budgets are often supported by blind appeals to loyalty. Often we hear that we should support $300 billion defense budgets because we should be loyal to the men and women who suffered or fought for us in past wars and allowed us to experience the freedom and prosperity we enjoy today. However, such appeals say nothing about the wisdom of the particular expenditures today. Do we need the particular items specified in the $300 billion budget? Is each one a quality product that will help us achieve our security goals? Is it not true that wasting money today, if the expenditures are not wise, would be an insult to the brave men and women who supported our country in the past? A tragic insult? Because these men and women suffered so that their children and their children's children would live in a society that spent more money on education and economic development.[1]

Although support for a military involvement is the most natural place for Appeals to Loyalty to occur, as the conclusion of the formal essence shows, Appeals to Loyalty can be used in as many ways as Appeals to Authority, Popularity, Traditional

[1] We should also keep in mind that we must distinguish between issues of public policy, how these policies should be decided in a democracy, and how they should be carried out. Once a policy is decided democratically, and the implementation of that policy involves military action, loyalty may be the most appropriate response. If every individual soldier second-guessed every democratically decided policy, we would not accomplish our policy goals.

Wisdom, and Provincialism. By the 1980s, the Japanese had become a major world economic power by producing quality products and exporting them to the United States and Europe. This occurred in part because they spent only a small portion of their capital assets on military defense and hence were able to invest these assets in ultramodern production techniques that used computer technology and robotics. While the United States poured an enormous amount of its assets into a military hardware cold-war race with the former Soviet Union, the best electronic products, such as VCRs, TVs, and stereos became almost all Japanese-made. And Japanese automobile manufacturing, previously an industry dominated by the United States, began to seriously threaten the profits of U.S. manufacturers. Some U.S. manufacturers responded with advertisements that appealed to loyalty to the country. Chrysler, for instance, began running ads that reminded United States citizens that they should buy its cars because they were made "in America."[1] For the most part, such ads did not work. People know that quality is the relevant and decisive issue in purchasing a product. Eventually, U.S. business and political leaders began to figure it out. The most loyal thing to do was to forsake fruitless attempts to shift attention away from the underlying problem, and instead invest more capital assets into retooling U.S. industry, which was seriously behind that of the Japanese in efficiency and modern equipment.

As we have noted several times in this book, most examples of bad reasoning are not reliable in the long run because they do not force us to test our beliefs. They shift attention away from relevant issues, perhaps making us feel good for awhile or relieving tension by offering alluring shortcuts to complex situations. Still reality has a way of eventually penetrating even our strongest psychological masks.

Two Wrongs Make A Right

Another way that attention is drawn away from testing the merit of a course of action is to shift attention to another action and question its merit. This method of reasoning is often persuasive because most people are morally sensitive to hypocrisy and to a basic principle of justice of equal treatment. If one person is doing something thought to be wrong, it hardly seems appropriate for that person to criticize someone else who is doing the same thing. But the inappropriateness of the criticism is a different matter and should be separated from the rightness or wrongness of the action itself. The inappropriateness of the criticism does not magically make the action being criticized right. The criticism and the action can both be wrong. For instance, if it is wrong for the average person to cheat on his or her taxes, then it is unethical for a government official to avoid just taxation. But unjust action on the part of a government official should not shift attention away from the general issue of the wrongness of cheating on taxes.

A *Two Wrongs Make a Right* fallacy confuses a charge of hypocrisy and injustice with the relevant discussion necessary to justify a course of action. This fallacy is committed when instead of providing the reasons why an apparently questionable

---

[1] Even though some of Chrysler's cars were manufactured in Canada, and some even had Japanese engines!

course of action is actually meritorious, it shifts attention to another, allegedly similar, questionable action. If, between two allegedly similar actions, one is considered acceptable and the other is not, this may be unjust or hypocritical, but the bare fact of the apparent similarity does not justify the claim that the actions are really similar or why either is acceptable.[1] Here is a classic example:

> **EXAMPLE 4-7**   Alcohol, valium, and many other harmful addictive drugs are legal in our society. Twenty thousand tons of aspirin are consumed each year by U.S. citizens, and many people die from its use. Carcinogenic chemicals can legally be placed in the food we eat. Therefore, marijuana should be legalized.

The conclusion of this argument is that marijuana should be legalized, but instead of discussing the marijuana issue directly and testing the merits of the legalization claim, attention is shifted in the premises to other "acceptable wrongs." That the consumption of marijuana is illegal and alcohol is legal could well be a constitutionally unjust situation. If it could be demonstrated that one group of people use marijuana and another group of people use alcohol, and if it could be shown that marijuana as a recreational means of personal relaxation is no more harmful physiologically and socially than alcohol, then a case could be made that the law discriminates against marijuana users. However, to demonstrate discrimination would be only the first step in a legalization argument,[2] and to do so would involve discussing why alcohol is an acceptable wrong followed by a comprehensive comparison of alcohol and marijuana to see if they are indeed similar. Once these issues have been investigated, a discussion could follow of whether marijuana should also be an acceptable wrong. In short, a better argument would involve a discussion of why these other drugs are legal and whether the results of this discussion apply also in the case of marijuana. The pros and cons of marijuana use should be discussed somewhere in the premises, otherwise we would fail to test the merits of the legalization claim.

Here is an analysis of the above argument, followed by the formal essence of all Two Wrongs Make a Right fallacies.

> **EXAMPLE 4-7A**
>
> | | |
> |---|---|
> | **Conclusion:** | Marijuana should be legalized. |
> | **Premises:** | Alcohol, valium, and many other harmful addictive drugs are legal in our society. Twenty thousand tons of aspirin are consumed each year by U.S. citizens, and many people die from its use. Carcinogenic chemicals can legally be placed in the food we eat. |

---

[1] A more technical way of putting this is to note the difference between an ethical and a logical judgment. If actions X and Y are both wrong, it may be unethical for one to be acceptable and the other not, but this is a different matter than the logically relevant reasons for judging X and Y to be wrong in the first place.

[2] There might be good reasons why our society has decided to discriminate. We might admit that marijuana is no worse than alcohol, but decide that we do not want to add one more legal drug to the pressures of our current social situation.

Label & Description:	Two Wrongs Make a Right. Other acceptable wrongs (the use of alcohol, valium, etc.) are cited in the premises that are not directly related to the issue stated in the conclusion.
Argument Analysis:	The reasoning is weak. Rather than referring to the specific issues related to marijuana legalization, the argument shifts attention to the fact that other wrongs allegedly similar to marijuana legalization are acceptable. A better argument would involve a discussion of why alcohol and the other drugs mentioned are legal and whether the results of this discussion apply also in the case of marijuana in the sense that laws should be consistent. In other words, somewhere in the premises the pros and cons of marijuana use (related health concerns and such) should be debated.[1]

EXAMPLE 4-7B

Conclusion:	X is okay or acceptable.
Premise:	Because Y is already considered to be okay or acceptable.
Label & Description:	Two Wrongs Make a Right. Cites an acceptable wrong in the premise.
Argument Analysis:	Reasoning. Even if it is true that Y is acceptable, a discussion is needed as to why Y is acceptable and then the results of this discussion applied to X by comparing X and Y. Are there good reasons why Y is acceptable? Do those same reasons apply in the case of X?

A version of Two Wrongs Make a Right that is often confused with Appeal to Popularity is called **Common Practice**. Appeals to Popularity have the tone, "Most people are doing X, therefore you should." Like Two Wrongs Make a Right, Common Practice has the tone, "Lots of people are doing X (even though it is wrong), so it is okay for me to do X also." Appeals to Popularity urge us to accept something as good because many other people apparently do. Common Practice urges us to accept or ignore a questionable action because others are doing it (the action is common). Similar to Two Wrongs, Common Practice shifts attention away from what should be discussed: the reasons for the acceptability of the actions of others, whether these reasons are good reasons, and so on.

In 1988, at the beginning of President George Bush's term, a major issue was made of his attempt to appoint former Senator John Tower as secretary of defense. Numerous allegations were made concerning Tower's character, such as his alleged alcohol use and attitude toward women. Of the many stories that surfaced in the

---

[1] Further arguments might involve the amount of money saved on law enforcement and redirected toward drug education, different ways of legalization, such as decriminalization, and the benefits of taxation.

news, one concerned Colonel Robert L. Moser, a former top-ranking staffer and aide to Tower at the 1985 United States–Soviet arms-control talks in Geneva. According to news reports, the Air Force Office of Special Investigations uncovered an "alarming pattern of disregard for security regulations" by Moser and raised "serious questions of Moser's integrity and conduct." The Air Force issued him a formal reprimand for adultery, sexually harassing his secretary, and having her perform personal business tasks for him at government expense. The Air Force further claimed that he had a two-year affair with an Austrian woman connected with Soviet agents. When Moser was confronted by investigators with these allegations, he did not deny them but responded with the defense that "others" in the one hundred-member Geneva delegation were conducting themselves in a similar manner. The nuclear armament situation between the United States and the then–Soviet Union was probably at an all-time low point during the mid-1980s. Thousands of nuclear missiles were targeted on each other's cities, then–President Reagan had called the Soviet Union an "evil empire" and was about to install more nuclear missiles in Europe, and there seemed no end in sight to the nuclear arms race and its negative effect on the economies of both countries. It was not a good time to get off track as to what was relevant to matters of "security regulations."

Ad Hominem Abusive and Circumstantial

It is a sad fact that most political debate involves personal attacks rather than an in-depth discussion of the issues. Although a candidate's character is clearly a relevant matter in voting for someone, too often character appraisal is used as a dangerous shortcut to understanding complex issues. Too often, many voters assume that if they like a person's character that person will support the positions they want supported. So, even though there is no necessary relationship between a person's character and their position on the issues, politicians will spend a lot of energy attacking their opponents personally. An ***Ad Hominem*** (to attack the person) fallacy is committed when the conclusion of an argument is directed at a person's stance on an issue, but the premises offer no relevant critique of the person's stance on the debated issue and substitute a personal attack instead. There are two types: ***Ad Hominem Abusive***, in which a person is attacked by simply being called creative names; and ***Ad Hominem Circumstantial***, in which a person's circumstances or motives related to holding a particular position are attacked. Here are some examples:

In the early 1980s, the Reagan administration proposed a massive legislative package that they said would implement a forward-looking "New Federalism." Consistent with Reagan's political philosophy, this legislation had the goal of decentralizing most government services, including welfare and education, by turning them over to the states. It was unclear to most members of Congress how the states would pay for these services, so the legislation received a very cold welcome in both the Senate and House. One of Reagan's responses to the rejection of his legislation was to refer to the members of Congress as just a "bunch of mindless dinosaurs." Some members of Congress, evidently stung by this criticism, responded that Reagan's proposal lacked substance and was only a political ploy to "direct attention away from the faltering economy." Reagan had called members of Congress a name, but did not respond to their substantive reasons for opposing his legislation; the congressional response attacked the Reagan administration's motives for introducing the legislation. In neither case were the main issues debated publicly for the voting public—the details of the New Federalism and the possible effects of implementing such a package. Here is the argument structure for both appeals followed by their formal essences:

EXAMPLE 4-8

Conclusion: Those who oppose Reagan's New Federalism legislation are wrong to do so.

Premise: They are just a bunch of mindless dinosaurs.

Label & Description: Ad Hominem Abusive. There is name-calling in the premise ("mindless dinosaurs").

Argument Analysis: The reasoning is weak. Attacking one's opponents by name-calling is not relevant for concluding that their arguments are poor or invalid. Attention has been shifted from the relevant details of the New Federalism to a personal attack of those who oppose its implementation. To know whether the arguments of congressional critics were poor or not, their arguments need to be discussed. To know whether the legislation had merit or not, a detailed analysis of the legislation and its probable effects was needed. Reagan might have been right that this legislation would be good for the United States, but only by discussing the issues would we know.

EXAMPLE 4-8A

Conclusion: X's argument or position on Y is invalid or poor.

Premise: X is a Z (name-calling).

Label & Description: Ad Hominem Abusive. There is name-calling in the premise (identify Z).

Argument Analysis: The reasoning is weak. Attention has been shifted from the relevant substantive issues related to X's argument and X is attacked personally by name-calling. To know whether X's argument is invalid, the details of X's argument need to be discussed. Finding a creative name with which to brand X's position does not mean that name is accurate nor is it relevant to the substantive issues raised by X's argument.

EXAMPLE 4-9

Conclusion:	Reagan's New Federalism is worthless and should be rejected.
Premise:	This legislative package has been introduced at this time simply as a political ploy to direct attention away from the faltering economy.
Label & Description:	Ad Hominem Circumstantial. A circumstantial appeal is made in the premise (Reagan's alleged motives).
Argument Analysis:	The reasoning is weak. Rather than analyze the details of Reagan's New Federalism proposal and describe substantive flaws in the legislation relevant to the conclusion, attention is shifted to the possible motives for the legislation. Even if the motives are suspect for the timing of the legislation, the ideas embodied in the legislation could still be worthy. To know that the legislation is worthless, it would have to be read, studied, analyzed, and debated. Poor motives are relevant grounds for suspicion that the ideas may not be sound, but not relevant to the conclusion that the ideas are indeed unsound.

EXAMPLE 4-9A

Conclusion:	X's argument or position on Y is invalid or poor.
Premise:	Because of the circumstances or motives related to X's defense or support for Y.
Label & Description:	Ad Hominem Circumstantial. Cites personal circumstances or motives in the premise rather than an analysis of Y.
Argument Analysis:	The reasoning is weak. Although the circumstances or poor motives may be true, this would not be directly relevant to concluding that the position supported is poor. Ideas, beliefs, proposals can be correct, acceptable, or worthy even if the motives for them are suspect. To know, the ideas, beliefs, or proposals need to be studied, analyzed, and discussed.

Note that Ad Hominems always involve two people or groups of people: An arguer A who has presented an argument, and an arguer B who attacks the character, motive, or circumstances related to A's argument instead of the issues raised by A. Arguer B is guilty of the Ad Hominem.

Ad Hominem Circumstantial appeals are very persuasive because as a shortcut in separating the reasonable appeals from all the myriad conceivable appeals, there is sometimes an underlying rationality to this type of thinking that we do accept. In a court of law, for instance, the character and motive of a witness is clearly relevant to the acceptance of testimony. Suppose a man has been charged with murder and I claimed to have witnessed it. If you were on the jury and I were the only witness in a murder trial, you would be justifiably suspicious of the truth of my testimony if you discovered through the defense's cross-examination that I had a criminal record, had failed lie-detector tests on numerous occasions, and not only knew the defendant, but had reason to hate him. Suppose you found out that years ago he had an affair with my wife, that she had left me for him, and that I had vowed publicly to get even someday. Clearly, if there were no other corroborating items of evidence indicating that the defendant was guilty, my testimony alone could not be used to convict the defendant. However, even in a court of law there is an important but major technical difference between being *suspicious of testimony* and the *truth of testimony*. I could still be telling the truth about this event even if I had lied many times before and even though my motives are clearly less than civic. But because in a court of law we are innocent until proven guilty, evidence related to a conclusion of suspicion is all important and appropriate for doubting the truth of testimony.

The underlying rationality of such circumstantial appeals is inductive. We reason that if a person lied previously, they are probably lying now. If there is enough doubt that a person is telling the truth, then it would be unwise to accept the truth of what they say. If a person's motives are less than ideal for a belief, proposal, or policy, then we suspect that the belief, proposal, or policy has been "conjured up" and must not be a sound solution to a problem. However, from a deductive point of view, circumstantial appeals present a problem of relevance. Evidence related to suspicion that a belief has just been conjured up is not necessarily evidence that the belief itself has no merit. Even conjured-up beliefs can be true! I might believe in God because it makes me feel comfortable, gives my life meaning, and offers hope that when I die I will not just be a terrifying "nothingness." That this belief makes me feel secure is not evidence that my belief is true, but it is also not direct evidence that my belief is false. My motives would be relevant to doubting that I had any real evidence that God exists, but my belief could still be true even if I had suspect motives, and I may have even "stumbled" upon some real evidence because of my need for security. The notion of a loving Supreme Being may have been invented by human beings to help face the harsh realities of existence. But a Supreme Being could still exist anyway. Only by looking at and criticizing what I consider to be evidence for a Supreme Being would we know if the evidence is poor or not.

Appeals to suspect motives and circumstances are often so persuasive because many times motives and circumstances are all that we have to go on. In autumn 1991 citizens of the United States witnessed a wrenching confirmation hearing for Supreme Court Justice Clarence Thomas. Unprecedented sexual harassment allegations were made against Thomas by Anita Hill, then a college law professor. The harassment allegedly occurred when she worked for Thomas in the Office for Equal Employment Opportunity during the Reagan administration. Political polls showed that most people were very confused over who was telling the truth. Hill seemed credible when she made the charge. Thomas seemed credible when he forcefully denied sexually harassing her. Because, as in most sexual harassment cases, there were no witnesses, supporters of Hill and Thomas shifted attention to possible motives and circumstances.

At the Senate hearings supporters of Hill and Thomas had witnesses corroborating their honesty and general character and attacking that of the other. Thomas obviously had a motive for lying. He wanted to be confirmed to the Supreme Court. Did Hill have any motive for lying? Yes, said Thomas supporters, to promote her career by being a brave hero for sexual harassment victims. Revenge: The claim was made that Thomas had rejected Hill's amorous advances, rather than, as Hill claimed, the other way around. Political differences: Hill was more liberal than Thomas, and she did not want him confirmed to the Supreme Court. Mental instability: The claim was made by supporters of Thomas that Hill showed previous signs of mental illness.

Because the evidential focus was now purely inductive, was there any corroborating evidence supporting either side? Yes, and no. Angela Wright also worked for Thomas and also claimed to have been sexually harassed. However, supporters of Thomas pointed out that she was fired by Thomas, so she had a motive for lying. Supporters of Hill claimed that Thomas had a history of being a consumer of pornographic videos, but the relevance of this claim was not clear, and it could not be corroborated. Supporters of Thomas cited the testimony of a former Hill boyfriend who claimed that Hill was very insecure and often made exaggerated, unstable advances to men. But when this former alleged boyfriend testified, he appeared to be a supreme egotist with a very conceited image of himself and his own appeal to women. Supporters of Thomas also cited the fact that Hill had waited more than two years to make these charges and that the timing of the charges being made then was suspect. Supporters of Hill countered that Hill's dilemma typified the situation of many harassment victims. Thomas and Hill were in an unequal power relationship; Hill needed Thomas to recommend her for future employment; there were no witnesses, and in a male dominated society charges against a powerful male are very dangerous.

As the Senate investigation proceeded day after day, broadcast live to a soap opera audience, the circumstantial allegations probably broke all political records. Thomas claimed that the hearings were a "high-tech lynching" and that the charges against him exploited "the most bigoted, racist stereotypes of the sexual prowess of black men." (Hill had alleged that Thomas bragged about the size of his penis and how much pleasure he was able to give to women.) Many women viewers became increasingly uncomfortable with the fact that all the senators on the select committee to investigate the charges were men, and that question after question to Hill targeted her motives and status as a woman.

The result of this matter was not uplifting and was hardly material for a civics lesson. There was neither enough time nor apparently the resources to corroborate either claim. So many undecided senators gave Thomas the benefit of doubt, apparently reasoning that we are innocent until proven guilty. But was this benefit-of-the-doubt argument appropriate? The Senate hearing was not a court of law, and a nomination to the Supreme Court was at issue. But then again, any candidate could be unfairly voted against simply with a charge that raised some unsubstantiated doubt about character.

The point of this digression is that in the Thomas-Hill case we did not have access to the relevant evidence, so circumstances and motives became the only rational focus; sometimes life is tough in an imperfect world and circumstances and motives are all we have to go on. But this acknowledgment should not distract us from the fact that direct, relevant evidence often is available and should be the primary focus. What was upsetting for many people about this whole affair was that so much attention was focused on character, especially character assassination, rather than obtaining direct evidence that the allegations were true or on the main issue. Was Thomas qualified to be on the Supreme Court?

Consider an example where direct evidence was available. In the 1980s, the issue of high cholesterol content in foods and its relationship with heart disease became very controversial. Although the medical community and the American Heart Association had reached a consensus that the scientific evidence warranted a dietary recommendation to lessen intake of high cholesterol and high-fat foods, there were conflicting theories and scientific findings related to the urgency of this recommendation. Within this context, the National Food and Nutrition Board issued a report citing evidence that an insufficient connection existed between cutting fat and cholesterol intake and heart disease to make a dietary recommendation. This report was condemned by many members in the medical community who were urging us to cut our fat and cholesterol intake. Furthermore, two scientists on the Board were found to be paid consultants to food companies with special interests in high-cholesterol foods. The chairman of the Nutrition Board received about 10 percent of his income from Kraft Inc. and Pillsbury, and another member was an adviser and speaker for the American Egg Board and the Dairy Council of California. Millions of dollars were at stake for these companies. The American Heart Association and other health groups claimed that the report was biased. However, note that if anyone would conclude from this that the Board's report had no merit, they would be guilty of an Ad Hominem Circumstantial fallacy. That two scientists on the Board had a potential conflict of interest seems clear, but even if these scientists were biased, the evidence the report cited could still substantiate the report's conclusion that the connection between high cholesterol foods and heart disease is weaker than previously thought. Unlike the Hill–Thomas affair, the relevant scientific studies cited in the report were available and could have been studied and analyzed. The conflict of interest of the two scientists was relevant to the charge of their being biased, but not relevant to the report's conclusion. They could have been right even if they were biased.

In terms of maintaining the proper focus in analyzing arguments, Ad Hominem Circumstantial appeals are particularly illustrative. If our intent is to conclude bias, then appeals to motives and circumstances are on track. However, if our intent is to conclude that a belief, policy, report, or political position lacks truth, merit, or value, then it is a direct analysis of the belief, policy, report, or political position that is important.

Many successful ideas have been the result of strange or questionable motives. During the Renaissance, both Copernicus and Kepler were committed to the then-revolutionary view of a heliocentric (sun-centered) astronomical picture of our place in the universe, in part because they held (by modern standards) a strange religious belief that the sun was a material manifestation of God and thus should be in the center of planetary motion and the universe. They were clearly biased and committed to this view prior to an analysis of much of the observational evidence, but today in introductory science books we celebrate these men because they were right about the Earth revolving around the sun.[1]

We are messy creatures, and our beliefs have many origins. Even if they have been "conjured up," what matters most is whether there is evidence that they will be reliable guides to the future. In the words of Harvard naturalist Stephen Jay Gould, "People may believe correct things for the damnedest and weirdest of wrong reasons."

As a final example for this important fallacy, consider another issue raised against Clarence Thomas during his Senate confirmation hearings to be on the Supreme Court. As a federal appeals judge, Thomas wrote an opinion overturning a large damage award against the Ralston Purina company in which Thomas's friend, mentor, former boss, and chief supporter, John Danforth (Republican senator from Missouri) and his family held millions of dollars worth of stock. Notice the different items that must be disentangled for proper focus: Was Thomas's legal opinion accurate and just? Was Thomas biased? Was it proper for Thomas to rule on such a case, and should he have disqualified himself from participating in the case? If he was biased and if disqualification was the proper thing to do, did this one instance say enough about Thomas's character to raise serious questions about him being on the highest court in the country? Seldom do we separate clearly the relevant issues in the heat of such a political battle. If the focus is on the details and legal accuracy of the legal opinion issued by Thomas, then his possible bias was not relevant. Only a detailed analysis of the opinion itself mattered. But if the focus was on the ethical propriety of Thomas's ruling on such a matter at all and not disqualifying himself from participating in the case, then his possible bias was clearly relevant.

Irrelevant Reason

With the exception of Questionable Cause, all of the fallacies we have looked at thus far are examples of irrelevant reasons—the premises are not on track and shift our attention away from what we should be thinking about. Our final fallacy for this chapter is in part a sort of a generic "grab-bag" label. When the premises are clearly off track, but, unlike Popularity, Authority, Provincialism, and others, do not involve a specific way of being off track, then we will use Irrelevant Reason as a label.

However, there is a psychological appeal to most Irrelevant Reasons that deserves comment. As an example, consider how often we watch the battle that takes place between investigative reporters and politicians. Both have been trained by their professions to be survivors in a world of half-truths. The reporter's goal is to get the "real story" behind a politician's actions and the politician has been trained to

---

[1] They were clearly wrong about the sun being the center of the universe. Our sun is simply a grain of sand in a single vast galaxy of 100 billion suns, and our galaxy only one of many billion galaxies.

put a positive "spin" on all his or her actions and not allow the reporter to negatively "slice up" (through editing) his or her comments for the evening news. For instance, if asked a pointed question regarding why a senator voted against a bill that would help the homeless, the senator might begin a long speech involving a narrative of his experiences during the Korean War, of how he personally witnessed homelessness on a broad scale, of how sympathetic this made him to the plight of people in need. And perhaps for good measure he will throw in a little story of his own sad childhood during the Depression. The trick is to talk about a general problem tangentially related to the issue of relevance, but to evade committing oneself to a specific solution for the problem. Everyone is in favor of solving the problem of homelessness, of eliminating poverty, of producing jobs, of maximizing freedom and equality, and of achieving world peace. But the specific issues revolve around specific proposals and actions to achieve these goals. Thus, many irrelevant reasons attempt to give the appearance of supporting a conclusion by discussing topics that may resemble a specific issue under discussion, but actually evade or shift our attention from a discussion of the specific issue. Consider this example, a letter to the editor toward the end of the Vietnam era:

EXAMPLE 4-10   Bill Noble was the last American to die in Vietnam before the cease-fire. He left behind four children and a dedicated wife. He believed in what he was doing. I wonder if those who have opposed our involvement in Vietnam ever consider such real flesh-and-blood commitments. Or, do they simply reason from an assumed dogmatic philosophy and morality? We should have stayed in Vietnam and finished what we started.

The Vietnam War was a great tragedy for the United States. Compared to the Persian Gulf War it was not a popular war, and from the point of view of the men and women who fought, and their families, the greatest tragedy concerned the lack of apparent meaning for their commitment and sacrifice. Human beings can suffer indignity, terror, hardship, and even death as long as the actions that lead to these consequences are meaningful, as long as their actions are seen as part of a righteous cause, a virtuous big picture.

Many who fought in Vietnam suffered numerous physical and psychological terrors. A friend of mine was in charge of a Navy patrol boat that was ordered to patrol the Mekong Delta. All of the men on his boat had been wounded within nine months of his duty period, and several sister boats had been totally destroyed and all the men on them killed, but he and his men never once saw their mysterious enemy. They often saw South Vietnamese, the people they were supposed to be fighting for, but these people never warned them of their impending doom. The Vietcong hid deep in the jungle and detonated mines along the river banks. These detonations sprayed shrapnel across the river, hitting the men as they patrolled up the river. They would return the fire and call in air strikes to rip up the jungle, but they never found any Vietcong bodies, not even one. When they returned to their base camp, my friend's commander would ask for a "body count," and it became clear to him that he was supposed to "estimate" how many Vietcong they had killed. Someone might not get promoted if it appeared they were not winning the war. My friend survived this ordeal physically, but when he returned to the United States he had a complete mental collapse. The hypocrisy and terror of his experience, mixed with others' lack of respect for his commitment that he faced upon returning were too much to bear. It was not the way it was supposed to be. It was not like a John Wayne movie where the battle lines and issues were clear, and the good guys, the honest guys, always won, and even if they died, their deaths were meaningfully mourned and respected.

There is, of course, something particularly tragic about being the last one to die in a war. If the war had ended just one day sooner, Bill Noble would have perhaps experienced a lifetime of Little League games, picnics, graduations, and birthdays with his four children, and many loving, tender moments with his wife. We can all relate to his tragedy and wish it were not so, but the issue of whether he should have been in Vietnam in the first place, whether the United States involvement was wise, remains. Staying in Vietnam and "finishing what we started" would produce more Bill Nobles, but would not necessarily make the war justifiable and hence meaningful.

Below is an argument analysis followed by the formal essence of all Irrelevant Reasons:

EXAMPLE 4-10A

Conclusion:	Our involvement in the Vietnam War was justified. ("We should have stayed in Vietnam and finished what we started.")
Premise:	Bill Noble was the last American to die in Vietnam, and so on.
Label & Description:	Irrelevant Reason. The premise is logically irrelevant to the conclusion.
Argument Analysis:	The reasoning is weak. The premise is logically irrelevant to the conclusion. Although it is unfortunate that a particular individual was the last United States soldier to die in Vietnam, and although his death is relevant to concluding that our involvement was tragic in some way, this is not relevant to the larger issue of whether the war was justified and whether Bill Noble should have been there in the first place.

EXAMPLE 4-10B

Conclusion:	Do X, believe X, or buy X.
Premise:	Because of Y.
Label & Description:	Irrelevant Reason. Y is logically irrelevant to the conclusion.
Argument Analysis:	Reasoning. Even if Y is true, the conclusion is not supported. Make a case that Y is only tangentially related to X, that Y is not directly relevant to X, that

although Y resembles X in some way, it would support another conclusion, but not X.

Irrelevant reasons are often found where the politically volatile issue of jobs is discussed. That a nuclear power plant, a military base, or the logging of a forest will produce jobs for people is of course relevant to the people who need those jobs. But if the nuclear power plant is to be built over a geologically unstable area such as an earthquake fault zone, and if the military base will use a large amount of tax dollars—money that could be used to build schools and train teachers better—and it is not needed for national defense, and if logging the forest would result in extinction of a bird species, then jobs are not the only relevant matter for the long-term best interests of the people who want those jobs. Often what may appear only short-term idealism is actually long-term pragmatism. If the nuclear power plant would someday suffer a melt down similar to that of Chernobyl, then building an unsafe plant would not be in the long-term best interests of the people who want the jobs, and the issue of jobs would be irrelevant to the issue of whether or not the plant is safe. If our country is falling seriously behind in educational competitiveness in a high technological world, spending priorities for tax money become the main issue, not jobs related to a military base. A particular military base might be needed for national security, but then jobs would be irrelevant to deciding this issue also. Finally, it may seem the height of idealistic folly to care about the extinction of a bird species at the expense of jobs. However, evolutionary biologists have discovered that life has evolved and flourished on our planet by a process of maximizing genetic diversity. Thus, genetic uniformity, which results from the actions of human beings destroying the habitats of animal and plant species that cannot survive anywhere else, threatens in many ways all life on Earth, including that of human beings. Biologists are also convinced that substances of great medical and educational value lie secretly locked within the recesses of every forest. Because each form of life is unique and nonrepeatable, each one we destroy is gone forever. These are complex issues, and it is often tempting to resolve complex issues by comfortable shortcuts. But if our shortcuts are not real solutions to difficult problems, our future may be even more uncomfortably complex.

This concludes our discussion of fallacies of relevance. Although there are many more irrelevant appeals than those covered in this chapter, the general message that emerges from this sample is that what is relevant in one context for a particular conclusion may not be in another. Whether or not an argument has a label is not as important as identifying the conclusion of an argument and then asking yourself what is most relevant to that conclusion.

Key Terminology

Logical fallacy	Bandwagon appeal
Informal fallacy	Snob appeal
Fallacies of relevance	Appeal to Authority
Appeal to Popularity	Traditional Wisdom

Provincialism   Ad Hominem Abusive
Appeal to Loyalty   Ad Hominem Circumstantial
Two Wrongs Make a Right   Irrelevant Reason
Common Practice

Concept Summary

Many persuasive arguments ought not to be so persuasive. A **logical fallacy** is an argument that is usually psychologically persuasive but is logically weak in some sense. **Informal fallacies** are relatively simple appeals that occur every day in the mass media, political exchanges, and common disagreements. Many mistakes in reasoning involve getting off track, of not thinking about what we should be thinking about. **Fallacies of relevance** are informal fallacies that shift our attention and focus away from what we should be thinking about when we are assessing whether we should accept the intended conclusion of an argument.

Examples in this chapter include appeals to aspects of human nature that are not intrinsically bad, but which play a distracting role given the conclusions and issues contained in these appeals. **Appeals to Popularity** shift our attention to the natural desire to want to belong with others, to have friends, to be a comfortable member of a culture, but they fail to help us decide if what others do is wise. **Appeals to Authority** play upon our natural tendency to look up to people we admire, to trust certain people for advice on how to deal with a complex world, but they fail to help us when experts disagree and they deprive us of a relevant foundation (the authority's reasons) for advice and endorsement. **Traditional Wisdom** reminds us that often the experience of past generations helps us in the present, but fails to help us separate those past ideas that still work from those that do not. **Provincialism** reminds us that we identify with cooperating subgroups, but fails to help us decide if the paths people like us take are good ones. **Appeals to Loyalty** remind us that sometimes we need to stick together and cooperate in achieving a common cause, but they do not help us decide which causes are just or wise.

There is a time and place for each of these appeals. Our species has evolved successfully by moderating our natural tendency toward selfishness, by cooperating with others as part of a group with knowledgeable leaders, and by learning from past generations. But we have also evolved by being able to allow individuals freedom to think and offer new ideas for groups to follow. To do so, we must be free to think about relevant matters other than mere expert opinion, group loyalty, and past wisdom.

Two Wrongs Make a Right reminds us that we expect people to be treated equally, but it does not help us decide if two similar-seeming courses of action that appear bad are good just because they may be similar. The **Ad Hominem** appeals remind us that beliefs are beliefs of people, of flesh-and-blood individuals that we may want to know something about when they offer us reasons to accept their conclusions. But they do not help us reasonably know if ideas and character are necessarily linked, because they shift our attention away from the evidential reasons people may have for their beliefs. Finally, as the fallacies discussed in this chapter illustrate, there are many ways that people can fall for simple gambits of getting off track. Most people don't read long, complicated arguments, or if they do, they do so carelessly. Without careful consideration of the matters relevant to a conclusion, it is easy to be fooled that evidence has been presented for a conclusion, when in fact you are being offered reasons that support another conclusion that merely resembles the issue being discussed. *Irrelevant Reason* fallacies are examples of this generic problem.

EXERCISE I

Indicate whether the following are true or false.

*1. From a logical point of view all appeals to tradition are bad.

2. In criticizing an Appeal to Popularity, you should always question the truth of the premise.

*3. Unlike appeals to improper or irrelevant authorities, appeals to proper authorities or relevant experts by themselves are considered valid and should be persuasive.

4. All irrelevant reason fallacies should be criticized by attacking the link (the reasoning) between the premise(s) and conclusion.

5. Loyalty to a cause is not evidence that the cause is just or wise.

6. Two Wrongs Make a Right is invalid because it uses an implied premise that is often false, that two courses of action are similar.

*7. Evidence that a belief has been made up to fulfill a motive is direct evidence that the belief is false.

8. All Ad Hominem Abusive fallacies have name-calling in the conclusion.

9. Provincialism fallacies are aimed at people who identify with being part of a particular group.

10. Irrelevant Reason is a generic label for any argument that uses premises that support another conclusion that resembles the issue being discussed.

EXERCISE II

Make a case that each of the following are fallacies by giving a complete written analysis similar to that provided in the text. Use the same format as that in the text: Conclusion, Premise(s), Label and Description, and Argument Analysis. Be sure to provide and develop the appropriate focus in the argument analysis section.

*1. Advertisement:

Johnnie Walker Red Label is the scotch for you. It is welcome everywhere. In California, more smart people enjoy Johnnie

Walker Red Label. (Photograph—handsome people having a cocktail sitting next to the first tee of an exclusive golf country club.)

2. A 1992 campaign pitch in favor of electing Bill Clinton president of the United States:

It is time for a change. It is time for a new generation to assume the leadership of this great country. It is our time now. It is time for the generation born after World War II to be in charge. For many decades we have had the same tired leadership like George Bush. It is time to stop living in the past and prepare for the twenty-first century.

3. Argument against allowing women to be priests in the Catholic Church:

Women who are clamoring for the priesthood just are not too observant. The mission for women in the church has always been, and always will be, a different service. If they would only look at the history of the Roman Catholic Church, they would realize that they traditionally have the greatest power in the church— the formation of minds and hearts in education, a ministry to the sick, and the teaching of the church to future generations.

4. Argument against building a space station:

The United States should not build Space Station Freedom. It may cost up to $180 billion over the long term. The majority of the nation's leading scientific societies and numerous Nobel laureates have spoken out against it.

5. Supporters of Space Station Freedom are wrong about there being any scientific justification for a permanently manned space station. What is really at stake for the fifty senators who have signed a letter endorsing the space station is pork-barrel projects for their states. Each of the senators has a number of businesses in their states with space-station contracts.

Note: "pork-barrel" is a political term that refers to questionable federal projects or funding a member of Congress is able to secure for his or her district. The projects or funding are said to be examples of wasteful spending because they are often not based on what is good for the country, but rather what is good for the particular congressional representative and his or her future reelection prospects. For instance, because by 1991 the former Soviet Union was no longer perceived as a military threat to the United States, the Pentagon tried to curtail the operation of a $700-million Maine radar system that had been designed to detect

Soviet bombers. However, the powerful senators from this state were able to continue the operation of the radar system, arguing that it could now be used to detect drug smugglers. Very few drug smuggling planes fly into Maine; the vast majority of cases of airborne drug smuggling are in southern states. These senators were able to "bring home the bacon."

6. Argument in favor of finding O. J. Simpson innocent during his 1995 trial on charges that he killed his former wife Nicole Brown Simpson and her friend Ronald Goldman:

   It should be clear that we should support O. J. Simpson. Consider what he has done for young black Americans. First, as a heroic football player, he proved that nice guys can finish first. Second, as an actor and articulate salesman for Hertz, he showed young black children that regardless of where they start in life (fatherless and living in a ghetto), through effort they can make something of themselves. At this crucial and dangerous time in America, when most young black males growing up in inner cities think that their only option is to be in a gang, black children and all America needs someone like O. J. Simpson to be free. The majority of people in this country think that O. J. is innocent. Even 70% of the trial lawyers interviewed said they believe that Simpson will not be convicted due to a hung jury. (Assume that "support O. J." means that he should be found innocent.)

7. Conversation:

   Jane: "I know your arguments are not worth listening to. Of course you're against legalizing abortion. I wouldn't expect anything else from a male chauvinist."

   John: "But you haven't even listened to my arguments concerning the life and viability of the fetus."

   Jane: "I'm sorry, John. I don't need to, I know what you are like."

8. Traditional argument supporting U.S. exports of military arms to other countries:

   There is nothing wrong with the United States being the leading exporter of arms to Third World countries. The Soviet Union and China are a close second and third, respectively. Almost all industrialized nations support their balance of trade bottom line with arms exports while simultaneously using these shipments as an instrument of influence and foreign policy. If we don't do it, someone else will.

*9. The fact that Linus Pauling and other researchers have had to fight tough uphill battles to get the National Cancer Institute to sponsor trials with vitamin C for human cancer patients is reason enough to conclude that vitamin C has a considerable potential usefulness in cancer therapy. The cancer establishment is heavily committed to surgery, radiation, and conventional chemotherapy. Many careers built upon these specialties would collapse if innovative cancer therapies were shown to be more effective.

10. Conversation:

    Don: "I can't see how you can believe in God."

    Jim: "I have thought about it a lot. It seems to me to make the most thoughtful alternative."

    Don: "Thoughtful?! How can an intelligent person like you continue to believe in such an outdated proposition as we move into the twenty-first century? Don't you know that most people with low IQs believe in God?"

11. Mayoral campaign advertisement:

    "Who are you voting for?"
    "Frank Fasi. Isn't everybody?!"

12. Conversation over how to vote on City Charter proposals:

    Alice: "How are you going to vote on the city charter proposals?"

    Donald: "I am going to vote for all of them."

    Alice: "What? But some of them deny initiative and abolish the neighborhood boards!"

    Donald: "Oh, I always vote for the proposals. They are too complicated to figure out. Besides, I figure the charter people must know what they are doing. State Representative Domingo and Senator Basil were on the committee that proposed the amendments, and they have been in government for many years."

13. From a speech to a local women's organization by their president:

    "Any modern woman ready to enter the twenty-first century knows that it is her right to have abortion on demand. The antiquated arguments of the antiabortionists are no longer worth considering. Arguments from natural law have been used for centuries to oppress women. It is our time now and time for a new philosophy of justice. Abortion is legal, should be legal, and will be legal in a new century of equality."

14. Argument against Republican economic proposal:

    "It is the same old tired argument (Newt Gingrich's Contract with America) we have heard from the Republicans before. They say that they want to produce an economic structure that will produce more jobs for average Americans by cutting taxes and government regulation. But it is no secret that the Republican party is the party of the rich, that the majority of Republican voters are wealthy and will benefit the most from tax cuts. The economic policies of the Republicans are simply a disguise to make the rich richer and the poor poorer."

*15. Argument in favor of the innocence of Lt. Col. Oliver North on charges of violating a law that restricted U.S aid to the Nicaraguan rebels and lying to Congress. Assume that the conclusion of this argument is that North is innocent.

"If you were on this jury, how would you vote? Well, I can tell you how I would vote. According to President Reagan, North is a 'national hero.' Do we want to continue to be a free country? Should we not continue to support the champions of the eternal vigilance that Jefferson and the founding fathers spoke of more than two hundred years ago? At a time when our country is being taken over by wimp liberals, homosexuals, and Gorbachev-lovers, I say support courageous men like Ollie North. He and his family have suffered enough."

16. Argument in support of continuing to give federal government subsidies to tobacco growers:

I don't see why people get all upset when they find out that the government continues to help tobacco farmers in this country make a living. We do the same thing for sugar growers in Hawaii and milk producers in Minnesota.

17. Argument against appointment of Clarence Thomas to the Supreme Court:

Can all these people be wrong? The Leadership Conference on Civil Rights, an executive committee of a coalition of 185 civil rights groups opposes this appointment, as well as the National Urban League, NAACP, and the Congressional Black Caucus. The groundswell of opposition rapidly building against this appointment must be telling us something about the fitness of this man to serve; that he does not, in the words of Alexander Hamilton, "unite the requisite integrity with the requisite knowledge" needed to serve on the highest court in the land.

18. News item:

"Yesterday, the U.S. Senate voted to block a survey of adult sexual behavior that supporters argued would provide information that could be used to prevent the spread of AIDS and other sexually transmitted diseases. According to Senator Jesse Helms

(Republican, North Carolina), the real motive for the survey was 'to support the left-wing, liberal argument that homosexuality is normal.' "

(Assume that Helms is offering a justification for why the bill should not have been passed.)

19. Argument by a U.S. senator against rushing into making deep cuts in U.S. defense spending after the disintegration of the former Soviet Union. (The proposed cuts included cutting more than one hundred thousand troops, the B-2 stealth bomber that would cost almost $1 billion each, and the Star Wars antimissile system.):

"How quickly people forget. Just a few short months ago our men and women were dying in the Persian Gulf. Just a few short blocks from where we debate this issue is a wall of granite with the names of men who sacrificed their lives for their country during Vietnam. Many of you in this chamber right now fought valiantly in Korea and World War II and left many devoted friends on bloody battlefields far from home and loved ones. I say we need a strong defense and that we should maintain our current troop levels, build the B-2 and our Star Wars nuclear shield. We owe it to dedicated American men and women, past and present."

20. Argument supporting Israel's treatment of the Palestinians:

The way the Israelis treat the occupied Palestinians is not wrong. The claim is that the Palestinians are made to live in virtual concentration camps, have no political rights, have little freedom of movement, and that their land is repeatedly confiscated for Israeli settlements. But look at what the United States did to its Japanese citizens during World War II. Their land was confiscated, and they were also made to live in concentration camps. And this is not to mention the U.S. treatment of American Indians and the confiscation of their land for pioneer settlements.

21. Conservative Republican advertisement supporting Clarence Thomas for the Supreme Court (paraphrased):

The arguments of those who oppose Clarence Thomas for the Supreme Court don't measure up. Who do you want to make the important decisions that will affect America for years to come? People such as Judge Thomas, or his opponents: Senator Ted Kennedy, who was thrown out of Harvard as a college student for cheating and left the scene of a terrible accident at Chapiquiddick; Senator Joseph Biden, who was found to have plagiarized a speech when he attempted to run for president; and Senator Alan Cranston, who has been implicated in the savings and loan scandal?

22. Argument aimed at getting African-American support for appointing Clarence Thomas to the Supreme Court:

African-American people need representation on the Supreme Court. If Clarence Thomas is approved by the Senate, he will be only the second African-American person in the history of the United States to achieve such a high office.

23. Argument by Rush Limbaugh attacking Tom Cruise for supporting concern for the environment:

"Some people care more about Bambi than they do about people. . . . Some people think that trees are more important than human beings. . . . Take Tom Cruise. Didn't Tom Cruise make a stock-car movie in which he destroyed thirty-five cars, burned thousands of gallons of gasoline, and wasted dozens of tires? Tom, most people don't own thirty-five cars in their life. Now you are telling other people not to pollute the planet? Shut up, sir!"

24. For further practice, analyze the Ice Cube–St. Ides malt liquor ad and the Marlboro and Virginia Slims cigarette ads on pages 133–134.

25. In 1994, former national security adviser Robert McFarlane published a book entitled *Special Trust*. In this book, McFarlane claimed that both President Ronald Reagan and then–Vice President George Bush knew that money received from arms sales to Iran was being given illegally to the Nicaraguan Contras. If true, this would have meant that Reagan knowingly broke the law and could have been impeached. Bush would have most likely not become president in 1988. According to McFarlane, Reagan "lacked the moral conviction and intellectual courage" to admit his involvement. McFarlane also claimed that Oliver North, the principal administrator of the illegal Iran–Contra deal, lied to him and Congress. North, who was then running for a senate seat in Virginia, responded to these allegations by stating that McFarlane's book was "a pitiful and mean-spirited attempt to glue his broken reputation back together again." Analyze North's argument.

ANSWERS TO STARRED EXERCISES:

I.   True and False

1. False. Appeals to tradition often reflect that we learn from experience and pass on what we learn to the next generation. But because times change, we need to know which traditions remain reliable by discussing the initial reasons for a tradition and whether those reasons still apply to the present.

3. False. Experts can disagree, and accepting expert opinion without knowing the reasons the expert has for endorsing a belief or a

course of action deprives us of important and relevant information.

7. False. A poor motive is relevant to a charge of bias, but bias is not direct evidence that a belief is false. Beliefs can be true, even if the people who hold them are biased.

II.   Fallacy Analyses

1. **Conclusion:**   Buy Johnnie Walker Red Label.
   Premise: It is welcome everywhere. In California, more smart people enjoy Johnnie Walker Red Label.
   Label & Description: Appeal to Popularity, if we emphasize the "welcome everywhere" appeal; Provincialism, if the in-group of "smart people" and country club elite is emphasized. (It is not necessary to quibble over which label best fits here. The argument analysis should be the same regardless of which label is selected.)
   Argument Analysis: The reasoning is weak. Even if it is true that Johnnie Walker is the most popular scotch, and even if it is true that intelligent and well-to-do people drink it, there is no discussion of why these people drink it, of how the scotch is made, or whether the ingredients make a superior scotch to other scotches. The premises are irrelevant to the conclusion because we are asked to buy a product without the quality of the product discussed.

9. There are actually two conclusions in this argument. The first one is part of an Ad Hominem Circumstantial fallacy and is then used as a premise for the second conclusion.
   Conclusion #1:   The cancer-treatment establishment is wrong about the uselessness of vitamin C for cancer therapy.
   Conclusion #2:   Vitamin C has a considerable potential usefulness in cancer therapy.
   Premises: The cancer-treatment establishment is biased. It is heavily committed to surgery, radiation, and conventional chemotherapy. Many careers built upon these specialties would collapse if inno-

vative cancer therapies were shown to be more effective. They are so biased that Linus Pauling and other researchers apparently have failed to get the National Cancer Institute to sponsor trials with vitamin C for human cancer patients.

Label & Description: Ad Hominem Circumstantial. The personal circumstances of the doctors—their jobs as physicians using traditional methods of cancer treatment—and their possible bias and suspect motives because of vested interests in traditional treatment are cited in the premises.

Argument Analysis: The reasoning is weak. Even if it is true that the medical establishment doctors are biased against any therapy that would destroy their careers, this is not direct and relevant evidence that the therapy works. These doctors could still be right about vitamin C having little positive effect on cancer. The premises are irrelevant to the conclusion that the establishment is wrong about vitamin C. Appropriate evidence would be to cite any controlled studies that have been done showing whether a group of people with cancer improved more than or as well as a group of people undergoing traditional therapy.

15. **Conclusion:** North is innocent of charges violating a law restricting aid to the Nicaraguan rebels and lying to Congress.

Premises: We should continue to want to be a free country and support the champions of eternal vigilance that Jefferson and the founding fathers spoke of more than two hundred years ago. We should continue to support national heros at a time that our country is being taken over by wimp liberals, homosexuals, and Gorbachev lovers. North and his family have suffered enough.

Label & Description: Irrelevant Reason. The premises are irrelevant to the issue of whether or not

Argument Analysis:

North is guilty of breaking a law and lying to Congress.

The reasoning is weak because the premises are irrelevant to the conclusion. The premises are mostly about the general topic of freedom and support for champions of freedom. But the issue revealed in the conclusion is whether or not North broke a law. In a democracy, no one has a right, including the president of the United States, to decide which laws to obey. (If Reagan had known about the diversion of money to the Contras, he would have been impeached.) North disagreed with the law passed by Congress making it illegal for our government to give military aid to the Nicaraguan Contras and even boasted about lying to Congress. That he and his family had suffered is irrelevant to whether or not he caused that suffering himself by breaking a law. In 1991, all charges against North were dismissed on the technicality that North's televised congressional testimony, in which he admitted breaking the law, made it impossible for a jury and witnesses to be impartial!

Chapter
5

INFORMAL FALLACIES II

Reasoning is the best guide we have to the truth . . . . Those who offer
alternatives to reason are either mere hucksters, mere claimants to the throne, or
there's a case to be made for them; and of course, that is an appeal to reason.
— Michael Scriven, *Reasoning*

Introduction

In the last chapter we examined one of the major causes of poor reasoning, getting off track and not focusing on the issues related to a conclusion. In our general discussion of reasoning (Chapters 1–3), however, we saw that arguments can be weak in two other ways:

1. In deductive reasoning, arguments can be valid, but have false or questionable premises, or in both deductive and inductive reasoning, arguments may involve language tricks that mislead us into presuming evidence is being offered in the premises when it is not.
2. In weak inductive arguments, arguments can have true and relevant premises, but those premises can be insufficient to justify a conclusion as a reliable guide to the future.

Fallacies that use valid reasoning, but have premises that are questionable or are somehow unfair in the truth claims they make, we will call *fallacies of questionable premise*. As a subset of fallacies of questionable premise, fallacies that use tricks in the way the premises are presented, such that there is a danger of presuming evidence has been offered when it has not, we will call *fallacies of presumption*. Fallacies that involve very weak, but often psychologically persuasive inductions, we will call *fallacies of weak induction*. We also noted in Chapter 4 that fallacies can be analyzed in terms of being weak (1) in the reasoning, (2) by having questionable premises, or (3) by having suppressed evidence. In the majority of cases the most important distinction is between that of weak reasoning and questionable premises. Fallacies of relevance and fallacies of weak induction are all weak in the reasoning. With the exception of the fallacy of Suppressed Evidence, fallacies of questionable premise and presumption are weak in the premise or premises.

It is important to remember what this distinction means. When we claim that the reasoning of an argument is weak, the focus of criticism should be on the inference *from* the premise or premises to the conclusion. In this case, we need not criticize the premises in terms of their truth claims, but instead should argue that *even if* the premises are true we do not have good grounds for accepting the conclusion. On the other hand, when we claim that the premise or premises are weak, the focus of criticism should not be on the reasoning but should instead center on a claim that the premise or premises are questionable (false or unreliable), presumptive, or unfair in some sense.

At this point, these distinctions and various categorizations may seem very abstract, but it is important to grasp them as part of our general theme and strategy of applying the proper focus when arguing. Like becoming an expert surgeon, it is important to learn not to wander around with one's knife. Proper analysis and criticism should stay on track; as we have seen, there is enough unfocused thinking in the realm of human discourse. If an argument involves a particular fallacy, and one makes a claim that one should not accept the conclusion, then the criticism should focus, first and foremost, on the standard problem of that fallacy. For instance, as we have seen in the cases of Appeal to Popularity and Questionable Cause, the standard problem in both of these fallacies is in the reasoning. Given any particular passage that may commit either of these fallacies, there might be a lot more going on that we could criticize. We might doubt the truth of the premises as well. However, if the charge is one of these fallacies, then the analysis should start by criticizing the reasoning, not the claim made in the premises. Furthermore, the goal of criticism is not criticism for its own sake, but to make our beliefs more reliable, and the proper focus is important to determining what would make these arguments better. In the case of Appeal to Popularity, because the premise is irrelevant to the conclusion, a follow-up discussion would involve identifying relevant beliefs that would support the conclusion if they were true. In the case of Questionable Cause, the premise is relevant to the conclusion but insufficient, and a follow-up discussion would involve identifying more evidence of the same type as that claimed in the premise.

Remember that identifying an argument as invalid or unsound does not mean the conclusion is false and must be banned from rational consideration forever. Similarly, calling an argument a fallacy may only mean that after identifying why the argument is weak, follow-up discussion will enable us to better support the conclusion. Like the cigarette research example and making follow-up controlled experiments better, criticism of an argument helps make the next argument better. To do this, we must focus our attention based upon the proper categorization.

Finally, as is the case in all formalizations and abstractions, proper categorizations save a lot of time and intellectual energy. Once we know the essence of a fallacy, how it "fits" with these distinctions and categorizations, we only have to think about it once, so to speak. To help make the major contrast clear between fallacies that are weak in the reasoning and fallacies that are weak in the premise, let's first contrast the fallacies of relevance covered in Chapter 3 with some examples of fallacies of questionable premise.

Fallacies of Questionable Premise

As noted above, fallacies of questionable premise are valid. If the premises were true we would be locked into the conclusion. Recall, however, from Chapter 2 that in passages of the form "X, because Y," to distinguish arguments from explanations, in arguments Y should be less controversial than X. In general, to rationally persuade someone to accept a conclusion we should start with relatively accepted or noncontroversial premises. Normally, in good reasoning we are attempting to move (infer) from what we think we know (relatively non-controversial premises) to what we do not know (a statement that is controversial or uncertain before the argument, the conclusion). Fallacies of questionable premise violate this guideline by offering what a little commonsense reflection will reveal as obviously controversial or unfair premises.

Slippery Slope

In our discussion of deductive and inductive reasoning, we noted that many of the premises used in deductive arguments are based upon inductive reasoning. In our discussion of inductive reasoning, we noted that we can never be absolutely certain of any belief that is the result of inductive reasoning. We are faced therefore with the reality of always living with an uncomfortable or insecure situation. The best we can do is be intellectually honest, admit our ultimate ignorance, and try to put together reliable, but tentative beliefs based upon stronger inductions. Slippery slope is a psychologically persuasive way of arguing that takes advantage of the fact that we live in an uncertain world and that many things are possible. In particular, it takes advantage of the fact that people naturally fear what might happen when we choose to do something different or new. A **Slippery Slope fallacy** is committed when a chain of events (one event causing another) is asserted in the premises with no supporting evidence offered that the chain of events is likely to happen.

Consider this example of a politician commenting negatively on an initiative proposed in San Francisco to abolish the vice squad:

EXAMPLE 5-1   If this passes, San Francisco will be the whorehouse of the nation. There'll be soliciting on the steps of city hall and lovemaking on Market Street, organized crime will profit, and residents and businesses will be affected. It will be a signal that "anything goes" in San Francisco.

The issue before the voters was whether the city needed a special division in the police department devoted to enforcement of laws related to prostitution, drugs, pornography, public decency, and so on, or if, given budget priorities, enforcement of these laws could be the responsibility of the entire police department. A related issue was the priority of law enforcement. Those who were for the initiative thought that the police department should spend more time and tax money on robbery and violent crime rather than interfering with what liberal supporters considered mostly matters of privacy. Actions, supporters argued, that infringed on the rights of others could be handled by the police in general; those that did not, either should not be dealt with at all or should be dealt with through education.

The more conservative politician is concluding that the abolition of the vice squad is wrong, and the evidence offered is that if the vice squad is abolished terrible things will happen. The terrible things he cites may happen. However, they also may not happen. He is offering us a slippery slope in his premises, that once a first step is taken a number of other steps are inevitable. Because the chain of terrible things he cites are offered as premises and no evidence is offered to support them, we would be justified in asking what evidence he has that these terrible things will actually happen. These premises are weak or questionable in the sense that they are controversial and no evidence is offered to support the claim that once the vice squad is abolished the chain of events cited is inevitable. The mere creative assertion of a possible chain of events cannot support the conclusion unless we have some reason to believe that the chain of events is likely to happen. We could just as well create a positive possible chain of events—also unsubstantiated—that abolishing the vice squad will save money and restructure police law-enforcement priorities, that this will free up more money for education, and that this will make people stop using drugs.

Here is a complete formal analysis of this argument:

EXAMPLE 5-1A

Conclusion:	(implied) Don't abolish the vice squad (Or, don't vote for this initiative.).
Premise 1:	If you do, San Francisco will be the whorehouse of the nation. There'll be soliciting on the steps of city hall and lovemaking on Market Street, organized crime will profit and residents and businesses will be affected. It will be a signal that "anything goes" in San Francisco.
Premise 2:	(implied) A signal that anything goes in San Francisco is not good.
Label & Description:	Slippery Slope. There is a slippery slope generalization in the premise.
Argument Analysis:	The premise is questionable and unfair. Although the reasoning is valid, the premise asserts a number of controversial possible causal connections for which no evidence is cited. There is also reason to doubt the truth of this premise. The police department in general will not be abolished, nor will the laws against prostitution, and such. So, it is unlikely that even if the vice

squad is abolished the police would not do something about an obvious violation of the law such as prostitutes soliciting on the steps of city hall.

Note that this argument is valid. If the premises are true, if we knew these things would actually happen by abolishing the vice squad, and if we do not want these things to happen, then the conclusion would follow—we should not abolish the vice squad. The focus for criticism and discussion, then, should be on the premise. The problem with the argument is that it violates the principle that in general we should start arguments with noncontroversial premises whenever possible. Because the slippery slope premise is a prediction of a possible causal chain of events, inductive evidence should be cited to support this premise. For instance, if it could be shown that another city had abolished the vice squad and this city had an increase in drug use and prostitution, probably as a consequence of abolishing the vice squad, then the argument would not be a Slippery Slope fallacy. This better argument might not be conclusive, but at least it would draw attention to what should be discussed: whether or not the chain of events predicted is likely to be true. The persuasiveness of a Slippery Slope lies not only in its valid reasoning, but in the scare tactics involved in the premises and the fact that we live in an uncertain world. The message of Chapter 3, however, is that with effort we can have reliable beliefs even in an uncertain world. The problem with Slippery Slope fallacies is that the effort is not made to establish the premise as a reliable belief.

Here is the formal essence of all Slippery Slope fallacies.

EXAMPLE 5-1B

Conclusion:	Don't do A. (Most often implied.)
Premise #1:	Because if we do A, then B will happen. If B happens, then C will happen. And if C happens, then D happens. (The chain here does not need to be exactly this long, but usually it at least involves an A, B, and C.)
Premise #2:	D is bad (implied).
Label & Description:	Slippery Slope. There is an unsupported slippery slope in the premises. One of the premises claims that once a first step is taken a number of other steps are inevitable.
Argument Analysis:	The premise is questionable and unfair. Although the reasoning is valid, an unsupported and controversial prediction is made in the premises regarding a possible chain of causal events. Without some evidence presented to support the connections asserted in the premise, the mere assertion of a possible chain of events cannot support the conclusion. Focus on whether the premise is a reliable belief. Argue that it is at least questionable by offering evidence that the chain is not likely to happen. Point out that the argument could be stronger if inductive evidence were offered to support this premise.

Note that a Slippery Slope fallacy involves causation. For this reason, students will often confuse it with a Questionable Cause fallacy. But Questionable Cause is a fallacy of weak induction, the causation claim is always in the conclusion, and some relevant evidence for the conclusion is always offered in the premise. Slippery Slope has the causal claim in the premise and offers no evidence at all for this claim. This is an important distinction, for it reminds us what to focus on if we want to make these arguments better.

Here are some famous examples of Slippery Slope fallacies.

EXAMPLE 5-1C   Bob Hope supporting U.S. military involvement in the Vietnam War:

"Everyone I talked to there (Vietnam) wants to know why they can't go in and finish it. And don't let anybody kid you about why we're there. If we weren't, those Commies would have the whole thing (the rest of southeast Asia), and it wouldn't be long until we'd be looking off the coast of Santa Monica (California)."

EXAMPLE 5-1D   Anita Bryant arguing against giving homosexuals the same constitutional protection against discrimination as that of other minorities.

"If we give homosexuals the civil rights issue, then next year we will have to give it to prostitutes. Then it will be for child molesters, then bestiality, then necrophilia. . . ."

Bob Hope's argument is a version of Slippery Slope called the domino theory. The role of U.S. military involvement in Vietnam was complex and controversial, and one of the issues was whether the conflict between North and South Vietnam was basically a civil war or part of a worldwide communist movement. Hope was assuming that what was at stake was stopping a worldwide communist movement, and that if we didn't fight in Vietnam and the communists won, other southeast Asian countries, such as Cambodia, Burma, and Thailand, would fall to communism. But this is precisely the issue Hope should have addressed before he assumed that the domino chain of events was inevitable.

Anita Bryant is addressing an issue that faced many states in the 1980s and early 1990s. Should homosexuals be protected against job and housing discrimination? Should they be allowed to marry? Take for instance a high school history teacher. Suppose this teacher has won awards for his teaching and he is very popular with his students. Suppose there is no question about his job competency. Suppose, however, that the high school principal finds out the teacher is gay, and because of pressure from the local PTA, he fires the teacher. Suppose also that when this becomes known, the teacher's landlord evicts him from his apartment. Should these actions be legal? Bryant is arguing that they should be and that homosexuals should not be protected, because this will set a precedent for giving similar protection to child molesters and people who have sex with animals and dead people! Both of these arguments will be exercises at the end of this chapter. In preparation for analyzing them, think of what kind of evidence both Hope and Bryant would need to support their slippery slope premises.

Questionable Dilemma

Black-and-white thinking is another way to avoid uncertainty. If the real world of confusing choices and tentative beliefs could be sliced up into just two nice categories of good guys and bad guys, how simple life would be. A **Questionable Dilemma** fallacy is committed when a major premise offers us only two choices, one of which is claimed to be bad, but a little commonsense reflection reveals that a good case can be made that there are more than just these two choices.

During the early 1990s, physician Jack Kevorkian was arrested in Michigan numerous times for assisting people in committing suicide. To clarify the law against assisted suicide, the voters of Michigan were asked to vote on new language that would strengthen the legal case against Kevorkian. In campaigning against the new law, Kevorkian stated, "It's as simple as this: Are you going to vote for your right, or are you going to vote it away?" Kevorkian has given us only two ways to look at the issue of euthanasia. It is a right we either accept or throw away. But opponents of euthanasia and assisted suicide were arguing for a third alternative: that the law should be strengthened because suicide is not an individual right. Kevorkian may have been right that compassionate and painless suicide should be our right in circumstances of excruciating pain and inevitable death, but in the above statement his appeal ignores the third alternative.

Here is a more in-depth example:

EXAMPLE 5-2   From Tim LaHaye, chairman of California for Biblical Morality:
"Either God exists and has given man moral guidelines by which to live, or God is a myth and man is left to determine his own fate. Your response to either position will determine your attitude toward such issues as abortion, voluntary school prayer, pornography, homosexuality, capital punishment, the priority you place on traditional family life, and many other social problems."

LaHaye is essentially claiming that either we believe in the traditional Christian God or we will likely be tolerant of abortion, pornography, and homosexuality; be against capital punishment; and place a low priority on traditional family life. But he has not made a case that these are our only choices, and a case can easily be made for other possible choices. Can't atheists, agnostics, or people who accept other religious beliefs (Hindus, Buddhists, Jews, Muslims) also place a high priority on traditional family life? It is also possible to believe in God and believe that we are responsible for our own fate. Religious humanism is such an alternative. The above argument assumes that humanists cannot be good persons, because humanists believe that human beings can be good without the direct help of divine intervention. There is also a philosophical-religious position known as deism, which states that God exists, but is more like a grand clockmaker who created human beings and the universe and then withdrew from day-to-day involvement with His creation. Followers of Eastern philosophies and religions, such as Hinduism and Buddhism, believe that although a personal God does not exist, we are not entirely abandoned to chaos, that there is a grand pattern or flow to all creation, and that we can mold our behavior positively with this pattern.

Here is an analysis of LaHaye's argument:

EXAMPLE 5-2A

Conclusion:	The Christian God exists and has given humankind moral guidelines to live by.
Premise #1:	*Either* this is true *or* God is a myth and we are left to determine our own fate.
Premise #2:	If you believe the latter, then (implied) this is obviously wrong, because you are likely to be tolerant of pornography and homosexuality, against capital punishment, and to place a low priority on traditional family life.
Label & Description:	Questionable Dilemma. One of the premises makes a questionable either/or claim.
Argument Analysis:	The major premise is questionable. The first premise contains a dilemma that reduces the number of possible alternatives to two. Because there are possible positions other than these two extremes, the premise is weak. The above argument ignores the existence of many other major world religious views, such as Islam, Buddhism, Hinduism, and Taoism. It is also possible to be a religious humanist and believe that even though God exists, we are left to make our own decisions about abortion, homosexuality, and capital punishment—that believing in God does not automatically endorse only one position on these issues. It is also possible to be an atheist or agnostic and accept traditional family values.

In attacking LaHaye's argument as a Questionable Dilemma we are not claiming categorically that LaHaye's premise is false. We are simply pointing to the fact that it would be premature to accept his major premise when there are other possibilities to discuss. It is possible that he could make a case that these are our only choices. But he has not done so, and thus we have good reasons to doubt his premise until he does. He has claimed there are only two possibilities. We need only point to a third alternative to draw attention to the questionable nature of his premise and place the burden on him to offer further support for this premise before we have a good reason to accept his conclusion.

Here is the formal essence of all Questionable Dilemma fallacies:

EXAMPLE 5-2B

Conclusion:	Do A, believe A.
Premise #1:	Because we must do (or believe) A or B.

Premise #2:	(implied) B is bad.
Label & Description:	One of the premises makes a questionable either/or claim; it restricts our choices to only two alternatives.
Argument Analysis:	The premise is questionable. The first premise contains a dilemma that reduces the number of possible alternatives to two. Because there are other possible positions other than these two extremes, the premise is weak. Make a case for an alternative C.

Notice that, as with Slippery Slope, the structure of Questionable Dilemma is valid. If it were true that we had only two choices and we did not want one of them, then it would follow deductively that we should make the other choice. Thus, to claim that this fallacy has been committed places the burden of discussion on the content of the argument, specifically the premise that asserts the either/or dilemma. To accuse someone of committing this fallacy, you must know something about the topic under discussion and must at least state a tentative case for a third alternative. You can't simply charge someone with committing this fallacy because he or she claims there are only two alternatives. There may be only two realistic alternatives, and unless we make a case that there is a third alternative, we can't merely assume there is one. By stating a tentative third case we halt the process of premature conclusion and focus attention on the proper place for further discussion.

As an example, recall from Chapter 4 that defenders of arms sales to foreign countries often phrase their arguments as a Two Wrongs Make a Right fallacy: It is acceptable U.S. policy to sell arms to third-world countries because the Soviets or Chinese are doing it. Suppose after charging someone with committing this fallacy by pointing out that they had not addressed the relevant issue whether or not arms sales by any country is right, the defender of arms sales responds by saying,

EXAMPLE 5-2C "Of course it is wrong for any country to do this. In an ideal world this would be categorically wrong, but we do not live in an ideal world. We live in an imperfect world of many necessarily evil responses to gain influence and defend our strategic interests. In the world we live in we have no other choice: either we sell arms or someone else will."

The focus of discussion has now changed from whether selling arms is right to whether we have any other choice. Without further discussion we should assume neither the existence of no other choice nor of a third choice. Both parties should now defend their positions by discussing whether there is a third foreign policy choice.

Straw Person

We cannot be knowledgeable on every issue, but a little reflection reveals that we should be cautious in accepting without direct testimony a person's interpretation of a position that person obviously rejects. Suppose I was very critical of the governor of my state for opposing a bill that would increase public school teachers' salaries, and to persuade you to agree with me, I described the governor's position for you. Until you can get direct testimony from the governor or a supporter of the governor's position, you should be cautious in accepting as accurate my description of the governor's position. Human nature being what it is, my description could be an exaggeration or distortion of the governor's real position. Suppose the governor's stated position is that the bill in question is a budget buster, that the state is too much in debt, that the economy is in such a precarious state that not enough tax revenue will be generated to cover the bill's costs. Thus, taxes will have to be raised. And although he is in favor of higher salaries for teachers, he believes this particular bill is flawed because it does not require accountability from teachers and it does not specifically target good teachers for higher salaries. Suppose the governor has summed up his position by saying that he has a responsibility to promote quality education, but he is responsible to all tax payers, not just those who send their children to public schools. Suppose in my description of the governor's position I claim that the governor is against this bill because he is more concerned with balancing the budget and saving tax payers money than he is about good education for the state's children. I have exaggerated the governor's argument. Although I attempted to describe his argument, and I did mention something related to a portion of his argument (a balanced budget, no new taxes), I twisted it to make the governor's reasons appear very weak.

A **Straw Person** fallacy is committed when a distortion or exaggeration of a person's position is used as a premise for concluding that the person's position is invalid. Like the little pig who made his house out of straw and had it blown away by the big, bad wolf, a straw person argument sets up one's opponent with a distorted or exaggerated description so that the position is easy to blow away.

Consider this famous argument against the 1950s antinuclear bomb activist and British philosopher, Bertrand Russell.

> **EXAMPLE 5-3**   "Mr. Russell believes that we (the United States) should stop making atomic bombs so that the Soviet Union can gain a strategic superiority over the free world and threaten peace throughout the world. I say better dead than red."

This was a distortion of Russell's true position. He argued in the 1950s that the psychology of the arms race was producing, or would soon produce, a situation in which there would be far more nuclear weapons than needed for defensive purposes. By 1970, the United States had thirty-six nuclear bombs per major Soviet city, and the Soviet Union had eleven nuclear bombs per major U.S. city—six thousand missiles for the Soviet Union, and ten thousand for the United States. By 1985, the numbers had increased to about eleven thousand for the United States and ten thousand for the Soviet Union. By 1991, when the Soviets and the United States finally announced a nuclear arms reduction agreement, the United States had 12,304 missiles, and the Soviet Union 11,626. These numbers represent missiles not bombs. Some of the missiles carried as many as ten independently targetable hydrogen bombs; in total, the arsenals of both sides added up to approximately fifty thousand bombs, just one of which could kill millions of people. Russell was arguing that no more nuclear weapons should be produced, but he was not urging that we let the Soviets get ahead and threaten world peace. Rather, he argued that the one thousand or so nuclear weapons then in existence already made nuclear war unthinkable. Here is how this argument would be analyzed as a Straw Person fallacy.

EXAMPLE 5-3A

Conclusion:	Mr. Russell's position on the production of atomic bombs is incorrect.
Premise:	He wants us to stop producing atomic bombs so that the Soviet Union will threaten world peace and freedom.
Label & Description:	Straw Person. The premise, although an attempted description of Russell's position, is a distortion of his true position.
Argument Analysis:	The premise is not true. The premise involves a distortion of Russell's true position. He argued that the arms race psychology was producing, or would soon produce, a situation where there would be far more nuclear weapons than needed for defensive purposes. Although he did advocate that the production of nuclear weapons cease, it was an exaggeration of his position to claim that he wanted the Soviet Union to gain a strategic advantage.

As in Slippery Slope and Questionable Dilemma fallacies, the focus of the discussion should be on the premise rather than the reasoning. This means that to participate in criticizing or defending the premise we must know something about the topic's details. We cannot charge someone with distorting a person's true position unless we can make a case for the person's true position. This entails knowledge, a robust world view, or the patience to research the position being criticized. If this is not possible, then one should at least be cautious in accepting a description of a position one knows the describer is against.

Here is the formal essence of the Straw Person fallacy:

EXAMPLE 5-3B

Conclusion:	X is wrong about A.
Premise #1:	An alleged description by an opponent of X's position on A.
Premise #2:	(implied) As described, X's position is weak or stupid.
Label & Description:	Straw Person. The description of X's position in the premise is a distortion or exaggeration of X's true position.

Argument Analysis: The premise is weak. Although if the main premise was true, it would offer good grounds for accepting the conclusion, the main premise is not true. Make a case that the main premise is a distortion or exaggeration of X's true position by describing X's true position.

Note that the conclusion of Straw Person is the same as that of the Ad Hominem arguments, in which a person is attacked. Because of this, students will often confuse the Ad Hominem appeals with Straw Person. Straw Person fallacies at least make an attempt in the premises to describe X's position, and thus these fallacies are at least on track in this regard. In comparison, the Ad Hominem fallacies do not describe X's position at all in the premises, but instead attack with irrelevant name calling or cite poor motives and questionable circumstances. Straw Person's premise is relevant to the conclusion. If the description were true, this would be a major consideration for accepting the conclusion. With Ad Hominem premises, even if the premises are true, they are irrelevant to the conclusion. In recognizing the possibility of a Straw Person fallacy we are directed to the premise and must discuss whether the description of X's position is accurate or not. In identifying an Ad Hominem fallacy we are reminded that a person's position on an issue could be worthy even if we don't think much of that person's character or motives, or suspect the circumstances related to the person's argument. We are reminded that although at some point motives may be a consideration, we should examine the details of the person's position before writing it off.

In the spring of 1995, as part of their Contract with America, the Republican controlled House Economic and Educational Opportunities Committee sent a bill to the full house that would repeal the National School Lunch Act. By this time 14 million schoolchildren received free or reduced-priced school lunches and another 5 million received free or reduced-priced breakfasts. House Democratic leader Richard Gephardt responded, "Now we see what the Republican Contract with America is really all about. It is a war against children. They would rather use this money to build highways." House Republicans responded that this was a "bizarre characterization" of their position. They claimed that the new approach, which would give block grants to states to run the program, would be more efficient, cut unnecessary audits and paper work, and lead to an increase in the amount of money spent on school lunches. The Republicans may have been wrong that the program would be more efficient and not hurt children, but Gephardt's statements were clearly not an accurate description of what the Republicans thought they were doing. A good case can be made that Gephardt was guilty of a Straw Person. On the other hand, an attack on the Republican proposals to the effect that they were "simply a political payoff to the rich, who supported Republicans during the 1994 elections and who want lower taxes," is best characterized as Ad Hominem Circumstantial. Gephardt was at least attempting to describe the Republican plan; whereas the latter attack did not mention any specifics of the plan, and referred to motives instead.

One final note on Straw Person fallacies. Sophisticated media techniques allow easy video distortions of an opponent's position. In 1993, Governor Carroll Campbell of South Carolina opposed President Clinton's proposals on health care. In a thirty-second commercial backing the Clinton plan, Republicans were depicted as dogmatically refusing to acknowledge how many people were suffering either entirely without health care or were financially drained by family health care costs. In the commercial, in a sound bite Campbell clearly states, "There's not a crisis." But Campbell's comment was edited from a much longer answer to a television interview. Here are Campbell's complete remarks:

"Number one, you shouldn't say there's not a crisis. There's a crisis for people that don't have health care. And there is a crisis in the financing. **There's not a crisis** in the whole medical system of America, and there's a different interpretation. . . . But there are areas that are in crisis that need to be dealt with."

A small piece of film was used to totally distort Campbell's opposition to the Clinton plan. Campbell was on record for opposing the forcing of all businesses to help pay for coverage as a destructive burden on small business, but had not taken the position that no legislation was needed.

Fallacies of Weak Induction

In the discussion of inductive reasoning in Chapter 3 we noted that although all inductive conclusions lack certainty, some inductive arguments give us a better basis for believing that some beliefs are reliable guides to the future. Based upon what we learned about the rigor and patience needed for good inductive arguments, we can classify some psychologically persuasive weak inductions as fallacies.

Hasty Conclusion

Although not all inductions are generalizations, most are. Every day we generalize from our experience. We walk into a room that we have walked into, say, only once before, and we generalize that if we sit in the same chair again, it will support us as it did previously. I eat a sandwich that I like at a restaurant, and I assume that every time I go to this restaurant the sandwich will be the way I like it. I see a bird of a particular color, and I assume that all the birds of this species have this color. Only when we are disappointed in our generalizations, when my sandwich does not come out the way I assumed it would, for instance, do we realize that such generalizations are actually hypotheses, not certain truths. Most of the time, we are being rational in making such generalizations because we are actually adding independent inductions and well-supported world views to corroborate our generalizations. When a restaurant makes a sandwich a certain way, I can reasonably assume that it will continue to be made this way. The employees are trained to make it this way, the owner wants to please customers and stay in business, and in general I know from past experience and common knowledge that this is the way most restaurants work.

However, the all-encompassing appeal of many of our generalizations, and the same wish to eliminate the complexity and insecurity in our lives that lead to black-and-white thinking, sometimes fool us into accepting quick generalizations we ought not accept. **Hasty Conclusion** is a quick generalization in which care has not been taken to see if a small sample warrants a generalization about the subject of that sample. In our apple example in Chapter 3 we saw that one rotten apple would not warrant a very secure inference about the entire barrel unless we brought some corroborating factors into the inference—knowledge of bacteria, the amount of time the apples have been in the barrel, and so on. A Hasty Conclusion fallacy is a generalization where no such corroborating factors have been offered.

Exaggerated, quick generalizations often occur when people try to adjust to a confusing and controversial event. In 1993, Lorena Bobbitt was declared innocent of malicious wounding by virtue of temporary insanity. She had cut off her husband's penis with a kitchen knife after he had allegedly raped her. Discussion of the pros and cons of the jury verdict made the rounds of all the TV talk shows, with views in general highly polarized, some people defended her action, and others condemned it. However, these talk shows generated extreme generalizations such as, "This is a feminist dream come true. What they're doing is licensing the feminists to come and slice our penises off. It will be open season on men."

Consider this example, a letter to the editor in my daily newspaper:

EXAMPLE 5-4   "I noticed yesterday that the Reverend Hoffman, a very vocal member of Jerry Falwell's Moral Majority, resigned from his post as minister to the New Faith Church, because of the considerable embarrassment over his admitted adultery with one of his parishioners. This characterizes the true nature of the fundamentalist movement; they are all just a bunch of frustrated psychopaths hiding behind the facade of moral righteousness."

From one example of adultery and hypocrisy it would be very weak induction to conclude that this characterizes the entire fundamentalist movement. Without a greater sample or corroborating independent inductions, this is a hasty conclusion. A more interesting argument would be the addition of corroborating psychiatric testimony to the effect that those who protest most vigorously against pornography and sexual tolerance tend to have sexual problems themselves, and the addition of more positive cases, such as the cases of televangelist ministers Jimmy Swaggert, Marvin Gorman, and Jim Bakker. Swaggert was caught several times with prostitutes, and Gorman and Bakker were involved in adultery scandals. Even with this additional evidence, the conclusion that this characterizes the entire fundamentalist movement would be too general. More reliable might be a tentative conclusion that a significant amount of hypocrisy exists in the fundamentalist movement.

Here is an analysis of the above argument followed by the formal essence of all Hasty Conclusions.

EXAMPLE 5-4A

Conclusion:	*All* those in the fundamentalist movement are a bunch of psychopaths.
Premise:	*One* member of the fundamentalist movement is found guilty of deviant and hypocritical behavior.

Label & Description: Hasty Conclusion. The conclusion asserts a considerable generalization given the small amount of evidence in the premise.

Argument Analysis: This argument is weak in the reasoning. Even though the premise is relevant to the conclusion—Reverend Hoffman is a member of the fundamentalist movement—this one case is insufficient to characterize the whole religious movement without other examples and independent corroborating inductions. Hoffman could be an atypical case.

EXAMPLE 5-4B

Conclusion: All As are Bs.

Premise: One A is a B. (or) A few As are Bs.

Label & Description: Hasty Conclusion. The conclusion asserts a considerable generalization given the small amount of evidence in the premise.

Argument Analysis: Reasoning. Although the premise is relevant to the conclusion, it is insufficient to support the generalization in the conclusion. Make a case that the argument is a weak induction.

From this formal essence you should note that the premise does not need to refer to only a single incident of the generalization. Consider these examples:

EXAMPLE 5-4C   Because nine out of ten marijuana smokers in the Harlem section of New York city later progress to heroin, it is obvious that smoking marijuana will lead to the use of heroin.

EXAMPLE 5-4D   Pine must be a liberal atheist, because in this book he criticizes conservative arguments for the existence of God.

For **5-4C**, although the sample cited in the premise is a relatively large percentage, given the sample, the acceptance of the generalization that this sample is typical is hasty. One could easily make a case that this is a very weak induction. The sample, even if true about Harlem, is not necessarily representative of other communities. Harlem could very well be an atypical case. Until a more representative sample is taken, the generalization is very weak. For **5-4D**, this argument is not only a very hasty conclusion—the arguments of liberals are also criticized in this book—but implies a failed understanding of the reason for criticism as well. I may be very religious, but find the arguments criticized embarrassingly poor. Hence, my goal may be to have students think of better arguments for the existence of God and not to convince them that they should be atheists.

Notice that like the fallacies of relevance covered in Chapter 4, but unlike the fallacies of weak premise discussed at the beginning of this chapter, a Hasty Conclusion is weak in the reasoning. However, also note that unlike fallacies of relevance, a Hasty Conclusion has a premise relevant to the conclusion. Recognition of these distinctions is important because in criticizing arguments with the goal in mind of making our arguments better, Hasty Conclusion, as is the case with all weak inductions, at least possesses the virtue of having evidence in the premise that is on track. So, we at least know what type of evidence we need more of to support the conclusion.

This latter point is also important for making another distinction. We should not mistake the fact that a Hasty Conclusion is a very weak induction with the directive that we should be so cautious about generalizing that we stifle all creative hypothesizing. As a fallacy to avoid, understanding Hasty Conclusion cautions us not to jump to conclusions about what to believe or accept as reliably true. This does not mean that we should not jump to conclusions in proposing new ideas or hypotheses. Hypotheses and new ideas are most often the result of creatively jumping to a conclusion from very little evidence. There is nothing wrong with having a lot of bold creative hypotheses around for consideration and testing. However, there is a difference between proposing a new idea and saying that the idea has been critically examined and tested. We should be cautious about *accepting* a new idea, but not necessarily in *proposing* a new idea for examination.

Historically, many currently accepted scientific ideas started as bold conjectures that had very little initial evidence for them. In the sixteenth century, the Copernican hypothesis that the sun was the center of all planetary motion was a bold and difficult to believe generalization with relatively little evidence for it. It was very hard for many rational people to accept because it implied that the heavy, solid earth was spinning at an incredible speed that we do not feel. Also, because of the geometry of a sun-centered universe, it implied that the stars would have to be an immense distance away, a distance that further implied a universe much larger and less congenial to our special role in the eyes of God. At first, no physics existed to explain why we would not fly away if the earth was spinning so fast. And the scientific evidence available at the time seemed to support the view that the stars were much closer to us than the Copernican system implied. Eventually, Newton produced the theory of gravity to explain why we don't fly away even though the earth is spinning at about one thousand miles per hour. And eventually, with more powerful telescopes, scientific evidence revealed that the stars are an immense distance away. Thus, it would have been a hasty conclusion that the Copernican system was not true, just because initially the evidence did not support it reliably!

Scientists have also learned to be cautious in accepting initial data as a falsification of a theory or generalization. Sometimes the data are wrong and the theory is used to correct the data. For instance, in the barrel full of apples example in Chapter 3, suppose we have pulled out ninety-nine clearly rotten apples. When we pull out the last one, initially it appears sound. It looks fine on the surface and appears to falsify the hypothesis that all the apples in the barrel are rotten. But because our hypothesis has been so well supported up to this point, we just cannot believe that this last one is not also rotten. So, we examine it inside and find that it is clearly rotten. The history of science includes many occurrences in which initial data, experiments, or observations appeared to falsify a theory, only later to be seen as corrected or confirmations. In short, when facts and a hypothesis conflict, the facts might be wrong!

Darwin's theory of natural selection and Einstein's theory of relativity have histories similar to the Copernican theory. When Darwin's theory was first proposed, the process of natural selection as described by Darwin did not appear consistent with the then commonly accepted age of the Earth. To explain the evolution of the variety of species that currently exist on Earth by natural selection requires a very old Earth. Today, geologists believe that there is reliable evidence that our Earth is approximately 5 billion years old, thus providing ample time for the process of natural selection to produce, in Darwin's words, the "grandeur . . . (of) endless forms most beautiful and most wonderful." Similarly, when Einstein's theory was first proposed, the boldness of the theory far outmatched the evidence. According to Einstein's special theory, time is not uniform or absolute, but "slows down" in one reference frame of high relative velocity compared to another. Today, time dilation (the slowing down of time) demonstrations by scientific experiment are routine. For instance, two high-precession atomic clocks can be synchronized, and one then placed on the space shuttle and the other left on Earth. When the two clocks are compared after the space shuttle returns to Earth, the one on the shuttle will have recorded less time, even though the two clocks will still be recording time at the same precise rate as when they were first compared. It follows from these demonstrations that if long, very high-speed space voyages are someday routine, twin sisters will no longer be the same age if one takes a long space voyage and the other remains on Earth!

Thus, in science there is a difference between **pursuing** a creative new idea and **accepting** it as a reliable belief, and a difference between a context of discovery and a context of justification. Our emphasis on the tentative and hypothetical nature of establishing reliable beliefs is consistent and actually promotes an attitude of the playful pursuit of and experimentation with bold new ideas.[1]

Questionable Cause

By way of introduction to structuring fallacies, we examined the fallacy of Questionable Cause in Chapter 4. However, this is such an important fallacy that further examination within the context of weak inductions is warranted. We can now see that Questionable Cause is a type of Hasty Conclusion. To assert that one event is **caused by** another is a bold generalization. When people assert claims such as Reaganomics caused the economy to get better in the 1980s, that Clinton's deficit reduction and economic stimulus policies of the early 1990s caused the economy to improve, that feminism caused an increase in crime, that sex education classes cause teenagers to want to have sex, which produces more teenage pregnancies and spreads venereal disease, they are making bold claims about very complex situations.

---

[1] For more on the difference between the pursuit and acceptance of belief, see Larry Laudan's *Progress and Its Problems* (Berkeley: Univ. of California Press, 1977), pages 108–14.

It is understandable why we make such claims and why we seek reliable beliefs about underlying causal connections that produce complex behavior and physical and social events. We want to know what causes what, so that we can control reality or at least adjust to it and live successfully. If it is really true that the 1980s economic policies of President Reagan produced the foundation for a better economy, then we want to sustain such policies. If it is really true that sex education classes for teenagers produces behavior that we don't want, then we should eliminate such classes.

However, great harm can also be done by assuming a causal connection where there is none, and because the world is complex at least on the surface, the amount of evidence needed to support a causal connection must be considerable. As we saw in our introduction of Questionable Cause in the previous chapter, many events take place at the same time. The fact that one event precedes another is very weak inductive evidence that these events are causally related. Let's look at another example:

EXAMPLE 5-5   Letter to the editor:

"Just why is everyone pushing this sex education in schools? Why is it necessary? The worn-out reason is that a lot of parents do not talk about it at home, so it must be taught in school. Yet, since this trend started, VD and pregnancy among teenagers and even preteens has sharply gone up. Why? I thought sex education was supposed to reduce it, not increase it. The answer is that it is a how-to-do-it course, nothing else. That is why the dirty books are being pushed in schools now to go with it. Sex education is stupid."

One interpretation is that the author's main conclusion is that sex education has actually caused what it was supposed to prevent. This is a rather bold and startling claim, considering that sex education classes were offered to educate teenagers about sex and lessen venereal disease and premarital pregnancies. Such classes were also offered based upon the higher-order induction that education is good, that it leads to rational choice and behavior modification.

Suppose we investigate this claim and find that the premise offered in this argument is true: That during the past ten years there has been an increase in sex education classes and simultaneous dramatic increases in venereal disease and premarital pregnancies for teenagers. As in the example of the initial correlation between cigarette smoking and lung cancer, this would be a good place to begin an investigation, not end one. Although this correlation is unexpected, there are so many other possibilities that a bare temporal correlation is by itself a very weak inductive premise. Suppose sales of acne lotion have increased at the same time sex education classes have increased? Although psychologically more persuasive, without further investigation the status as evidence of this unexpected correlation between sex education classes and an increase in teen sex problems is no better than having a door slam shut immediately after a sneeze.

Many other factors could cause the increase of teenage venereal disease and pregnancy. An increase in the population of teenagers could cause every activity related to teenagers to increase: automobile accidents, purchasing of acne lotion, or particular types of clothing and music CDs, for example. There could be an increase in the population of particular types of teenagers, those in an area of the country where sex education is not taught or where early experimentation is encouraged by various social or family pressures. Temporal association and correlation do not prove causation. Like the initial correlation between an increase in the use of cigarettes and lung cancer, a correlation between sex education and teen sex problems does not reasonably substantiate a causal connection, and by itself, it does not give us a clear indication in which direction there may be a connection. For all we know at this point, an increase in teen sex problems has led to an increase in sex education classes!

As with the cigarette example, to establish a more reliable belief concerning a possible connection between sex education classes and teen sex problems, a controlled study should be conducted. Recall that in a controlled study the goal is to control as many variables as possible, so that given two populations of teenagers these populations will be about the same except that only one group will have had sex education in school. In this way, if we found a much higher percentage of pregnancies and VD in the group that had sex education, we would then have a more reasonable basis for claiming a causal connection between sex education classes in public schools and subsequent teen sex problems. On the other hand, if we found no significant difference between the two groups, it would be reasonable to conclude that a population increase or some other causal factor was involved. Or, if we found that the group that had sex education classes actually had a much lower percentage of teen sex problems, this would be reasonable support for the conclusion that the sex education classes were fulfilling their purpose of alleviating teen sex problems.

Such studies have been done. In 1991, Girls Inc. released a report on studies they had sponsored. Controlled studies had been conducted over a three-year period in Texas, Tennessee, Nebraska, and Delaware, in cities where girls were at higher than average risk of becoming pregnant. Girls in the sex education programs ages 15–17 were half as likely to have sexual intercourse for the first time as girls who did not participate in the program. And girls having sex reported having sex without birth control about half as often as girls in the control group. In other words, the sex education program appeared to work. Girls in the sex education programs abstained from having sexual intercourse more often than those not in the program, and if they did have sex they were far more likely to use contraception. The programs promoted parent–daughter communication for the purpose of delaying sexual activity, assertiveness skills for saying "no" while remaining popular, motivation for avoiding pregnancy by helping girls set educational and career goals, as well as responsible decisions about contraception. Girls Inc. estimated that it could provide a program containing these components for about $116 per year per girl.[1] If further studies conducted using the techniques of this program corroborate this result, then a more reliable belief would be that a well-run sex education experience for teenagers is a causal factor in lessening teen sex problems, not increasing them.

---

[1] Leticia T. Postrado and Heather Johnston Nicholson, "Effectiveness in Delaying the Initiation of Sexual Intercourse of Girls Aged 12–14: Two Components of the Girl's Incorporated Preventing Adolescent Pregnancy Program," *Youth and Society*, Volume 23, March 1992, pages 356–379.

Here is an analysis of **5-5** followed by the essence of all Questionable Cause fallacies.

EXAMPLE 5-5A

Conclusion:	Having sex education in schools has caused an increase in VD and pregnancy among teenagers and preteens.
Premise:	Since sex education has been introduced at schools, VD and pregnancy among teenagers and preteens have increased dramatically.
Label & Description:	Questionable Cause. There is a causal connection between two events asserted in the conclusion, and only a temporal association between those events asserted in the premise.
Argument Analysis:	The reasoning is weak. Although the premise is relevant to the conclusion, the reasoning is weak because even if it is true that since there has been sex education in schools, VD and teen pregnancies have increased, this is not sufficient to know whether there is a causal connection between these events. The connection could be purely coincidental and other factors might be involved, such as an increase in the population of teenagers, or teenagers living in a social situation that promotes teen sex problems. The correlation between these events would be a good place to begin an investigation (with a controlled study), not to end one.

EXAMPLE 5-5B

Conclusion:	A causes B.
Premise:	A happened, then B happened.
Label & Description:	Questionable Cause. A causal connection is asserted in the conclusion and only a temporal association in the premise.
Argument Analysis:	Reasoning. Although the premise is relevant to the conclusion, it is insufficient to support the conclusion. Make a case that this is a weak inductive inference by indicating that other factors were also happening at the same time as the alleged cause.

Note that in criticizing a Questionable Cause you do not have to make a case that one of the other events happening at the same time was the real cause. In the above example, we need not prove that some other factor actually caused the increase in teen sex problems. Because the person committing the Questionable Cause has cited only an event happening just before or at the same time as the alleged effect, once we point out that other events were also happening at the same time, the burden of proof is on the person committing the fallacy to show why we should believe the proposed event, rather than the others, is the cause. It is identical to the situation of a sneeze followed by a door slamming. Once we reflect that any one of many other things could be the cause because each also happened prior to the door slamming, the mere fact of sneezing before the door slammed cannot be used to separate it as a cause from the other events. At this stage of investigation, any other event that happened at the same time has an equal chance of being the real cause. A temporal sequence of events is only a good place to begin an investigation into causal connections.

Appeal to Ignorance

Another common hasty reasoning jump involves inferring that, because we can't prove something true, it must be false; or, that because something can't be proven false, it must be true. It would be an unreliable induction to conclude that ghosts exist because they have not conclusively been proven to not exist.

To understand the weakness in this mode of reasoning, but also to understand the psychological persuasiveness, consider the following example. Suppose I have a new religious belief that somewhere in the universe there is a pearl on a planet that is exactly ten times larger than any other pearl on Earth and that this pearl is special because once it is found by the human species we will have reached a "doorway" to God, and all the suffering that we experience on this earthly plane will be over—we will all go to heaven. Suppose I gain followers who are so faithful that they build rocket ships and begin to crisscross our galaxy in search of this pearl. If my followers do not find any such pearl within a given time, I can say that it is a big universe and that we just need to have faith and keep looking. If, to my amazement they do find a pearl that is exactly ten times larger than any other pearl and no state of heaven appears, I have ready responses. They either found the wrong pearl—there must be another one exactly ten times larger than this one—or it was the right pearl, but our state of mind was not right for God to reveal Himself; He did not think we were worthy yet.

The logical problem with my belief is that I have made it *irrefutable*, it cannot be proved false by any conceivable experience. If we find another pearl and nothing happens, I can again claim that there must be another one exactly ten times larger than this one. My belief is such that I can take any conceivable experience and twist the interpretation in such a way that the result not only does not disconfirm my belief but actually (I would claim) confirms it. "After all," I might say, "we must be on the right track; at least we found a pearl exactly ten times larger than any known on Earth." My belief may be true, but there is no way to show it to be false.

There are many irrefutable beliefs, and many could be true. However, if there is no way to test these beliefs when they clash in the claims they make, there is no way to separate them in terms of which one to believe. There is no way to judge which belief from a pair of irrefutable beliefs that clash is likely to be a reliable belief, because no matter what happens, an interpretation is possible that supports both irrefutable beliefs. For many people the security offered by irrefutable beliefs is enough for them to jump to the conclusion that such beliefs must be true. The mistake in reasoning is: ***my belief will never be proven false, so it must be true***. No matter what anyone says or discovers, my belief will stand the test of time and be able to handle anything new. I need never revise my belief. I need never be challenged to think of a new belief or modify my old belief.

We call this form of reasoning an appeal to ignorance because it appeals to what we don't know to mistakenly conclude that this is evidence for something we do know. An ***Appeal to Ignorance*** fallacy is committed when the premise of an argument appeals to what we don't know—specifically that we don't know if a belief is true or false for sure—and is used for concluding a belief that we are sure of. In her book, ***Only Words***, Catherine A. MacKinnon cites a premise for an Appeal to Ignorance fallacy when she states, "There is no evidence that pornography does no harm." Even if this is true, it is an unwarranted quick generalization to conclude from this fact alone that evidence exists that pornography does harm.

Consider this example:

EXAMPLE 5-6   Letter to the editor concerning AIDS and letting a child with AIDS attend school:

Since science cannot prove that breathing the same air as an AIDS victim will not result in the spread of the virus, children with AIDS should not be allowed to attend public schools.

If the arguer intends the conclusion that it must be true that breathing the same air as an AIDS victim will result in the spread of AIDS, classification as an Appeal to Ignorance is appropriate. It might be unfair on my part to interpret this argument in such a strong sense to make it into an Appeal to Ignorance. The arguer might just be urging caution, that because science cannot eliminate absolutely such an important possibility, prudence demands that we be cautious and not let AIDS-infected children come to school. However, we can use the strong interpretation as a guide for criticizing the weaker or more cautious interpretation as well.

As we noted in Chapter 3, the conclusions of science are never absolute, but are the result of a critical process and are based upon a great deal of carefully gathered evidence. After many years of scientific study, all that is known about AIDS at present points to the conclusion that the virus cannot survive outside of the body and must have sexual or other intimate contact for transmission, such as an exchange of semen or blood. This belief could be wrong, but to have a reasonable basis for believing it wrong, direct evidence must be shown that atypical AIDS transmission has actually occurred in some cases. So, in addition to accepting the poor conclusion that we know breathing the same air as an AIDS victim will result in the spread of the virus, unreasonable harm could be done to AIDS-stricken children—children already suffering physically and psychologically—by not letting them come to school. We must take risks with everything we do, but if we have a reliable belief, to be too cautious and not act on it could be harmful and irrational.

Here is a possible interpretation and analysis of the above argument, followed by the formal essence of all Appeal to Ignorance fallacies:

EXAMPLE 5-6A

Conclusion:	It must be true that breathing the same air as an AIDS victim will result in the spread of AIDS.
Premise:	Science cannot prove that it is false that breathing the same air as an AIDS victim will result in the spread of AIDS.
Label & Description:	Appeal to Ignorance. Concluding that something is true, because (premise) it has not been proven false. There is an appeal to what we do not know in the premise and this is used to conclude what we (allegedly) do know.
Argument Analysis:	The reasoning is weak. As a type of hasty conclusion, even though the premise is relevant to the conclusion, it is insufficient (weak induction) to accept the conclusion. Although it has not been ruled out categorically that breathing the same air as an AIDS victim may result in the spread of AIDS, and this is relevant to the possibility of this way of getting the disease, it is a very hasty inductive jump to conclude that because something has not been proven false, it must be true. That it may be true that breathing the same air spreads AIDS, because we can't show it to be impossible, does not mean that it is true. Direct evidence of a link between breathing the same air and the spread of AIDS, plus corroborating evidence that we need to change our theories about how AIDS spreads, would be needed before this conclusion can be considered a reliable belief. For instance, better evidence would be a documented AIDS case in which the person worked around AIDS patients occasionally and there was no sexual contact or exchange of blood or other bodily fluids.

EXAMPLE 5-6B

Conclusion:	X is true. (Or, X is false.)
Premise:	Because X has not been proven false. (Or, because X has not been proven true.)
Label & Description:	Appeal to Ignorance. Concluding that something is true because (premise) it has not been proven false (or false because it has not been proven true). There is an appeal to what we do not know in the premise, and this is used to conclude what we (allegedly) do know.
Argument Analysis:	Reasoning. A type of hasty conclusion and hence a weak induction. The premise is relevant to the conclusion, but is insufficient to support the conclusion as a

reliable belief. Make a case that this form of reasoning is a weak induction because it appeals to what we do not know as a justification for what we allegedly do know, and explain what kind of direct, positive evidence would be more convincing to establish the belief as a reliable belief.

Fallacies of Presumption

Our last group of fallacies we will call fallacies of presumption because, as do the fallacies of questionable premise, they show that we should be cautious in accepting or presuming the premises of an argument as always true or fair. However, unlike the fallacies of questionable premise, they will not always have valid reasoning nor will the issue always be whether the premises are true.

Begging the Question

Begging the Question is a fallacy with valid reasoning, but the problem is not whether the premise is true, but whether any evidence at all has been given by the premise to support the conclusion. Another name for this fallacy is a *circular argument*. The reasoning is valid, but it goes in a circle: the premise surreptitiously and unfairly assumes an answer to or a position on the very question that should be defended in order for us to accept the conclusion. Here is a famous example:

> **EXAMPLE 5-7** God exists because the Bible says so. And we know that the Bible is an authoritative source of information, because it was divinely inspired.

The conclusion of this argument is that God exists. One of the premises asks us to believe this because God's existence is a clear message of the Bible. But there are many other religious texts reflecting different cultural beliefs about different stories and types of Supreme Beings, so how do we know that what the Bible says is more reliable than what these texts assert? The argument anticipates this objection, and the next premise claims that we can separate out the Bible as correct because its authors were "divinely inspired." But to say that the Bible was divinely inspired—that the people who wrote it had some special communication with divine thoughts—is already to say that some sort of divine creature exists. And because the arguer wants to assert that there is only one God, that divine creature must be God. So, the argument goes in a circle. It essentially claims that God exists, because the Bible says so, and the Bible is right because God inspired it. Essentially no more is said than "God exists, therefore God exists."

This argument is no different than my trying to persuade you that classical music is better than rock and roll, because intelligent people listen to classical music, and you can always tell which people are intelligent, because they are the ones who listen to classical music. Or, if I were to attempt to persuade you that a sure sign that intelligent life must exist elsewhere in the universe is that we have lots of problems on Earth (pollution, species extinction, overpopulation, and ethnic wars) and no life form has been stupid enough to visit us. The premise, that we have not been visited by smart ETs who want nothing to do with messy human beings, assumes an answer to what is at issue in the conclusion: Do ETs exist?

Here is an analysis of **5-7**, followed by the formal essence of all Begging the Question fallacies.

EXAMPLE 5-7A

Conclusion:	God exists.
Premise #1:	Because it says so in the Bible.
Premise #2:	What the Bible says is reliable because it was divinely inspired.
Label & Description:	Begging the Question. The argument assumes or prestates in one of the premises a position that we are asked to accept in the conclusion.
Argument Analysis:	The premise may or may not be true, but presuming it is true would be a mistake because it merely restates the main claim made in the conclusion. There is no independent evidence offered in the premises to support the conclusion that God exists. To say in a premise that the Bible was divinely inspired is already to commit to the claim that God exists. The argument goes in a circle, and no independent evidence is offered for the conclusion. The argument is valid, because if the premise about divine inspiration is true, then the conclusion that God exists would be true, but because the conclusion and the premise say and commit us to the same thing, we should not presume the premise is true.

EXAMPLE 5-7B

Conclusion:	X is true.
Premise:	Because Y is true (but Y already says and commits us to X).
Label & Description:	Begging the Question. The argument assumes or prestates in Y a position on X.
Argument Analysis:	The premise is presumptive. The premise may or may not be true, but because it is just a surreptitious restatement of the conclusion, it would be a mistake to accept it without some evidence for it. Describe what the premise implies, and make a case that it already assumes the conclusion to be true.

Although all deductive valid arguments "contain" their conclusions in some sense, Begging the Question offers little but the conclusion. The conclusion is asserted in a disguised way in the premises. Substantive valid deductive arguments correctly stay on track by carrying out the implications of information and evidence. Such arguments involve a very creative process of putting together the right premises and finding the right chain of reasoning that goes somewhere interesting. They bring to our attention what we have committed ourselves to when we accept the beliefs stated in the premises. To see this, think of the difference between the blind man's (Chapter 1) and Eratosthenes's (Chapter 3) accomplishments and the following example.

Suppose I am the chief executive officer of a local electric company, and I am trying to get government approval to build a nuclear power plant. Opposing me is an environmental group that not only fears all applications of nuclear energy, but opposes this particular plant. They claim that it will be extremely dangerous because of the region's geological instability and earthquake potential, and the predominant wind patterns that would bring radiation over most of the city if a nuclear accident occurred. Suppose the environmental group calls themselves Life of the Land. One day, when being interviewed by a local TV station regarding this controversy, I say:

EXAMPLE 5-7C "I am not really against organizations such as Life of the Land, but I wish they would concentrate their efforts on helping people."

In making this statement, I have "begged the question" at issue. The issue is whether or not building the nuclear power plant is beneficial to the people of the city—whether or not the trade off of cheap electricity is worth the potential risk. My position is that the potential risk is minimal and building the plant will benefit the people of the city. But rather than offer evidence for this, my statement just assumes it by claiming that anyone who is against my project opposes what is good for the people of the city. My statement amounts to no more than this: "The project is a good project because it is good for the people of the city, and it is good for the people of the city because it is a good project."

The deductive accomplishments of the blind man and Eratosthenes are much different. Both discover and draw out for our attention the implications of accepted information. They are not just repeating an assumption hidden in the premises.

Complex Question

In Chapter 2 you were introduced to a rhetorical question, a question that implies its own answer by the context. If I were to say to my students a week before a big exam, "You don't want to fail to do your homework do you?," I am really saying "It is very important to do your homework this week." We often find such rhetorical questions as appropriate parts of arguments because they are really declarative statements in disguise. However, there are apparent rhetorical questions that do not imply one declarative statement, but actually are complex in the sense that they are asking two questions disguised as one. When such complex questions become a premise in an argument a **Complex Question** fallacy is committed. The premise then involves a complex question that cannot be answered with a straightforward yes-or-no answer without implying a mistaken conclusion.

Consider this example and its analysis, an argument between a father and son during the Vietnam War:

EXAMPLE 5-8   Son: "Dad, I've decided to fight the draft and oppose the war in Vietnam."

Father: "What?! Aren't you going to be a man and support your country?"

Son: "Yes, I mean no!"

Father: "What a coward."

EXAMPLE 5-8A

Conclusion: The son is not a man (coward).

Premise #1:   "Aren't you going to be a man and support your country?"

Premise #2:   "No."

Label & Description: Complex Question. Premise #1 involves a complex question that cannot be answered with a straightforward yes-or-no answer.

Argument Analysis: The premise is presumptive. It involves more than one question: "Are you a man?" and "Are you going to support your country?" The son is trapped into accepting a conclusion regardless if he answers yes or no. If he answers 'yes,' this implies that he has changed his mind and will support the war effort. If he answers 'no,' this implies that he is a coward. In either case, the conclusion does not follow, because the question needs to be divided. The Vietnam War was very controversial and divided the country on whether the United States should have participated in what many claimed was a civil war. Many young people at that time thought that our involvement in Vietnam violated the principles of democracy and self-determination our country was supposed to represent. The son's position was that he was not a coward, and that the war was not justifiable.

Some other examples of Complex Questions are:

Prosecutor to defendant: "Have you stopped beating your wife?"

Traditional husband to wife: "Aren't you going to stay home and take care of your kids?"

Religious advertisement: "Where will your kids spend eternity?"

Orange juice ad: "Have you had your orange juice today?"

Here is the formal essence of all Complex Questions:

EXAMPLE 5-8B

Conclusion:	You accept, believe, or want to do Y (or Z).
Premise #1:	Don't you accept, believe, or want to do X? (But X actually involves Y or Z.)
Premise #2:	Yes-or-No answer.
Label & Description:	Complex Question. Premise #1 is a complex question.
Argument Analysis:	Presumption. It would be unfair to accept the Complex Question premise and hence the conclusion. The premise involves a Complex Question that cannot be answered with a straightforward yes-or-no answer without implying a mistaken conclusion. Make a case that the premise is a Complex Question by specifying what the two questions are (Y and Z) and showing that the question needs to be divided into two independent questions.

Ambiguity-Equivocation

In Chapter 2, we saw that mistakes in reasoning can occur when the words used in premises are not precise in their meaning. In particular we saw that ambiguous phrases such as "the highest rating" can have two different meanings in a given context, and that it was a mistake in the Dunlop tire advertisement, first discussed in Chapter 1, to presume that only one meaning was intended. We saw that it was a mistake to infer from the premise—that Dunlop had been given the highest rating (it placed in the highest category)—to the conclusion that the Dunlop tire was better (another meaning of highest) than all major competitors. The following is an example of one version of an *Ambiguity-Equivocation* fallacy. In this version, a word or phrase occurs in both the premise and the conclusion, but with different meanings:

EXAMPLE 5-9

Conclusion:	The Dunlop SP-4 tire was the best (highest-rated) radial tire of all major brands in the *Car and Driver* tire test.
Premise:	Because it had the best (the highest category) rating in the *Car and Driver* tire test.

A lot of humor is based on deliberate equivocations, where language is used that is subject to two or more interpretations. Here are some examples from cartoons:

Peanuts	Charlie Brown:	"Lucy, if you miss one more fly ball, you've had it!"
	Lucy:	"If I'd had it, I wouldn't have missed it!"

Shoe	Senator Battison D. Belfrey:	"And I'd like to introduce Trixie, my wife of twenty-three years."
	Editor:	"The Senator's been married twenty-three years?"
	Cosmo:	"*She's* 23, not the marriage. . . ."

And an unintended humorous example from a politician:

Dan Quayle: "What a waste it is to lose one's mind, or not to have a mind is being very wasteful."

In the *Peanuts* example, the phrase "had it" has two meanings. When Charlie Brown uses the phrase he means that if Lucy misses another fly ball, she is probably going to be kicked off the team or something else drastic. When Lucy uses the phrase she means catching the ball. In the *Shoe* example, the editor assumed that the phrase "my wife of twenty-three years" means to his amazement that the sleazy Senator has been married twenty-three years, but we find that the Senator meant a twenty-three-year-old wife. Like cartoons and humor in general, such deliberate equivocations are harmless and add to the gaiety of life.[1]

The same probably cannot be said about the Dan Quayle case. Because he was vice president of the United States when he made this statement, and a supposed role model of leadership and accomplishment for other U. S. citizens, the confused equivocatory nature of his statement is regrettable. Quayle was trying to recall the United Negro College Fund's motto, "A mind is a terrible thing to waste," and probably trying to say something about the importance of self-actualization and the development of potential, of developing the full potential of the human mind regardless of race. However, the meaning of the word *mind* in the phrase "lose one's mind" usually refers to someone being very upset or forgetful, or in extreme cases to a mental breakdown. On the other hand, the word *mind* in the phrase "have a mind" is often used in deep philosophical discussions about whether human beings are more than just a physical body, whether they have a mind or soul that will continue to exist after the death of the physical body. Quayle also seems to be equivocating on *waste*. It is sad (a waste) to lose one's mind (go nuts), and bad (wasteful) not to develop one's potential??? Go figure.

However, our focus in this section is on arguments that contain ambiguous or equivocal words or phrases. Specifically, where an argument is presented using a word or phrase inconsistently, where a word or phrase occurs in both a premise and the conclusion, but with different meanings. In such cases we should not presume that evidence has been offered in the premise.

---

[1] Equivocations are also used to get our attention, such as in the advertisement from the Environmental Defense Fund supporting recycling, which shows a beautiful picture of our fragile Earth in space with the caption *IF YOU'RE NOT RECYCLING, YOU'RE THROWING IT ALL AWAY.* "A little reminder from the EDF that if you're not recycling you're throwing away a lot more than just your trash."

Let's look more closely at the Archie Bunker example given in Chapter 2 (pages 54–55). Archie commits two back-to-back Ambiguity-Equivocation fallacies. In his first argument he equivocates on the word "cheating," and in the second he uses two senses of "honest." Here is an analysis of his first argument followed by the formal essence of all Ambiguity-Equivocation fallacies. The second argument will be left for a student exercise.

EXAMPLE 5-9A

Conclusion:	I'm (Archie) not cheating (by using hidden notes on the exam).
Premise:	I'm supposed to give the right answers on the test, and it would be cheating if I don't, because when you are supposed to give something to someone and you don't it is cheating.
Label & Description:	Ambiguity-Equivocation. Inconsistent use of the word "cheating" in the premise and the conclusion.
Argument Analysis:	Presumption. We should not presume that Archie has offered any evidence for his conclusion that he is not cheating, because the word "cheating" is not used consistently in the premise and the conclusion. The use of "cheating" in Archie's premise does not have the same meaning that Gloria is using. He has not responded to Gloria's charge of cheating (the issue in the conclusion) in the sense that using hidden notes on an exam is cheating. In denying Gloria's charge, his use of "cheating" in the sense of not giving something to someone does not answer Gloria's charge of the inappropriateness of using hidden notes on an exam.

EXAMPLE 5-9B

Conclusion:	X is the case (using a particular word or phrase with meaning A).
Premise:	Because Y is true (using the same word or phrase with meaning B).
Label & Description:	Ambiguity-Equivocation. Using the same word or phrase (identify) in the premise and the conclusion, but with an inconsistent meaning.
Argument Analysis:	Presumption. Because the word or phrase used in the premise does not have the same meaning as the word or phrase used again in the conclusion an illusion of evidence is created, but no evidence is actually supplied by the premise. Explain why it would be presumptive

to accept the premise as evidence by identifying the word or phrase and explaining the two different meanings used.

As a final, and perhaps more serious, example of Ambiguity-Equivocation consider the following, which also uses sexist language.

EXAMPLE 5-9C   It is not true that the use of robotic space craft is better than manned space flight. Only manned space flight could accomplish the data-gathering flexibility exhibited by the Apollo-Moon program. Therefore, only men should be astronauts.

The word *man* has different meanings in our language. It sometimes refers to a single male, as in "The man was very tall," and sometimes to the entire human race, as in "Man has accomplished so much in recent years with new technology." In recent years a revolution in language use has taken place alongside changing attitudes toward women and minorities. A much more inclusive attitude has developed regarding the wisdom of allowing all people to develop their potential. The above example reminds us that sexist language should be avoided because it leads to sexual stereotyping. Words such as *mankind, businessman,* and *chairman,* are being replaced with *humankind* (or the *human race*), *businessperson,* and *chairperson.* The fallacious nature of the above argument can easily be seen if we replace the phrase *manned space flight* in both premises with *human-piloted space flight.*

Questionable Analogy

Analogical reasoning is a type of inductive reasoning that is often strong enough to provide us with reliable beliefs. If a friend of mine has a Honda Accord LX, and she tells me that it has the particular internal mechanical features X, Y, Z, then it is a fairly reliable inference on my part that the next Honda Accord LX I see will also have these features. When I see the next Honda Accord LX and note its external features, I infer that it must be "just like" my friend's Accord and must also have the internal mechanical features X, Y, Z. There might be a few things that are different between the two cars, but it will be a reasonable inference on my part to conclude that basically the cars are built the same. The strength of this inductive inference is based on a higher-order induction—my background knowledge of how cars are manufacturered.

However, analogies can also be used as hasty premises in the sense that they are offered as a creative comparison of two things with little or no evidence that the two are comparable. When a weak analogy is used as a premise, a **Questionable Analogy** fallacy is committed. Consider the following letter to the editor during the Reagan presidency.

EXAMPLE 5-10   President Reagan's economic policies, which are supposed to solve our economic problems, are just like ancient practices of human sacrifice.

Selected individuals were slaughtered at the altar to appease the gods and enhance the well-being of the majority. There is no difference between this and Reagan's policy, which gives large tax cuts to the rich and cuts government programs for the poor.

The implied conclusion of this argument is that Reagan's economic policies of the 1980s were bad for the country. The alleged goal of Reaganomics was to stimulate the economy by cutting government expenditures and taxes. This was supposed to put more money into the hands of the rich and the middle class, but it was also supposed to make more money available for investment, creating more jobs for the middle class and the poor. In short, supporters of Reagan's policies argued that this approach would help the poor, not sacrifice them. The author of the above argument may be right that it did not help the poor, but instead sacrificed their needs in favor of the rich. However, no evidence was offered to support this claim, instead only a creative analogy was used to compare the treatment of the poor with the primitive practice of sacrifice. A better argument would cite statistics showing that the number of people classified as poor increased dramatically during the Reagan presidency, that the rich tended to keep their money and use it for other purposes than investment in new industries, and that budget deficits also increased, which placed a heavy burden on the current economy and future generations.

The analogy used could help us understand an argument by way of introduction or summation, but the bare analogy should not persuade us. If we claim that X is like Y, we should be able to justify that this is so because they share some characteristics, A, B, C. Thus, it is the substantiation that both X and Y do have characteristics A, B, C that should persuade us. Did the poor get poorer under Reagan's policies? Was Reaganomics the cause? Did more people become poor because of Reaganomics? Did the rich prosper? Was there a conscious effort to sacrifice the poor for the betterment of some and their niche in the economy?

Here is an analysis of the above argument followed by the formal essence of all Questionable Analogy fallacies.

EXAMPLE 5-10A

Conclusion:	Reagan's economic policy is wrong.
Premise #1:	Reagan's policy is just like ancient practices of human sacrifice. It sacrifices the poor to enhance the well-being of the majority. There is no difference between the ancient practice of human sacrifice and Reagan's policy, which gives large tax cuts to the rich and cuts government programs for the poor.
Label & Description:	Questionable Analogy. A weak analogy was used in the premise; Reaganomics was compared to ("just like") the ancient practice of human sacrifice.
Argument Analysis:	Presumption. A creative analogy was used in the premise, but it would be unfair to presume that this was evidence for the conclusion. Although the above anal-

ogy can be used to help others understand the author's feelings about the Reagan economic policy that cut funding for government programs for the poor, it did not provide evidence that Reagan's policy was not working. Reagan claimed that his policies would benefit all the people. Evidence should have been introduced to show that this was not true. Furthermore, there are at least some obvious differences between the two things being compared that weaken the above analogy. No one was literally murdered (slaughtered) under Reagan's policies. And no evidence was introduced showing that a conscious policy existed of sacrificing the poor for the betterment of the rich and the middle class.

EXAMPLE 5-10B

Conclusion:	X is bad (Or, X is good.)
Premise #1:	X is just like Y.
Premise #2:	(implied) Y is bad (Or, Y is good.)
Label & Description:	Questionable Analogy. A weak analogy is used in the premise (indicate what is being compared to what).
Argument Analysis:	Presumption. Just because a creative analogy is used in the premise we should not presume that evidence is offered for the conclusion. By way of summation or introduction, creative analogies can help us understand an argument, but unless the characteristics of similarity are discussed and justified, bare analogies should not be considered evidence. Critique the specific analogy. Indicate what evidence is left out that would make the analogy stronger. Indicate how the two things being compared are dissimilar.

As noted above, the use of an analogy as a conclusion (rather than a premise) is a type of inductive argument, and not all analogies are inappropriate or weakly supported. A strong case can be made that two things really are similar by demonstrating in detail a representative sample of the characteristics they have in common. But we should not presume that such a demonstration has been made just because a creative comparison is offered.

Jerry Falwell, one-time leader of a group called the Moral Majority, once claimed that giving teenagers free condoms was like giving "cookbooks to people at fat farms." We can imagine what might happen to people who are trying to lose weight being given cookbooks that detail, complete with sumptuous pictures, how to cook gourmet meals. We can imagine that their dieting discipline will suffer. But imagining that their discipline will suffer is not evidence that their discipline will suffer. Furthermore, Falwell has given no evidence to support his comparison that if we give free condoms to teenagers, their discipline will similarly suffer, or that the benefits of giving free condoms to teenagers will not far outweigh the potential problems. He has not cited any actual studies that show giving free condoms increases teenage sexual activity beyond what it would be anyway. His creative analogy is another example of a "write-off" excuse not to test an important claim. Are the two things being compared truly similar? Are all teenagers sexually obsessive to the same extent as problem eaters? Isn't there a major difference between adults with a specific problem and at specific facility (a fat farm) and teenagers in the general population with many different interests and involvements? Perhaps not. Maybe Falwell's basic claim is correct. But do we know? His claim could be tested and the analogy possibly supported as a reliable inductive generalization, but only if we let the world of experience speak. But if we presume his analogy is an acceptable premise offering evidence, we would be cutting ourselves off from exactly the type of reasoning necessary to gain a reliable belief about an important matter.

Besides being another excuse for not testing beliefs, weak analogies are persuasive because many people accept only what they understand, and analogies do help us understand a person's argument. But understanding an argument should be separated from judging that argument. Although we should understand an argument before we judge it, we should be able to understand many arguments and beliefs that we do not accept. Falwell's analogy helps me understand his claim, but I don't have to accept his conclusion just because I understand his argument.

Furthermore, there may be many things that are true that I have a hard time understanding. For instance, as a human being with goals and consequent purpose to my actions to achieve these goals, I may have a hard time understanding the possibility that the universe may be here for no reason, that it may be here for absolutely no purpose, that it and our existence are just accidents. I may have a hard time imagining this, but it could be true. Because we think this way, we are easy prey for someone who offers us the argument: "Believing that God doesn't exist and that the universe is here for no reason is like believing that a building collapsed for absolutely no reason." Understanding this analogy is not proof that there is a God or a purpose to life.

Perhaps anyone who is not a theoretical physicist will have a hard time believing that the universe was created from absolute nothingness. The human mind seems to naturally believe that all events must have a cause. But reality need not conform to the way our minds work. Today, modern physicists have constructed well-supported scientific theories that deny that every event has a cause and propose that the universe could be created from a kind of nothingness. Below, the noted science author Jonathan J. Halliwell attempts to describe this situation in a *Scientific American* article. Note his liberal use of analogies (tunneling, fuzz) to help us understand these difficult concepts.

"Perhaps then, the universe has tunneled from 'nothing.' The evolution described by inflation and the big bang would have subsequently occurred after the tunneling. . . . The picture that emerges is of a universe with nonzero size and finite (rather than infinite) energy density appearing from a quantum fuzz."[1]

---

[1] Jonathan J. Halliwell, *Scientific American*, 265, no. 6, Dec. 1991, 84, 85.

Suppressed Evidence

Our last fallacy is a special case. The fallacies that we have looked at thus far involve persuasive tricks either in the reasoning or the premise. It is also possible to select evidence in a special way such that although the evidence is true and offers a relevant supporting reason for the conclusion, we still should not accept the conclusion because not all the available information associated with accepting the conclusion is given. A *Suppressed Evidence* fallacy is an argument that is valid or inductively reliable and offers a relevant, true premise for its conclusion, but omits an important fact that when known makes a major difference in how we ought to view the conclusion.

To take a simple example, a commercial urges us to buy Crest toothpaste because it has fluoride. It is true that Crest has fluoride, and based on reliable scientific studies that is a good reason to buy it, because fluoride prevents tooth decay. What is missing in this commercial is the simple fact that many other toothpastes also have fluoride in them. So, we have not been offered a reason to buy Crest rather than one of the other toothpastes. Information is omitted that would make a significant difference in how we view the persuasiveness of the argument.

Many advertisements use this technique. They do not make false claims, but instead slant the reasons offered in such a way that the premises are true and the reasoning valid or reliable in isolation (the premises in isolation do give a good reason to buy the product) but omit a key fact that would change our perspective on the force of the premises.[1] A popular approach in early TV advertising was to use what is called the **Brand X** technique. For instance, in attempting to use the rational force of a controlled study, two sets of identical shirts would be washed in separate washing machines. One would have Tide detergent, and the other Brand X. After washing, the Tide shirt would of course appear cleaner, thus giving us a good reason to buy Tide rather than Brand X. But what was Brand X? Was it one of the major competitors of Tide? Laundry soaps are not all the same, and they do get improved. What this commercial did was compare Tide to an older detergent that was no longer on the market. Similarly, an old Shell gasoline commercial urged us to buy Shell gasoline because it contained Platformate, a special ingredient that would give greater gas mileage. This commercial showed two cars being given the same amount of gasoline, one with Platformate and one without, and then being driven down a road at the same speed. It ended in a dramatic fashion with the car without gasoline containing Platformate running out of gas and the Shell car continuing until it broke through a paper barrier with the Shell logo on it. What we were not being told is that the other car also contained Shell gasoline but with the Platformate removed, and that all the gasoline competitors of Shell had similar mileage ingredients added to their gasolines.

Suppressed evidence is also a favorite technique of politicians, because it does not involve lying but does require knowledge (to understand the slanting) that the average person will not have. In the 1988 presidential campaign, then–Vice President

---

[1] Other common examples are: Buy Sanka coffee; it's decaffeinated. Bank at Bank of America, because they offer equity loans.

Bush attacked the economic record of Michael Dukakis, then governor of Massachusetts, in a speech. As part of that speech, Bush charged with dramatic emphasis the following:

EXAMPLE 5-11   Under my opponent, Massachusetts has lost—lost—twenty-six thousand jobs since 1983, more than any other state in the country.

In this speech Bush was attempting to counter the Dukakis strategy of pointing to the economic turnaround of the state of Massachusetts during his tenure as governor. Dukakis claimed that if he could lead Massachusetts through an economic miracle from a backward economic focus with reliance on out-of-date manufacturing jobs to a high-tech, competitive, new industrial structure based on robotics and computer technology, he could do the same for the nation. Bush was claiming that no such Massachusetts economic miracle had occurred and that Dukakis was not fit to be president. So how could Dukakis claim that he had been a good governor and that the economy had improved in his state, if his state had lost twenty-six thousand jobs? Was Bush lying? Bush was not lying, but awareness of an important fact suppressed by Bush makes a very big difference in the persuasive force of his premise. The state had lost approximately twenty-six thousand *old* industrial manufacturing jobs. But according to federal figures, Massachusetts had gained more than 230,000 new industry jobs in the same time period!

Here is a complete analysis of Bush's argument followed by the formal essence of all Suppressed Evidence fallacies.

EXAMPLE 5-11A

Conclusion:	Michael Dukakis has not been a good governor for the state of Massachusetts.
Premise:	Since he has been governor, his state has lost twenty-six thousand jobs.
Label & Description:	Suppressed Evidence. Although the premise was true and would have been a good reason for accepting the conclusion in isolation, an important fact was suppressed.
Argument Analysis:	Presumption. We should not presume that the premise offers a good reason to accept the conclusion because this argument involved suppressed evidence. Although the premise was true and in isolation offered a good reason for accepting the conclusion, the argument omitted the fact that during the time Dukakis was governor his state had gained 230,000 new industry jobs. Because Dukakis was claiming that he had helped turn his state's economy around with policies that encouraged the establishment of such new industries, this fact would now seem to support Dukakis's claim. Bush had unfairly slanted his argument by selecting a fact

that appeared to support his conclusion while omitting a fact that would change our perspective on the persuasiveness of his argument.

EXAMPLE 5-11B

Conclusion:	Accept X.
Premise:	Because Y is true.
Label & Description:	Suppressed Evidence. Although Y is true and would be a good reason for accepting X in isolation, Z is suppressed.
Argument Analysis:	Presumption. Argue that although Y is true and offers a good reason to accept X in isolation, an important fact, Z, has been omitted. Describe Z and explain how it should change our perspective on the persuasiveness of the argument.

Note that in charging suppressed evidence it is not sufficient to simply claim that there is another side to the story or that some fact may be suppressed. Whatever fact is being suppressed must be known and described, and then used as part of a counter argument showing how our perspective of the persuasiveness of the original argument should change. Also, the Suppressed Evidence fallacy is reserved for cases in which a single fact or a small set of facts has been omitted. Every discussion that involves a complicated situation can of course produce different points of view. An argument that supports only one point of view is not necessarily suppressed evidence. A charge of Suppressed Evidence fallacy is reserved for the relatively focused cases where a few facts have been omitted. Because every argument is guilty to some extent of not providing all the possible available evidence—if for no other reason than we never know all the facts in an imperfect world, and it is never the case that all the facts are "in," so to speak, it would not be fair to charge an argument with suppressed evidence simply because there may be something not considered by the argument. Given any situation, there is always more to discuss, and every argument would be a fallacy if we did not acknowledge that this is appropriately so. That we live in an uncertain world and that all arguments are fallible, that we must place our bets based only on reliable evidence, should not be confused with specific cases of identifiable suppressed evidence.

Key Terminology

Fallacies of Questionable Premise
Fallacies of Presumption
Fallacies of Weak Induction
Slippery Slope
Questionable Dilemma
Straw Person
Hasty Conclusion

Questionable Cause
Appeal to Ignorance
Begging the Question
Complex Question
Ambiguity-Equivocation
Questionable Analogy
Suppressed Evidence

Concept Summary

In this chapter we covered fallacies that are valid but have questionable premises in terms of their truth status, fallacies whose premises are relevant and probably true but are weak by inductive standards used to establish reliable beliefs, and fallacies in which the premises may not be false but are presumptive in some way.

The fallacies of questionable premise play upon the natural fear and uncertainty generated by living in a complicated world. *Slippery Slope* offers a chain of maybe-events in one of its premises, but cites no evidence for why we ought to accept such a controversial premise. *Questionable Dilemma* simplifies our choices to a black-and-white situation in its premise, but a little reflection usually reveals that the world and our choices are more complicated than that offered by this premise. *Straw Person* also appeals to our desire to have uncomplicated choices by offering a simplified, but distorted and exaggerated description of an opponent's position.

The fallacies of weak induction take advantage of the fact that many of our beliefs are generalizations from particular experiences, but they also take advantage of the fact that human beings have a tendency to be epistemologically lazy. That is, we want "quick" beliefs and often do not have the patience to submit our generalizations to the inductive rigor of establishing them as reliable beliefs. *Hasty Conclusion* jumps to a generalization from only one or a few positive particular cases of the generalization and does not offer the kind of evidence that would give us a reliable reason to believe the cases are representative. *Questionable Cause* infers a very important and strong conclusion—a causal connection between two events—but offers only a basic minimum of evidence for such a conclusion—a temporal connection between those events. *Appeal to Ignorance* takes advantage of the fact that if we have not shown a belief to be false (or true), it could possibly be true (or false). But instead of inferring from this lack of knowledge to the appropriate conclusion that more investigation is needed, it hastily jumps to the conclusion that we know something to be true, because of our lack of certainty that it is false, or that we know something to be false, because of our lack of certainty that it is true.

Finally, fallacies of presumption have premises that may be true but the premises are presented in a slanted or unfair way. *Begging the Question* surreptitiously prestates the conclusion, so it offers no evidence for the conclusion and only argues in a circle. *Complex Question* offers two or more questions in its premise disguised as one, such that the question cannot be answered with a yes-or-no answer without implying an unfair or unintended conclusion. One version of *Ambiguity-Equivocation* uses a word or phrase in the premise that also occurs in the conclusion, but the meaning of the word or phrase in the premise is not consistent with that of the conclusion. *Questionable Analogy* takes advantage of the fact that analogies are often used as one of the inductive strategies for arriving at reliable beliefs. But there is a difference between an analogy presented as a complete argument and an analogy used as premise only. Analogies as complete arguments present a representative sample of similarities between two things in the premises and then generalize reliably that the two things being compared are similar. Analogies as persuasive fallacies present a quick creative comparison of two things without offering a detailed or representative sample of similarities of the two things being compared. In the latter case, we should not presume that evidence has been offered just because the analogy may help us understand the argument's claim. Evidence still must be presented that the claim is true. Finally, **Suppressed Evidence** shows that we should not presume that a good argument has been presented even if the premises are true and offer a good reason for accepting the conclusion. The premises may be true and in isolation offer a good reason for accepting a conclusion, but our perspective on the persuasiveness of the premises can change if we find out that a specific fact has been omitted.

EXERCISE
I

Indicate if the following are true or false.

1. Logicians have identified twenty informal fallacies, and any bad argument can be identified as one of these fallacies.

2. All fallacies of weak induction have premises that are at least relevant to the conclusion.

*3. Once an argument has been identified as a fallacy, the belief contained in the conclusion should not be accepted by any rational person, because we know that the conclusion is false and that it cannot be strengthened with another argument.

4. The fact that we should avoid hasty conclusions does not mean that we should avoid all creative generalizing and hypothesizing.

5. The Slippery Slope fallacy is similar to the Questionable Cause fallacy. Both make a causal connection claim in the conclusion.

*6. It is not fair to charge someone with committing a Questionable Dilemma fallacy without at least making a tentative case for a third alternative.

7. A Complex Question fallacy is any argument that has a question in its premises, and it is a bad argument because premises should be declarative sentences, not questions.

8. The main problem with the Begging the Question fallacy is that the argument is always circular.

*9. Analogies are helpful in understanding arguments, but we must be careful in how we use them as evidence. Understanding an argument is not evidence that the argument is strong.

10. A Suppressed Evidence fallacy is committed whenever all points of view of an issue are not discussed or when all facts have not been found.

11. The problem with irrefutable beliefs is that no way exists to test them when they clash with each other.

12. The only purpose for criticizing arguments and identifying fallacies is to make sure that we don't believe in false beliefs.

EXERCISE II

Give a complete written analysis similar to that provided in the text
for the following fallacies. Use the same format as that in the text:
Conclusion, Premise(s), Label and Description, and Argument Analysis. Be sure to provide and develop the appropriate focus in the
argument analysis section.

1. Conversation:

    Bob: "If it wasn't for the demonstrations and protests by college
    students and liberals against the Vietnam War during the
    late 1960s and early 1970s, we would have won the war."

    Sam: "How do you know that?"

    Bob: "Before the demonstrations and protests against the U.S.
    military involvement in Vietnam, we were winning the
    war. After they started, we began to lose."

2. Argument by Daniel Brent, U.S. attorney opposing a program
that would give sterile needles to drug addicts as a measure to
control the spread of AIDS:

    "We should not support this program, because if drug addicts
    who have not been mainlining (because of their fear of AIDS)
    get clean needles they will try intravenous use. If they get addicted
    to intravenous use, then when they can't get clean needles, they
    will use dirty ones and put themselves at risk for AIDS. Supporting
    this program will backfire, actually leading to an increase in
    intravenous drug use and the spread of AIDS."

*3. During the early years of the Reagan administration, troops were
sent to Lebanon to help establish peace between the Islamic and
Christian factions fighting for control of Beirut. Reagan vowed
that the United States would not be intimidated by terrorists,
but after 3 truck bombings, one of which killed 250 soldiers, the
United States withdrew its troops. Here was President Reagan's
response to a question regarding why we did not have better
security by the time of the third bombing, the bombing of the
U.S. embassy:

    "Anyone who has ever had their kitchen done over knows that
    it never gets done as soon as you wish it would."

    Assume that Reagan's reference to having a kitchen remodeled
    was a reference to security plans that did not get into place as
    fast as the Reagan administration had hoped. Assume that Reagan
    was offering a reason for the conclusion that poor security that
    led to the success of the truck bombings was understandable
    and defensible.

4. Advertisement (1991) for Lean Cuisine frozen dinners:

Of all the things we make, we make sense! There are some things we skimp on: Calories. Fat. Sodium. With less than 300 calories, controlled fat, and always less than 1 gram of sodium per entree, we make good sense taste great.

Note: The salt content in food is normally referred to in milligrams. One gram is equal to one thousand milligrams. In response to criticism from the Federal Trade Commission, Lean Cuisine dinners now contain no more than six hundred milligrams of sodium. Shortly after this advertisement appeared in the early 1990s, the Food and Drug Administration issued rules on package labeling in compliance with the National Labeling Education Act. For salt content, "low salt" was to mean "less than 140 milligrams of sodium."

5. Argument by Archie Bunker on the sex drive of retarded people:

Archie: "Retarded people have an abnormal sex drive."

Mike: "That's a myth!"

Archie: "Oh yeah, then why are you on automatic all the time?"

Although Archie's final argument is a hasty conclusion—that because one person he knows (Mike) is (allegedly) mentally retarded and has a abnormal sex drive ("on automatic all the time"), therefore all retarded people are this way—Archie seems to be using another fallacy against Mike: That he is mentally retarded because he has an abnormal sex drive and he has an abnormal sex drive because he is mentally retarded. Analyze the latter argument.

6. Letter to the editor defending Lt. Colonel Oliver North during his trial for breaking government laws by diverting money from arms sales to Iran to the Nicaraguan Contras:

It seems to me we must either support the causes of dedicated men like Lt. Col. Oliver North, who are champions of the principles of freedom and equality, or allow our country to be taken over by the wimp liberals, homosexuals, and Gorbachev lovers. I am confident that Americans know the evil consequences of the latter choice. So, let's support North and the other defendants in the Iran–Contra trials.

7. During the 1988 presidential campaign, then–Vice President Bush attempted to portray his opponent Michael Dukakis as weak on military defense issues, as one who would dismantle the U.S. military at a crucial time in our struggle to end the cold war. According to Bush, one of the reasons that Dukakis should not

be supported was because he has "opposed every new weapons system since the slingshot."

Note: As a Democrat, it is true that Dukakis favored cutting military spending in favor of better domestic programs and reducing the trillion-dollar deficit produced by the Reagan administration. Dukakis was also against the Star Wars nuclear shield to be placed in outer space. But he supported the Stealth bomber, the D5 sea-launched ballistic missile and Trident II submarine, the M1 tank, and the F15 and F16 jet fighters.

*8. In September of 1991, a consumer group charged that the Food and Drug Administration made a mistake in approving the drug Prozac, widely prescribed for treatment of depression. The group claimed that it was a dangerous drug that sparked suicidal impulses and should be banned. Suppose as evidence of this claim the group cited the case of a woman who had terrifying nightmares, headaches, forgetfulness, and anxiety three days after she began taking the drug. That during that period she tried to kill herself with a gun, but her husband stopped her before she could pull the trigger. She was on the drug for three weeks, and all suicidal impulses disappeared when she stopped taking the drug. Suppose the consumer group was concluding from cases like these that all people risk suicidal impulses if they take this drug.

Note: By this time, the drug Prozac had been prescribed for more than 3 million people in the United States, and 5 million people worldwide. Also, 15 percent of untreated depressed patients commit suicide, and depression is a major cause of suicide.

9. Science cannot prove categorically that the theory of evolution is true. Therefore, any scientist who accepts it is doing so on faith alone.

10. As a prelude to the 1992 presidential election, the Democrats charged that George Bush was insensitive to domestic issues. As evidence of this they cited Bush's threatened veto of legislation to extend jobless benefits for as many as twenty more weeks to unemployed Americans. The Democrats claimed this bill was badly needed because every month three hundred thousand unemployed Americans were seeing their jobless benefits expire due to a long recession.

According to Bush, he vowed to veto the bill because its $6.1 billion cost was too high and the recession was almost over. At a Republican fundraiser, Bush said that the Democrats' bill was "garbage," and that "We want to help people, but also we want to see that what we do is fiscally sound."

House Democratic leader Richard Gephardt of Missouri responded, "This is the president's domestic agenda: mingling with the millionaires, raising campaign funds, calling the recession no big deal, and referring to unemployment benefits as garbage. Unemployment compensation isn't garbage, Mr. President, and neither are the people who need it."

Analyze Gephardt's argument.

11. Reverend Jerry Falwell, founder of Moral Majority on abortion: "Calling the abortion issue a question of freedom of choice is ridiculous. 'Freedom to kill' is more appropriate. It's like a bank robber saying, 'I have freedom to break safes, and I want my freedom.'"

12. Analyze Archie's second equivocation fallacy ("being honest") Chapter 2, p. 55.

13. In Richard Wright's novel, *Native Son*, a black man has been arrested and is on trial for the murder of a daughter of a wealthy Chicago family. The evidence against him is purely circumstantial: As the family's chauffeur he was allegedly the last one to see the girl alive when they dropped off her boyfriend at his apartment. The girl's burnt bones were found several days later in a basement furnace, and the black man's other job was to keep the furnace stoked with fuel to heat the family's mansion. Here is a key exchange between the prosecutor and the boyfriend at the trial:

    Prosecutor: "What time did they pick you up?

    Boyfriend: "About 7:30."

    Prosecutor: "The negro chauffeur was driving?"

    Boyfriend: "Yes."

    Prosecutor: "And what time did you get dropped off by the drunken negro chauffeur?"

    Boyfriend: "About 11:30."

    Prosecutor: "So you left her alone with the drunken negro chauffeur about 11:30?"

    Boyfriend: "Yes, but . . ."

    Prosecutor: "Thank you, that will be all."

    Hint: How many questions is the prosecutor asking in his third question?

14. Argument in the 1970s in favor of legalizing marijuana: "No one has been able to show that marijuana is harmful, therefore it must be safer to use than alcohol."

15. Letter to the editor arguing in favor of legalized abortion:

    Abortion should be legal, just like any other medical procedure. As a human being I have a basic right to control what takes place in my own body. If you had cancer or a parasite growing in your body, it would be absurd to think that the government could prevent you from having it removed.

16. Give a complete analysis of the Bob Hope and Anita Bryant Slippery Slope fallacy examples (**5-1C** & **5-1D**) on page 166.

17. Analyze the Shell Platformate commercial discussed on page 195.

18. Letter to the editor on Christianity and the gay rights movement:

A few weeks ago, the gay rights movement held its annual parade in Washington, D.C. I was appalled to see the shameless promotion of sexual immorality on the streets of our nation's capital. Through this event and also through the public blessing of the gay and lesbian community by President Clinton, enough evidence has been shown to prove that our nation has forgotten God.

*19. From a 1993 Democratic fundraising letter, asking for money to help pass President Bill Clinton's economic proposals:

Ever since Bill Clinton first proposed his New Direction economic stimulus and deficit reduction package in his State of the Union address, the lights have been burning late in the plush offices along "Gucci Gulch" as the "Me First" crowd—the special interest establishment and its lawyers and lobbyists—plots and schemes. We need your help. This is an epic struggle between the special interests and the public interest, between the old Republican road and Bill Clinton's New Direction.

Some background: According to Republican opponents of Clinton's economic proposals to reduce the federal deficit and stimulate the economy, they involved mostly raising taxes and spending large sums of money on inefficient bureaucratic programs. They claimed that Clinton's proposals were essentially the same tax-and-spend tactics of former failed Democratic programs. The Republicans claimed to be in favor of cutting federal spending as the best way of stimulating the economy and decreasing the deficit. Supporters of Clinton's plan claimed that the increased taxation would fall mostly on the rich.

20. The following is part of a campaign pitch in favor of the 1992 reelection of then-President Bush:

Of all of the president's accomplishments, there should be little doubt in anyone's mind that President Bush's war on drugs has worked, that his policies of rigorous interdiction and police law enforcement efforts have caused a dramatic decrease in the use of drugs in our society. Before Bush was president, the use of marijuana and cocaine were skyrocketing. Now, since he has been president, the use of both drugs has dropped dramatically.

21. Argument against President Clinton's 1993 proposal to lift the ban on homosexuals in the military:

President Clinton should not lift the ban on homosexuals in the military. If he does, this could severely weaken the recently refurbished armed forces and irreparably harm public confidence in the new president. Furthermore, if homosexuals are officially welcomed into the military, it won't stop there. This will produce the argument that if homosexuals can fight and die for their country, the state should then sanction same-sex marriages. Then soon pedophiles (those who sanction adult–child relations) will be knocking on our cultural door asking for legitimacy.

22. Argument in favor of biological evolution:

Imagine that through some magic, all we could see of a real tree were its individual leaves distributed in space. Would we suppose that somehow those leaves had just sprung into existence where they were? Surely not! We would suppose that they were supported by an unseen trunk, branches, and stems, dividing and subdividing, and that the leaves hung at the end of the finest, final stems. And this is what evolution is, just like a tree. The leaves are the species of plants and animals we observe today. The trunk, branches, and stems are the extinct species of the past. It would be just as stupid to not believe in evolution as it would to believe tree leaves could miraculously come into existence without a trunk and branches. So, believe it or not, evolution is true.

23. Argument in favor of the notion that paying federal taxes should be voluntary:

In a democracy we are supposed to think for ourselves. Well then, I should be able to think for myself on whether or not I pay my taxes this year.

24. People who believe in God are ignorant. You can always tell who these people are: They actually believe there is a heaven and an afterlife!

25. Argument in favor of the absolute freedom of the press:

In a democracy a free press is in the public interest. Therefore, tabloid newspapers, such as the *National Inquirer* should not be sued for distorting or exaggerating the intimate details of famous personalities, because this is what the public is interested in.

EXERCISE III

Read the letters to the editor in your daily newspaper, look critically at advertisements in magazines and on television, until you find at least five different examples of some of the fallacies covered in Chapters 4 and 5. Supply a complete analysis in terms of Conclusion, Premise(s), Label and Description, and Argument Analysis.

EXERCISE
IV

Make a case for any fallacies you think I have committed in this book! Note that the next chapter is considered very controversial by people who do not endorse what I have called the world view of modern science. In making your case, supply a complete analysis in terms of Conclusion, Premise(s), Label and Description, and Argument Analysis. (Feel free to discuss them with me via one of the e-mail addresses listed in the Introduction.)

ANSWERS TO STARRED EXERCISES:

I.   True and False.

3. False. To identify an argument as a fallacy is only to identify it as a weak argument, not to offer proof that the conclusion is false. Furthermore, one of the purposes of identifying fallacies is to understand what kind of evidence is needed to make arguments stronger and conclusions better supported.

6. True. Questionable Dilemma is a valid argument, so the focus of discussion should be on the either/or premise. This argument claims there are only two realistic alternatives. To show that the conclusion is not supported, you must be able to show that the two alternatives are not the only alternatives by describing a tentative case for a third alternative.

9. True. There is a difference between analogies as complete and reliable arguments, where the premises offer a comparative representative sample of two things being claimed as similar in the conclusion, and simple creative comparisons offered in the premises with no detailed justification that the two things compared really are similar.

II.   Fallacy Analyses.

3. **Conclusion:** Poor security that led to the truck bombings in Beirut is understandable and defensible.

Premise: Constructing security barriers against terrorism is just like having a kitchen remodeled when it doesn't get done as soon as you wish it would.

Label & Description: Questionable Analogy. A weak analogy is used in the premise. Constructing security barriers against terrorism is compared to remodeling kitchens.

Argument Analysis: Presumption. Although Reagan's analogy may serve as an introduction to a defense of why we could not protect U.S. personnel better, we should not presume that he has given any evidence for his conclusion yet. U.S. personnel were put into a very dangerous situation, and Reagan gave us no reason to believe that there was a systematic security plan. Furthermore, the analogy is a poor one, because as president of the United States, surely he has more power over the implementation of security measures than the average person who is trying to work with slow carpenters to remodel a kitchen.

8. If we focus on the claim that Prozac caused the woman's problems, a case could be made for questionable cause. The fact that her symptoms increased after taking the drug could be a coincidence. Presumably she was taking the drug in the first place for treatment of depression and it is possible that the underlying source of this depression caused her symptoms, the increase in her problems being a natural progression of the underlying problem.

However, because the conclusion of the consumer group was that such cases prove that all people risk suicidal impulses if they take this drug, a case can be made for Hasty Conclusion as follows.

Conclusion: In general, Prozac is a dangerous drug that produces suicidal impulses.

Premise: Some people who have taken this drug have attempted to commit suicide.

Label & Description: Hasty Conclusion. There is a considerable generalization in the conclusion, given the small amount of inductive evidence in the premise.

Argument Analysis: Reasoning. Although the case cited in the premise is relevant to the conclusion, it is insufficient to support the generalization in the conclusion. The case cited does not appear a typical reaction to the drug. Because millions of people take this drug, most with apparent success and no adverse symptoms, more evidence should be given that the alleged suicidal reaction is common. For a genuine test of this claim, a controlled study should be cited—two groups of people suffering from depression, one group given the drug Prozac. Generally, for the FDA to allow for the prescription use of any drug, such controlled studies are mandatory. That a few people have had suicidal tendencies after taking the drug does not prove that the drug caused this reaction. The people were in depression, so their suicidal tendencies may have surfaced anyway. Like most drugs, perhaps Prozac did not work for them. Furthermore, even if it is true that some people do have adverse reactions to the drug, which is most important for the charge of hasty conclusion, this would not prove that all or most people will.

19. Although there are some Ad Hominem attacks in this letter, I would argue that the following Questionable Dilemma interpretation is the strongest;

Conclusion:	We should support Bill Clinton's economic proposals.
Premise:	Either we support Clinton's New Direction economic proposals and the public interest or we support special interests, the "Me First" crowd along "Gucci Gulch," and the old Republican road.
Premise:	Supporting the special interests is bad for the country.
Label & Description:	Questionable Dilemma. There is a questionable dilemma in one of the premises. This premise restricts our choices to only two possibly extreme choices.
Argument Analysis:	The major premise is questionable. The first premise contains a dilemma that reduces the number of possible alternatives to two. Because there are possible positions other than these two extremes, the premise is weak. Although the Clinton supporters could be right that these are our only choices, at least a third possible choice is not being discussed. Being against adding more government programs does not make one automatically a fan of special interests and the rich. One could argue for reducing the

deficit and stimulating the economy by getting very serious about government waste. Was the only way to stimulate the economy a program of government projects? Were all the projects good projects, or were some merely pork barrel? Is spending money on government projects the most efficient way of stimulating the economy?

Chapter 6

LOGIC AND HOPE

Man is a rational animal.

—Aristotle

Logic does not exclude madness.

—Erich Fromm

Insanity in individuals is something rare—but in groups, parties, nations, and epochs, it is the rule.

—Friedrich Nietzsche

Somehow I am not distressed that the human order must veil all our interactions with the universe, for the veil is translucent, however strong its texture.

—Stephen Jay Gould

Comfort women, rape camps, the new racism, ethnic cleansing, Sudan, Somalia, Rwanda (see Figure 6-1)—is this the beginning of the twenty-first century? An essential tenet in Aristotle's philosophy (384–322 B.C.) is that the human species is a rational animal. For Aristotle, implied in this statement is the claim that because we have the power of reflective thought, we can learn about the world and make progress, that we can not only improve our physical lot technologically but make moral progress as well. But does history support this claim?

In this short transitional chapter between informal and formal logic, we will discuss two questions. (1) Why be logical in an illogical world? (2) Is our reasoning ability a tool for making a better world, or is it actually responsible for violence, destruction, and exploitation—a tool in the service of our wretched nature? Obviously, I believe there are very good, positive, and hopeful answers to these questions, or I would not be writing this book. However, because these questions are of such great consequence, it is important that we be as honest as possible and confront directly the most negative possibilities.

FIGURE 6-1

In the 1990s, the cold war gave way to dozens of ethnic, religious, and sectional conflicts. Millions of people were tortured, starved, and slaughtered. Many were innocent children. This photo of a vulture standing near a starving Sudanese child won Kevin Carter the 1994 Pulitzer Prize for photography. A few months after receiving this award, Carter, who was only thirty-three years old, committed suicide. The war in Sudan was between an Arab Muslim–dominated government in the north, and black Christian insurgents from the south. © Kevin Carter/Sygma.

The first question poses an immediate, personal pragmatic question. Why should you be logical if success in modern society involves the expert use of rhetorical, psychological, and inferential tricks? Why should you attempt to persuade others with logic when alternative means often work better?

The second question raises a deep philosophical issue concerning the nature of human existence. From an evolutionary standpoint, the human species seems to be a "messy" result of many past successful survival strategies. Territoriality, tribalism, xenophobia, the herd instinct, aggression, wishful thinking, deception and self-deception, rationalization—all behaviors based on self-preservation and personal well-being—can be seen as natural behaviors that once and probably still have survival value given certain situations. If eons ago an evolutionary ancestor of yours and its family were attacked, and no effort of cooperative defense and territorial bonding was made, you would probably not be reading these words now. Today, however, with modern technological advances, these natural behaviors have many threatening, inappropriate expressions in terms of the overall well-being of our species. We have the weaponry to instantly annihilate the groups we fear. We have the multimedia technology to make our acts of deception so persuasive that the distinction between reality and fantasy becomes virtually nonexistent for many people. And, with enormous computer databases available to the rich and powerful, and their lawyers, just about anything can be known about you as a demographic bit of information in seconds—known in preparation for manipulation and control.

The essential answer to the second question that we seek is whether our reasoning ability, a relatively new evolutionary characteristic, can be used to control, help control, or at least not encourage the inappropriate and destructive expression of past survival strategies, or whether our reasoning ability is but a subservient tool to make the inappropriate expression of these strategies even more destructive and tragic.

Concerning the first question, many students often become quite cynical after covering the material in the first five chapters of this book. They tell me that they now see fallacies "everywhere," how resistent to change and the critical testing of beliefs most people are, how most people use the fallacies described in this book as excuses not to think, and how many people in power and in advertising take advantage of this situation by using the same psychological and rhetorical tricks to control and manipulate the masses. One young man once told me that after taking my class he got into an argument with his father over the wisdom of a presidential policy. After a few minutes discussion his father responded with, "Well, the president must be right. He is the president. Why else do you think he is president?"[1] When the young man accused his father of offering a very weak justification of his position, of offering an excuse not to think critically about the decision the president had made, his father responded with, "Kid, you wouldn't know your ass from a hole in the ground!" The young man was also grounded and was not allowed to use the family car for more than a month.[2] After this experience, my student was very disheartened. How can a democracy work, he thought, if his own father thinks like this. He had just experienced the joy of critical thinking and was now learning, he thought, that it was not valued. Other students have told me that attempting to be logical gets them absolutely nowhere with their bosses, and even worse, that relationships with their spouses or significant others become dangerously imperiled if they try to be too logical. For many students who would consider themselves lucky to have a family car and a father who would ground them, life just seems to be one of basic survival in a very complex mix of subcultures, where manipulation

---

[1] Depending on the interpretation, there are several possible fallacies implied. Begging the question is one possibility: The president must be right. To be president you must be smart, and you can only be smart if you are right about the issues. Another possibility is Appeal to Authority: The president is right, because he is the president.
[2] This stance can also be interpreted as a fallacy, one that we did not cover, called Appeal to Force, summarized as follows: "I'm right, because I have the power to make it right." For a relativist, there is nothing illogical about this mode of persuasion. Given the antirelativist position adopted in this book, however, this argument would be a version of irrelevant reason. Having the power to force an action is logically irrelevant to that action being right.

is the rule interspersed occasionally with some shallow short-term contracts, "I'll help you with that, if you help me with this."

So, why be logical? If we can't beat them, shouldn't we join them? Recall Protagoras's formula for a successful life (Chapter 2). The game of life is not one of logic and truth, he said, but rather one of artistic persuasion. Not one of reasoning to or describing the truth, but rather one of creating reality, of molding a perspective and persuading people to live in that perspective. For Protagoras, as a relativist, everything is a matter of interpretation and *any* interpretation can be defensible if you are creative enough. Beliefs are not true or false, and behaviors are not good or bad. Like a good lawyer it is simply a matter of making the best "case" for one belief or another, one action or another. For Protagoras, the game of belief is all a matter of presentation, perspective, and rationalization. What matters most is SUCCESS, success in getting people to change their minds to your way of thinking. So, perhaps what you should be learning is how to use fallacies rather than avoid them!

My answer to this challenge is, of course, to first reject relativism and then remind you of the major themes of this book. Logic can be seen as a defensive tool that allows us to defend ourselves against the onslaught of persuasive appeals that bombard us daily, and it also forces us—as a society, as a culture, as a species—to test our beliefs to see if they are reliable.

Concerning relativism, I have thought about this question most of my adult life, and I would not be telling you the truth if I told you that relativism was not a challenging philosophical position that must be taken seriously. However, you must ask yourself the same question I ask myself after examining the reasoning trail of relativism as fairly as I can: Is there not an objective world of some sort that intrudes upon our most cherished beliefs? Even after acknowledging that every human belief is a result of some perspective or other, even after acknowledging that it is impossible for us to get completely outside our minds and cultures and examine reality firsthand without the filters of our biases, is it not really true that some beliefs are better than others? That some perspectives are better than others? The conclusion I come to is that some beliefs just do not work, and some beliefs are better than others, that some beliefs are more successful in long-term reliability, even though certainty for any belief is impossible, and all are fallible, tentative, and potentially revisable.

When considering the conceptual and empirical problems of some beliefs, the evidence is overwhelming that these beliefs will not be reliable and are not likely to help us be successful in the future, and that dogmatic adherence to them in the face of overwhelming evidence is like beating your head against a wall. Sometimes, perhaps much more often than we desire, reality "kicks back"; it responds and overwhelms our most cherished beliefs, often in a very forceful way, and resists what we want to be true. To paraphrase Stephen Jay Gould's opening quote above, the veil of our biases and perspectives are translucent, however strong their textures.

In March 1993, a religious cult in Waco, Texas, ambushed and killed several federal marshals when the marshals attempted to confiscate an arsenal of illegal arms (automatic assault weapons, grenades, and homemade bombs) being held by this group of radical millenialists. Calling themselves the Branch Davidians, this sect believed that the end of the world would take place at the turn of the century, and that before the return of God's rule there would be an apocalyptic war between

Christians and the secular army and police, representatives of evil secular political forces. They believed that they were the shock troops of the Apocalypse, that their use of force was biblically determined, and that the raid by federal officials was part of biblical prophesy.

For months after the ambush, police and marshals surrounded the heavily fortified ranch of this group of more than 130 men, women, and children, and a tense standoff ensued with their leader, David Koresh, a self-appointed prophet who at various times claimed to be Jesus Christ. At one point, Koresh announced that he would come out peacefully if he were allowed to do a radio broadcast explaining the mission of his group. Koresh was allowed the broadcast, but at the appointed time for the surrender he announced that he had changed his mind because God told him to wait for further instructions.

As the story unfolded, it became clear that Koresh really believed that he was Christ and that he was in direct contact with God. As a millenarian, he was able to convince his followers that because the world would end soon, whatever they had been taught about society's standards of sinfulness or immoralality was no longer so, because he was operating with God's direct authority. Koresh admitted that he routinely had sex with many of his female followers, including married women and even some mother-daughter pairs. According to some of Koresh's former followers, the women were honored to produce a race of warriors who would rule in the new kingdom to come.

Obviously, Koresh was able to **persuade** many people to accept his vision of himself and the future, motivating these people to cooperate in an elaborate venture. In addition to amassing a small fortune in arms, the group had purchased a seventy-seven-acre farm and reconstructed a former farmhouse into a labyrinth of rooms, secret tunnels, and observation towers. But is persuasion equal to truth? Was Koresh's problem only that he failed to persuade enough people? Would the world have come to an end in the year 2000 if enough people believed this to be true? Or, was Koresh's vision of the future not reliable, just wrong? Conceptually, was this version of Christianity logically consistent with a religion that believes in a loving and forgiving God, urges us to "turn the other cheek," and generally asks us to follow the nonviolent life-style of its founder, Jesus Christ? Empirically, how likely was it in 1993 that the world would come to an end by the year 2000?

It seems to me that unfortunately for the followers of Koresh, reality does intrude upon our most deeply held beliefs. Although with a little Protagorean creativity and rationalization the tragic results could be made consistent with Koresh's belief system (more than seventy men, women, and children died in a fire set by Koresh's followers when federal officials unleashed tear gas to get them out), at some point the message from reality becomes relatively clear. Like the tobacco industry's position on smoking and lung cancer, at some point a reasonable person knows that it is not likely that this belief system is going to be a reliable guide to the future.

The British philosopher Bertrand Russell (1872–1970) once remarked that if everyone in the world were required to take at least one logic course, our world would be a more pleasant place in which to live. From the point of view of this book, Russell's comment means that much of the world's suffering is caused by the power that selfish, self-serving, manipulative leaders are able to amass by exploiting the insecurity, confusion, and suffering experienced by many people. The remedy, simply put, is to be aware of the manipulative tricks discussed in this book, giving each individual an arsenal of defensive, critical tools, and thus undermining the potential power base of these leaders.

In other words, this book does not claim that by being more analytical and critical one will automatically be more ethical. In fact, many evil leaders are very logical in planning their manipulative adventures. As Eric Fromm noted, "Logic does not exclude madness." As we have seen, the direction of a logical trail must be guided by the values one starts with in the premises. This does not mean that values cannot be objectively argued about, but it does mean that no automatic or necessary relationship exists between being logical and being ethical. But a large part of the problem is the acceptance of unreliable beliefs and the power that unethical leaders have. Logical analysis and critical thinking are effective tools for testing beliefs and diffusing this power. In short, my answer to the first question is that we should be logical, because if more people were aware of the defensive and epistemological stances taught in this book, the manipulative ploys of the powerful would not work any longer.

However, another question arises that leads to the second issue mentioned above. It would seem that the smartest thing to do would be to learn these defensive stances so that you could avoid having them used against you, but also so that you could use them to successfully persuade people to accept your conclusions! Shouldn't you watch out for the tricks used against you, but use them whenever you can against others? The point here is that the above interpretation of Russell's claim does not explain why we should be ethical. Or put another way, whether or not human beings are capable of being ethical, whether being logical helps in any way or is simply a neutral tool subservient to our baser nature.

Intelligent people can be very violent. In fact, a very persuasive case can be made for the claim that the greatest acts of violence have been made by people who have had the very best educations. This is the point of the Nietzsche quote at the beginning of this chapter. Think of the informational, mathematical, and logical capabilities needed to create "smart" bombs, nuclear bombs, and most recently, biological weapons. Consider a few examples:

Toward the end of World War II, British and United States military strategists carefully planned the firebombing of the German cities of Hamburg and Dresden. Special bombs were dropped in strategic locations in both cities so that a fire storm was created (the air caught fire). The German people were used to air raids, and because they were all in air raid shelters underneath the city when the bombs were dropped, many were roasted in an enormous oven, approximately seventy thousand men, women, and children in Hamburg and eighty thousand in Dresden. Rescuers found grotesque dissolved corpses. The bodies were not charred as if they had been on fire; the flesh of most of the bodies of these people had simply melted like wax around their skeletons. Almost every one of the stripped skulls showed gaping, contorted jaws, no doubt a reflection of the trapped agony in which they died. Regardless of the possible justification of the Allied goal of the ultimate destruction of Hitler's Germany, the fact remains that the fire storm was not an accident; *it was intelligently planned*.

Forty million people were killed in World War II. By the end of the 1980s, our intelligence and knowledge in military matters had "progressed" to the point that the combined firepower alone of the former Soviet Union and the United States equaled six thousand World War IIs. We are speaking, of course, of nuclear weapons. A common measure of destructive firepower during this time was the megaton, a million tons of TNT. The total firepower of World War II was three megatons. By the end of the 1980s, the combined arsenals of the two superpowers was more than eighteen thousand megatons. A single United States Trident submarine carried twenty-four megatons, enough to "roast" every major city in the Soviet Union. A single nuclear bomb falling on a city would subject millions of people to deadly radiation, blast waves, and incendiary levels of heat, such that they would be irradiated, crushed, or burned to death. But both sides targeted major cities and military installations with at least ten nuclear bombs.

By the 1990s, both superpowers began to get smart. The economic exhaustion experienced by both countries of attempting to fund this gruesome arsenal of guaranteed extinction, and the dissolution of the Soviet Union, prompted serious efforts to cut total megatonnage to six thousand, or two thousand World War IIs. Although this was euphorically promoted as the end of the cold war and the opening up of a safer and saner new world order that would replace military competition with peaceful economic competition, cynics claimed that a significant factor was simply a change in military strategy from virtually useless nuclear weapons to cheaper and potentially more manageable biological weapons. Making use of our knowledge in biological matters, it is possible that viral or bacterial agents could be manufactured that would kill people more selectively. For instance, by taking advantage of the fact that certain races are more susceptible to certain diseases, it is possible to design and deliver a fatal disease to, say, only Jews or Arabs.[1]

Whether the cynics were right in this matter or not, the euphoria of the new world order was short lived. Soon the front pages of our newspapers were covered with a plethora of ethnic wars. In the former country of Yugoslavia, Orthodox Christian Serbs, Catholics, and Muslim Croates were attempting to "ethnically cleanse" each other from the face of the Earth. They put each other in concentration camps, gruesomely tortured and then killed captured soldiers, systematically raped each other's women, shelled innocent children trapped in cities, and attempted to starve each other to death. In the former Soviet Union, Georgian Christians and Abkhazian Muslims began a bloody battle over prime tourist real estate, Armenians and Azerbaijans fought over Nagorno-Karabakh, and in Tajikistan a bloody battle between north and south Tajiks for power and economic resources soon involved Uzbekistanians, Turkmenistanians, and Afghanistanians as religion and ethnicity were used to rationalize the hatred and slaughter. Elsewhere in the world, to give but a small sample of ethnic warfare and other group against group violence, Turks and Kurds fought each other, and Iraqis tried to terminate Kurds and Shiites, Jews and Arabs continued their relentless cruelty to each other, Irish Catholics fought

---

[1] See David Suzuki and Peter Knudtson, *Genethics: the Clash between the New Genetics and Human Values* (Cambridge, Mass.: Harvard University Press, 1989). Perhaps, regrettably, we will soon hear our military commanders speak of "collateral biological damage."

Irish Protestants, Indonesians attempted to cleanse their country of Timorese, Hindus and Muslims fought each other throughout much of India, Hutus killed at least two hundred thousand Tutsis in Rwanda, and in the inner cities of the United States large ethnic gangs with firepower to equal some third world countries emerged to kill each other. By 1993, CNN reported that $600 billion in weapons were sold worldwide in one year, there were 200 million hand guns in the United States, and ninety thousand gang members in Los Angeles alone.

The one thing that stands out about all these tragic rivalries is how well human beings cooperate in groups to fight other members of their species. It is time to remind you of a theory first discussed in passing in a note in Chapter 1, p. 7). According to Harvard naturalist Richard Alexander, we can only begin to understand the moral and immoral behavior of human beings by honestly reflecting on the historical evolutionary conditions of our species' social behavior. For Alexander, evolution is true, and ignoring it is like trying to ignore the law of gravity. We can build planes to defy gravity, but not without knowledge of gravity. We can also adjust to evolution, but we must understand it to do so, just as we must understand gravity and other laws of nature to design our planes. So, what should we understand?

Other animals also sometimes kill members of their own species, but only human beings do so at such an unprecedented level. The cause is obvious—our intelligence! Other animal groups spend much of their time competing against the hostile forces of nature, but because of the special trait of intelligence—our self-consciousness, our ability to reflect, plan, and reason—for the most part many human beings no longer have to worry as much about the hostile forces of nature. So, we have replaced competition with nature with an "ominous group-against-group within species competition."[1] And, according to Alexander,

> . . . in no other species do social groups have as **their main jeopardy** other social groups of the same species—therefore, the unending selective race toward greater social complexity, intelligence, and cleverness in dealing with one another.[2]

In other words, our intelligence not only eliminates much of our competition with nature, fuels complex forms of deception and exploitation as we compete with each other for resources (more and more lawyers, greater and greater expertise in media persuasion), but allows us to create massive forms of destruction with which to deliver our brutality to each other. According to Jonathan Schell, in a powerful book written in the early 1980s on nuclear war and human extinction,

> . . . the fundamental origin of the peril of human extinction by nuclear arms lies not in any particular social or political circumstance of our time but in the attainment by mankind as a whole, after millennia of scientific progress, of a certain level of knowledge of the physical universe.[3]

And Alexander again,

---

[1] Richard D. Alexander, *The Biology of Moral Systems* (New York: Aldine De Gruyter, 1987), page 228.
[2] Alexander, page 80. Emphasis added.
[3] Jonathan Schell, *The Fate of the Earth* (New York: Knopf, 1982), page 48.

Humans alone have been equipped by their evolutionary history with traits and tendencies that, *as a consequence of their normal functioning*, can bring about human extinction.[1]

From a broad evolutionary perspective, we see that ancestor species to modern Homo sapiens were greatly "out-gunned" by competing creatures in terms of superior physical characteristics conducive to immediate survival. The human species is an evolutionary irony. Our initial physical inferiority was an important accidental condition that led to our present success on this planet. Many creatures have always had better eyesight and hearing, some can directly experience electromagnetic fields or experience chemicals at long range, and some even have sonar. Many creatures can run faster and are physically much more powerful, with razor sharp claws and teeth. Our ancestors survived because they cooperated against these outside threats and because they had the ability to symbolize, to abstract, to reason, plan, and reflect. The vast majority of creatures on Earth have what can be called a *direct survival interface* with their environment. They do not need to think. The *indirect survival interface* of reflective symbolic processing is not needed. Initially, and up to the present, the human way has served us well. We dominate the Earth. With our intelligence and science, we can duplicate with machinery the direct sensing abilities of other creatures. Now, however, the very characteristic responsible for our initial survival threatens our extinction as a species.[2]

This is not an unusual situation in the evolution of species. Often a characteristic that initially plays a positive role in a species's survival, with time and environmental change will play a very detrimental role, often being the cause of extinction. Consider the Irish deer. The male of this species gradually evolved larger and larger antlers. By the time of its extinction, some antlers extended horizontally to twelve feet and had huge palmlike spiked lobes at each end. As is often the case in nature, this characteristic served as a sexual selection sign for the female of the species. The male with the largest antlers had the most mates. But as this trend continued, the large antlers became dysfunctional, and coupled with other environment changes, the species became extinct. The modern peacock is another example of this process. If not protected in zoos by humans, the beautiful male of this creature would easily follow in the footsteps of the Irish deer. Is human intelligence just another version of large antlers and beautiful feathers? Initially a characteristic of great value, is it now a characteristic that will guarantee our extinction?

Or is our problem fundamentally different? Could the very characteristic that has contributed so much to our uncertain current state be the source of a solution for us as well? Could the late Isaac Asimov have been right when he said, "The dangers that face the world can, every one of them, be traced back to science. . . . (but) The salvations that may save the world will, every one of them, be traced back to science"? Or, according to Alexander, "This means that the problem before us is an absolutely stupendous one: we have created it with our intellects and now

---

[1] Alexander, page 232. Author's original emphasis.
[2] For further elaboration of this theme, see Ronald C. Pine, *Science and the Human Prospect* (Belmont, CA: Wadsworth, 1989), particularly Chapters 9 and 10.

cannot relax until we have used those same evolved intellects to resolve it."[1] Can we use our intelligence to not only be aware of the destructive tendencies of our intelligence, but to make our intelligence work for us rather than against us?

In 1932, Albert Einstein asked this same question. No doubt envisioning the dangerous combination of a rise in ethnic nationalism and resentment in Germany and Italy and the destructive power of science and technology, he wrote a letter to Sigmund Freud, asking, "Is it possible to so guide the psychological development of man that it becomes resistant to the psychoses of hatred and destruction?" Freud's response was not encouraging. "It seems," he wrote on the systematic killing of members of our own species, "to be a natural occurrence, biologically well founded and . . . scarcely avoidable."

Such destructive tendencies may be biologically well founded, but I think Freud was wrong about successful avoidance. Success is possible, and at a very fundamental level that connects the concepts we have studied in this book, it will involve two levels of understanding: (1) A humble understanding of the power of our intellect, an understanding of the constructive and uplifting, but fallible nature of our scientific, deductive-inductive reasoning abilities; (2) A thorough understanding of the wonderful product of centuries of application of this reasoning ability, the world view of modern science. Sections of this book thus far have been concerned mostly with the first level. If you understood the message of Chapters 1–3, you are aware that logic does not automatically produce truth and certainty. At their best, the tools of deductive and inductive reasoning produce tentative, relatively reliable beliefs that a reasonable person ought to accept at a given time as a practical guide to the future, given the evidence and alternative beliefs then available. An understanding of this epistemological result should produce a playful, open-minded, hypothetical, tolerant, experimental attitude toward most beliefs and traditions. It should produce a willingness to accept fallibility and self-correction.

This understanding should also underscore the importance of values. As we have seen, logic is viciously neutral. Like a computer, it produces a product that is contingent on what you start with. If you want to blow up a city and kill millions of people, science and logic will show the necessary reasoning trail to accomplish this goal. If you want a child to grow up sane and happy, science and logic will help accomplish this goal. Conclusions follow from premises. Hence, the premises we use must be based, not only on our most reliable knowledge, but value commitments as well. And we know what kind of commitment is needed for a better world. We know that given our present predicament, to fuel the cooperative efforts necessary to control population[2] and better distribute the world's resources, we need a greater commitment from all currently competing groups to the human species as a family. We need to recognize the human species as one race of people whose differences are trivial compared to their similarities, as one family whose real threat is not other members of this family but our fragile, contingent existence together on this planet.

---

[1] Alexander, page 237.

[2] Human species population growth took tens of thousands of years to reach the 2 billion mark in 1927. It will take just seventy years to triple that number. Every twenty-four hours enough people are added to the Earth to fill a city the size of Newark, New Jersey.

I am advocating what is usually viewed as a very idealistic perspective. How can competing groups see themselves as a family after centuries of mutual brutality and cruelty, in a world of growing population and shrinking economic resources? The response is that, given what modern science tells us about human nature, this is not an unrealistic idealism, but rather long-term pragmatism. What do we know? According to Alexander, we know ". . . that there is one basic functional substrate for trends toward *cooperative* group living and that is active and cooperative defense against some common extrinsic threat or uncertainty . . ."[1] And what is our most common extrinsic threat or uncertainty? It is not each other, but rather our fragile, contingent role in an awesome universe.

Let's call this view of life the ***cosmic perspective***. Here are some of its details:

Our Earth is but a defenseless grain of sand, a small speck of some relatively recent leftover star stuff, floating in a rather hostile cosmic ocean that is 12–15 billion years old. How big is this ocean?

Our sun is approximately 93 million miles away and a million Earths could easily fit within it. Trillions and trillions of miles farther away are stars so large that millions of suns could easily fit within them. Our sun and billions of other stars make up a single galaxy. This galaxy is currently whirling around like a gigantic pinwheel or merry-go-round. Our Earth spins around at one thousand miles per hour at the equator, revolves around the sun at a speed of approximately sixty-six thousand miles per hour, and sails around our galaxy with our sun at more than five hundred thousand miles per hour. As you read this sentence, our apparently stable Earth will rush through 150 miles of space! Trillions and trillions of miles farther away are as many as 10 billion other galaxies, many much larger than ours. Some whirling their stars around at more than a million miles per hour. But so large are galaxies that even at these great speeds, millions of years pass before one star can complete one circuit around a galaxy. Our sun requires 250 million years to revolve once around our galaxy.

The stars within galaxies have a life cycle of birth and death. The largest stars die with cataclysmic explosions, called supernovas, and seed the galaxy with heavy elements that could become new stars and new planets. Surrounding our one sun and nine planets are several layers of cometary debris left over from the formation of our solar system from a collapsing cloud of galactic dust; the collapse was initiated by an exploding star. Pieces of this debris periodically get gravitationally kicked into our solar system. In July 1994, pieces of a disintegrating comet slammed into Jupiter with a force estimated at 40 million megatons of dynamite, 500 times more explosive power than all the world's nuclear weapons detonated at once.[2] We know that in the past pieces of this debris slammed into the Earth causing explosions and environmental destruction equal to full-scale nuclear war. Today, we believe that after a successful reign on Earth of more than 150 million years, the dinosaurs were destroyed by such an event. Several comets slammed into the Earth 60 million years ago, destroyed the environment of the dinosaurs, and made the current dominance of mammals possible. We know we are bound to be hit by this cosmic debris again.

---

[1] Alexander, page 65.
[2] Stephen Jay Gould, "Jove's thunderbolts," *National History*, vol. 103, No. 10, October 1994, p. 8.

Given enough time, enough debris (there is plenty), and gravity, like strangers in the night our Earth and some of these cosmic visitors will meet. Some scientists believe that our Earth takes significant hits every 25 million years or so. But even without these cosmic hits, our Earth as we know it will perish. Our sun will die 5 billion years from now, and in just five hundred thousand years it will be too hot to sustain life on Earth. Cometary collisions and the death of our sun may be far in the future, but the only reason life on Earth appears calm and secure from cataclysmic outside forces is time.

If this is too remote for you in terms of the immediate present, think of your life this way. Human infrastructures that seem so durable and impressive reside on relatively thin shells or plates of solid rock. These plates are slowly, but relentlessly, moving and crashing into each other. Constant earthquakes that can flatten our infrastructures in seconds and slosh the ocean around to produce massive tidal waves are the inevitable result of these moving plates. For perspective, consider that one hundred thousand years from now Los Angeles will be next to San Francisco, because Los Angeles is on a plate sliding toward San Francisco.

Beneath these relatively thin shells are massive amounts of molten rock—99.9 percent of our Earth is an inhospitable hell. Inevitably, the plates move and crash which produces pin pricks and fissures that allow some of this fiery mass to explode to the surface without regard to where humans reside. We are speaking, of course, of volcanoes. Along with the molten lava, which is impossible to stop from inundating human habitats, gas, ash, and debris are thrown into our atmosphere, sometimes disrupting worldwide weather patterns for years. Some scientists believe that massive past volcanic eruptions produced major extinctual disruptions for all life on Earth.[1]

Think also of the fact that just a few miles above your head is a relatively thin blanket of ozone protection from deadly ultraviolet radiation from the sun and cosmic radiation left over from the inevitable march of exploding, dying stars. Only within this thin layer then, a small slice of oasis, sandwiched between a bottomless molten hell and an endless deadly cosmic ocean, can humans live unprotected. We live on a small, momentary slice of biological heaven, spinning dizzily through inhospitable astronomical space, surrounded by an inevitable procession of violent and destructive events, and look what we do with this fragile, momentary gift! With all the group-against-group violence and destruction mentioned above, we are like two men on a raft headed for a deadly waterfall, who are too busy fighting and trying to kill each other to notice where they are or to cooperate to save themselves. And even within this biological heaven, we must remember that it is also heaven for many other forms of life, that this life is constantly changing and evolving, and inevitably some of these changes will be great threats to us as the recent AIDS epidemic tragically underscores.[2]

As Alexander points out, people are capable of incredible acts of cooperation and sacrifice when faced with an outside threat. Can billions of people be educated

---

[1] For a summary of nature's cataclysmic control over our lives, see Charles Officer and Jake Page, *Tales of the Earth: Paroxysms and Perturbations of the Blue Planet* (London: Oxford University Press, 1993), and Bruce A. Bolt, *Earthquakes* (New York: W.H. Freeman and Co., 1993).

[2] The AIDS virus is thought to be a mutation of a monkey virus that was previously harmless to humans.

of this cosmic perspective, of the real threat to their children's children's children? Cynics will slam this view as a hopeless romantic idealism. Teach people about the stars, and they will be nicer to each other? Well, my response is that we have not even tried yet. Less than one tenth of one percent of humankind is aware of what I have described in the last several paragraphs. Is it not at least an idea worthy of test?

Perhaps not all people will even have the reaction we desire upon learning of the cosmic perspective. Many may become so afraid that their natural selfish aggression may be enhanced.[1] But hopefully many will have the reaction of a former student of mine. She had already raised a family and was now coming back to finish a college degree that she had postponed for thirty years. She told the young people in our class that when she thinks of the cosmic perspective, when she is able to put aside all the mundane, daily commotions that call for her attention, she feels "like family." She has an urge to hug and educate children, to hold her neighbor's hand, to lecture people not to take their families for granted, to contemplate, and to prepare emotionally and intellectually for the greatest threat the human race will ever face—our survival within this cold, apparently loveless and heartless universe.

This last statement may give the appearance that I believe that we are alone in a godless universe. This book has made no commitment to whether God exists or not, and you should not hastily generalize from this lack of commitment that you know how I personally feel about this issue. My views on God will remain private. There are many religions and traditions, and it would not be fair to you for me to give my opinion. However, I do claim that the cosmic perspective sketched above constitutes not only a set of our most reliable beliefs, but is consistent with all enlightened religious views. It is consistent with all religions *except* those who wish to deny our true predicament, who feel that God is on their side and will help them kill other members of the human species; it is consistent with all religions except those that feel that they need not be responsible or fear the consequences of their actions.[2]

There are also many similarities between the cosmic perspective and traditional attitudes toward God. We should fear and respect the universe. It will always be in charge, regardless of how smart and powerful we become. But we should also love it. In the words of physicist Heinz Pagels, the universe is an "unfathomable beautiful ocean of existence." Science has not destroyed the religious feeling of mystery and awe, but has enhanced it. As very lucky guests the universe is not only our home, but the natural universal processes that threaten us also have produced us through a grand evolution and given us a chance at life.[3] In other words, science has revealed the same religious sense of living at the mercy of a wondrous foreboding power. With each deductive and inductive scientific trail followed, with each solution to

---

[1] Some of my students get mad at me after I have told them of the cosmic perspective. They say I frightened them, and knowing how big the universe is has ruined their lives.

[2] It is especially not consistent with those views that assert that the end of the human race is fine, because then members of their select group will go to heaven. See W. Martin, "Waiting for the End," *The Atlantic*, 249, June 1982: 31–37.

[3] There were an astronomical number of contingent events that had to be just right to produce the Earth in its current state. To name but a few lucky results: The Earth is just the right size, just the right distance from the sun, and our sun is in just the right place in our galaxy.

a previous mystery, new mysteries and wonders are revealed. With each new mystery, we are humbled by our informed ignorance. With each new wonder we are encouraged to have an unpretentious view of and reverence for existence. With such reverence we can express the best in ourselves: Our ability to cooperate for the common good, to be tolerant of and promote diversity, and to be true cosmopolitans—citizens of the universe—who see the universe as both a sovereign threat to our significance and welfare, and a magical playground for unending learning.

So, my answer to the second question posed above is to take seriously certain reliable beliefs produced by science. To take seriously the results of testing our beliefs, and to have a reasonable faith in the thinking tools covered in this book, the same tools used by Eratosthenes to follow the trail that first shattered humankind's egocentric complacency.

Chapters 7 through 12 may not be easy for you. You will be made to experience firsthand the challenge of symbolic reasoning. However, for your motivation, steadfastness, and discipline, keep in mind that you will be made to rigorously exercise the same symbolic reasoning faculty of mind that has helped produce such a startling view of our lives in the universe.

EXERCISE   Indicate whether the following are true or false.
I
1. According to this chapter, people who are logical are never violent.

*2. According to this book, relativism is wrong because it is possible to achieve absolutely certain beliefs.

3. According to this chapter, we should use logic to defend ourselves against the persuasive appeals of others, but use fallacies as persuasive techniques to get what we want from others.

4. According to this book, relativism is still wrong even though it is not possible to achieve absolutely certain beliefs.

5. According to this chapter, being more logical will automatically make us more ethical.

*6. A major theme of this book has been that an accurate account of the power and limitations of our reasoning ability implies a playful, open-minded, hypothetical, tolerant, experimental, and tentative attitude towards most beliefs and traditions.

7. According to this chapter, given the nature of evolution we are doomed as a species.

8. According to Alexander and Schell, the possibility of human extinction is due primarily to the natural functioning of human intelligence.

9. By the end of the 1980s, the former Soviet Union and the United States targeted each other's major cities with at least ten nuclear bombs apiece.

*10. According to this chapter, acceptance of scientific reasoning and results such as the cosmic perspective are inconsistent with a belief in God.

11. Soon it may be technologically possible to biologically engineer viruses or bacteriological agents that will selectively kill only people of a particular ethnicity.

12. According to modern science, our universe is only a few thousand years old and a few trillion miles in circumference.

EXERCISE II   Essays

1. One of the themes of this chapter is that our most common extrinsic threat or uncertainty is our fragile and contingent situation as a human family in an apparently indifferent, very old, and very large universe. Furthermore, it is important that people have this cosmic awareness. Write a two- to three-page essay analyzing this theme and its implications. Do you agree with this perspective? Is a cosmic awareness relevant? Will it help change the world? Should we be sure that children learn the cosmic perspective on how old and vast the universe is?

2. Write a two- to three-page essay explaining the interpretation given in this chapter of the Stephen Jay Gould quote from the beginning of this chapter. Hint: This will involve summarizing the theory of reliable beliefs and my arguments against relativism presented in this book.

3. This chapter presents some very controversial opinions. I hope I have made it clear that in my opinion the world view of modern science gives us a set of the most reliable beliefs we have. However, according to A. B. Alcott, "The true teacher defends his pupils against his own influence." With this in mind, write a critical essay making a case that I have committed fallacies in this chapter. Here are some possibilities:

A. Straw Person. My presentation of relativism may be a Straw Person. Do some philosophical research and see if I have distorted the relativist's position.[1]

B. Hasty Conclusion. One reviewer of this book from Georgia claimed that I have unfairly generalized from the facts of modern science—that the universe is old and vast—to a

[1] For some challenging, but relatively accessible reading on this topic, see Paul Feyerabend "How to Defend Society Against Science," *Scientific Revolutions* (1981), Ian Hacking, ed. (New York: Oxford Univ. Press), pages 156–167, and Larry Laudan, *Science and Relativism: Some Key Controversies in the Philosophy of Science* (1990)(Chicago: Univ. Chicago Press).

"metaphysical" interpretation—that humans are insignificant and unimportant. Did I make this claim? Suppose I did. Make a case that this is a hasty generalization from the results of science by making a case that the results of science are consistent with human beings being very important in the universe. Here are some possibilities to think about.

1. Einstein once remarked that the most incomprehensible thing about the universe is that it is comprehensible. His point seems to be that it is amazing that we are able to figure out how vast the universe is if we are so insignificant. Perhaps we are not.

2. There is also an old saying, "Astronomically speaking, humans are insignificant, but astronomically speaking, humans are the astronomers." In other words, just like Einstein's statement, how are we able to figure all this out, if we are not special in some sense?

3. During the Renaissance, several scientists of that time saw no inconsistency between believing in a big universe and God. In fact, that the universe is big and that we can figure it out was to them almost a proof that God existed and that God has given us a special status with a reasoning ability to figure out this great work. Copernicus, Kepler, and Galileo believed this.

C. Questionable Dilemma. On page 213 I talk about only two ways of viewing our reasoning ability, and then later in the chapter reject one of these alternatives. Is there a third alternative that significantly weakens my premise and hence my argument?

I, of course, don't believe I am guilty of these fallacies. But I would be very interested in reading what you come up with; these are very important issues. Again, feel free to correspond via one of the e-mail addresses listed in the Introduction.

Answers to Starred Exercises:

I.

2. False. It has been argued throughout this book that just because we can't have certainty, it is not true that anything goes. We can obtain reliable beliefs given alternatives and the weight of evidence for those alternatives. I have argued that what I call the world view of modern science is true, in the sense that we possess an overwhelming amount of evidence that the universe is very old and vast, that our Earth is a small grain of sand in comparison, and that life on our planet has evolved over billions of years.

6. True.

10. False!

Chapter 7

SYMBOLIC TRANSLATION

Introduction

By now you should have an appreciation for the practical nature of formal analysis. In addition to saving a lot of time by being able to see the essence of an argument, formal analysis is also valuable when arguments and inference situations are complicated and a way is needed to carefully follow the details of a reasoning trail. Much of our technological society is based on the results of people sitting down at desks and following, in one form or another, long logic trails. The computer programmer who analyzes a portion of the millions of lines of symbolic code that guide the space shuttle's computers, the engineer who follows the physical implications of a new engine design, and the astrophysicist who from a few facts attempts to reason back 15 billion years to the details of the origin of our universe, are all following the same tradition of Eratosthenes: they take premises and assumptions and "play" with them in a disciplined way to see where they lead. And like Eratosthenes, all use a form of symbolic reasoning, because it is much too difficult to follow the complex trail of reasoning if expressed in ordinary language.

The goal is the same. The trails must be valid or they will be worthless, or in cases such as the space shuttle, even dangerous. Thus, symbolic reasoning is used because even though we may have an understanding of what to look for in ascertaining a valid argument, it is often very difficult to see if complicated arguments stay on track and meet the validity standard. For instance, consider the following discussion between two friends over some of the beliefs contained in the Christian religion. John considers himself a Christian, and Dan, an agnostic, is challenging John on what Dan considers to be a major inconsistency in the literal or fundamentalist version of Christianity.

Dan: So God exists and is compassionate and forgiving?

John: Yes.

Dan: But your God is also omnipotent, all powerful, and there are rewards and punishments for our behavior. For many Christians this means that we go somewhere when we die, either to a heaven or a hell.

John: Yes, although there are many versions of this, all Christians believe that ultimately God is in charge so to speak, and that He has created a situation where there are consequences for our thoughts and actions; that we create one way or another a spiritual heaven or hell for ourselves. Christians are united in the belief that there would be no basis to morality otherwise.

Dan: Well, let me prove to you that if you are a good Christian, concerned with justice, compassion, and forgiveness, you should not believe in any place called 'hell.' If God's existence is necessary for the foundation for morality, would you agree, hypothetically, that if God does not exist, then there will be neither a heaven nor a hell for us when we die?

John: Yes, I would agree hypothetically. As I have said, for a Christian, reward and punishment, good and bad, right and wrong have meaning only if there is a God.

Dan: And you also agree that although human beings suffer greatly while on this Earth—that whether we are rich or poor, intelligent, gifted, well-educated, or intellectually dull, mentally disabled, or uneducated—we all suffer in one way or another, and experience pain, doubt, absurd misfortune, and disappointment?

John: Yes, but for a Christian the absurdity is only apparent; everything that one experiences is meaningful from God's perspective.

Dan: Yes, exactly. So, all human suffering that exists must contribute in some fashion to fulfilling God's purpose.

John: Yes, all suffering is part of His plan, no matter how absurd and unfair it may seem, because God is good and His compassion is limitless.

Dan: Ah, but you see here is my problem. If God is supposed to be good, infinitely compassionate and forgiving, then if there is to be a kind of human suffering that is eternal—and this by definition is what hell is, an eternal, endless suffering—then this cannot contribute to fulfilling a compassionate and forgiving God's purpose. If there is a hell for some human beings when they die that includes human suffering and eternal suffering, and if eternal suffering is inconsistent with a Christian conception of a forgiving and compassionate God, it follows for a Christian that a literal place of eternal suffering, a hell, cannot exist!

Dan's basic argument is that the literal interpretation of the concept of an eternal hell is inconsistent with the basic postulates of the Christian religion. So, if a good Christian wants to believe in a good, infinitely compassionate and forgiving God, he or she must reject the conception of a literal eternal hell. Here is a formalization of Dan's argument in terms of premises and a conclusion.

Premises:

1. If God does not exist, then there will be neither a heaven nor a hell for us when we die.
2. If He does exist, then there should be human suffering only if this suffering contributes to fulfilling God's purpose.
3. However, if there is to be human suffering and eternal suffering, then this cannot contribute to fulfilling God's purpose (because God is supposed to be good, forgiving, compassionate).
4. There will be human suffering and eternal suffering, if there is a hell for us when we die.

Conclusion:

It follows that there will not be a hell for us when we die.

Can you follow Dan's argument? Regardless of your religious orientation, which would influence your judgment on the truth or falsity of the premises, can you use the skills we have learned thus far to evaluate the reasoning of the argument? For the average person it is not at all clear whether we would have to accept the conclusion if we accepted the premises. The reasoning of this argument may "sound good," seem to flow and stay on track, but there are many arguments that psychologically give the appearance of staying on track that are not valid.

I often tell my students that the logic course they are taking is structured in such a way that they can pass the course only if they pass the final exam. Some will mistakenly infer from this that if they pass the final they are guaranteed to pass the course. Some even invalidly infer from my statement that they can skip quizzes and other exams, because only the final exam is important. In making this inference they are committing the *fallacy of affirming the consequent*, an argument form that seems to stay on track but does not. This inference form is persuasive because it is close to a valid form of reasoning called *modus ponens*, which we will study later.

Dan's argument is valid (although it may not be sound, a Christian could object to any one of the premises—but most likely the third premise).[1] But how can I prove this to you—that you should accept the conclusion that a literal eternal hell is impossible if the premises are accepted? It is very hard to keep the implications of all the premises in mind to see if we are locked into the conclusion. The average

---

[1] John might also respond with something like, "Your logic is excellent, but it has only uncovered the pain of being a human being, the anguish that will always result if reason and logic alone are relied upon. As Pascal said, 'The heart has its reasons which reason knows nothing of.'"

person will forget the implications of the first premise by the time he or she reads the second or third premise.

Like mathematics, symbolic logic was invented to enable us to follow trails that would be practically impossible (or at least take a very long time) using our normal language-based reasoning tools. As in the case of mathematics, logicians have discovered that our common sense can be systematized symbolically and then valid mechanical techniques used to follow difficult trails without getting lost in complex distracting and irrelevant details. In other words, we can take the smallest, most obviously valid and agreed-upon pieces of our common sense and methodically and objectively check or produce a trail of complex reasoning. This discovery, combined with advances in electronics and philosophy, gave birth to our current computer revolution. Computers are essentially symbolic logic machines that know how to do only one thing, stay on track with a vengeance, following the trail of an initial starting point with a single-minded purpose, similar to a Komodo dragon following a meal.[1] Our goal for the next several chapters will be to learn how to be symbolic Komodo dragons, to work on arguments like the one above and understand why they are valid, and learn to appreciate the process of symbolic reasoning.

Logical Connectives

Unfortunately, even though symbolic logic is merely organized common sense, the first step in the learning process is usually the most difficult for students. We must learn to *translate* arguments from our normal language into a symbolic notation. We will approach the learning of this translation process as if learning another language. Although the vocabulary of this new language is small, as when learning any language, lots of practice is necessary. Do not expect to be an expert right away.

The new language you will learn is part of the first stage of symbolic logic. This first stage is often called *propositional logic*, because it deals with the manipulation of the logical implications of linked propositions or statements, such as in "Either John passes the final or he will not pass the course." The parts of this statement—"John passes the final" and "John passes the course"—are statements that are linked by the words *or* and *not*. These linking words are called *logical connectives*, and we will see that the validity of propositional inferences depends necessarily on the arrangement of propositions by logical connectives. For instance, once you become proficient at manipulating logical connectives symbolically, you will see that from this statement about John's situation it logically follows that passing the final is a necessary condition for John to pass the course, or, put another way, that it is not possible for John to pass the course and not pass the final. But it does not follow that passing the final is sufficient for John passing the course, or, put another way, it is still possible for John to pass the final, but not pass the course. (Stay calm, it is easier to follow this in symbols.)

---

[1] The Komodo dragon is a giant lizard that reaches the length of a midsize car and lives on the isolated islands of Komodo, Gillimontang, and Rintja, part of the southeastern edge of the Indonesian archipelago. As a primitive reptile, it is famous for its ability of relentless pursuit of usually much faster prey. According to native stories, some will stay on track of their prey for days or weeks until the prey dies of hopeless fright!

Because this is only the first level of symbolic logic, and because we will only be interested in manipulating and analyzing inferences of whole statements, the new language to be learned is very simple. In a sense, there are only five key vocabulary terms for the whole language.[1] The five new terms are symbolic representations of five logical connectives as follows:

not	and	or	if . . . , then. . .	if and only if
~	•	∨	⊃	≡

Any statement that contains one or more of these connectives we will call a **compound statement**. A **simple statement** will be any complete sentence in ordinary language that does not contain any logical connectives. Simple statements will be represented symbolically by capital letters, such as:

A = *Alice* is going to the party.

C = The students passed the essential *competency* exam.

F = John passed the *final* exam.

W = The United States has pledged to *withdraw* all of its landbased nuclear weapons from the Korean peninsula.

There is nothing absolute about which capital letter is used to stand for each simple statement. The capital letters are merely abbreviations. Consistency, convenience, and agreement are the only requirements. Because we are interested in analyzing the implications of connected simple statements, if we agreed we could have used a U rather than a W for the last statement above.

With these preliminaries sketched, we are now ready to translate our first compound statement from ordinary language to a symbolic notation. If A stands for "Alice is going to the party" and B stands for "Barbara is going to the party," how would we translate the compound statement: "Alice and Barbara are both going to the party"? If you answered A • B, then you are well on your way to learning symbolic translation.

This first step may seem overly simple to you, but for psychological, motivational, and pedagogical reasons it is important to reflect here that the essential ability of mathematical and logical analysis is no more difficult than being able to understand A • B, of being able to see the simple pieces that make up a complex representation or reasoning trail. Remember that in the reasoning of Eratosthenes (Chapter 3) and the example of counting the number of atoms in the entire universe (Chapter 4), an apparent complex trail was just a series of commonsense steps. Similarly, soon we will be dealing with symbolic statements like:

$$\{\sim[(A \vee B) \supset \sim C] \equiv (D \bullet G)\} \supset \sim[(A \bullet D) \vee \sim B]$$

---

[1] This is certainly simpler than learning a foreign language, in which case more than five new words would be introduced on the first day of class. Our symbolic logic will be very simple because we will be learning to map for the most part only the **syntax** of our language.

Most students tend to suffer from "sensory overload" when confronted by such complex statements. So, it is very important to remember that, in a sense, all such complex statements are actually just a bunch of A • Bs in disguise, and that a calm, disciplined analysis will show that all complex statements are made of simple parts.

Keeping this in mind, let's continue with some basics. Here is how the other logical connectives would be used in basic compound statements.

Alice is going to the party *or* Barbara is going to the party.

$$A \lor B$$

If Alice is going to the party, *then* Barbara is going to the party.

$$A \supset B$$

Alice is going to the party *if and only if* Barbara is going to the party.

$$A \equiv B$$

Alice is *not* going to the party.

$$\sim A$$

For the most part, language contains rich nuances that propositional logic ignores and also some others that it addresses but that we will ignore in the beginning. Let's elaborate on the simple mechanics of these translations and touch on a few of the complexities.

Statements such as the last statement above, which contain a *not,* we will call **negations**. There is nothing absolute about the placement of the ~, or negation, symbol in the above translation. We could have placed it after the A. But it is important to have one way to translate negations so that we can communicate symbolically. As Humpty Dumpty points out to Alice in Lewis Carroll's *Through the Looking Glass,* there is nothing absolute about the words we use to describe things; using the words we do is a matter of convention. A table, for instance, does not have a sign on it saying, "Call me *table.*" If we all agreed or were part of a culture that reared its children this way, we could call a table a *dog,* and a dog a *table.* We would not think that this was strange, because we would be accustomed to it. But to communicate we do need a standard way of referring to objects. Likewise, in symbolic logic we need a standard way of translating. So we will adopt the basic rule that when a statement involves a negation, we will put the negation in front of the capital letter that stands for the same statement if it were not negated. If F is to be used for "John passed the final exam" then "John did *not* pass the final exam" would be translated as ~F.[1]

Compound statements that use <u>*and*</u> are called **conjunctions**. Sometimes conjunctions will not explicitly state the word *and,* as in "<u>Even though</u> Alice is going to the party, so is Barbara." From a propositional point of view, we would still translate this statement as A • B, because the bottom-line claim is that they are both going, even though semantically the use of the phrase *even though* implies something

---

[1] Incidently, Lewis Carroll's real name was Charles Lutwidge Dodgson (1832–1898), and he was a professor of mathematics and logic at Oxford University. There is much more than excellent children's stories involved in his *The Adventures of Alice in Wonderland* and *Through the Looking Glass.* Political criticism and theories on the nature of logic and language lurk behind the scenes.

different from a straightforward, unqualified *and*. A similar situation occurs when we reflect on the semantic and possible cultural differences between "Mary became pregnant and married John" and "Mary married John and became pregnant." In some cultural contexts, say, the 1950s in the United States, there would be a very big difference in the implications attached to the meanings of these two statements. The first would have implied a very embarrassing situation for Mary and John, whereas the latter would have been cause for celebration. Propositional logic ignores this semantic difference and reflects only the minimum description that Mary is married and pregnant. But in propositional logic we are not interested in capturing the full meaning of statements, because a considerable amount of logical analysis can be accomplished by simplifying.[1]

Logicians refer to compound statements that use <u>*or*</u> as **disjunctions**. Sometimes a disjunction will use the word <u>*either,*</u> as in "Either Alice or Barbara is going to the party," but this will not change the translation. Sometimes *or* statements will imply the use of a "strong" or **exclusive** *or,* as in "With your dinner tonight you may have cream of asparagus soup or crab salad." At other times *or* statements will imply a "weak" or **inclusive** sense, such as in "(Well, I'm not sure, but I think) Either Alice or Barbara is going to the party." In the first case, we know that the context implies that we can't have both the soup and the salad, unless we want to pay extra. In the second case, we know that it is possible that both Alice and Barbara may be going to the party. The claim is made only that at least one is going, leaving open the possibility that both may go. An exclusive *or* makes the claim that one thing or another is true, but not both; an inclusive *or* makes the claim that one thing or another is true, possibly both. After we become more comfortable in translating, we will have to address these differences in the use of *or*.

Statements that use or imply an <u>*if*</u> . . . <u>*then*</u> hypothetical situation are called **conditionals**. Notice that although two words are used, only one symbol is used (⊃). This symbol looks like a horseshoe turned on its side. The open part of the horseshoe will always face left. In a straightforward *if* . . . *then* statement, that part of the statement that immediately follows the *if* is called the **antecedent** and is placed on the left side of the horseshoe when translated. The part of the statement that follows the *then* is called the **consequent** and is placed on the right side of the horseshoe. So, in the translation A ⊃ B, A is the antecedent and B is the consequent.

Statements with an <u>*if and only if*</u> phrase are called **biconditionals**. This phrase is used a lot in contractual situations such as, "An employee gets a day off during the week if and only if the employee works on Saturday." Beginning logic students usually have a difficult time feeling fully comfortable with *if and only if* phrases because in everyday communication we don't say, "I'll help you with your homework tonight *if and only if* you buy me a soda." However, such statements are found in logic, science, law, diplomacy, and any field where precise communication is very important. In 1979, the government of Iran told the U. S. government, "The U. S.

---

[1] In this context, it is important to remember the message of the Huxley quote at the beginning of Chapter 2 and the discussion of its implications for logic. Although we will be "cutting up" our normally rich experience into a logical skeleton, our task will be no different from creating a map. We will be "mapping" reasoning, and although maps should not be confused with reality, they can be used as useful guides.

hostages will be freed *only if* the United States returns the Shah of Iran and his assets to the Iranian government." (The Shah was living in the United States at the time, and his considerable assets were in U. S. banks.) It was important for U. S. leaders to know the difference between this offer and "The U. S. hostages will be freed *if and only if* the Shah and his assets are returned." The Iranian statement implied no guarantee that the hostages would be released, even if the Shah and his assets were returned. On the other hand, the use of *if and only if* would have been an implied guarantee.

Perhaps if we spoke this way more often in everyday exchanges there might be less quarreling. Consider the following statements made by a mother to her son.

1. "You do not go out tonight with your friends unless you clean your room."
2. "You do go out tonight with your friends if you clean your room."
3. "You go out tonight with your friends only if you clean your room."
4. "You go out tonight with your friends if and only if you clean your room."

There are major differences in the meanings and logical implications of these statements. Even the simplest things in life are potentially much more complex, rich, and interesting than they seem, which is why there will always be so many points of view on any issue. Logic, properly understood as a set of tools to deal with complexity, is not intended to destroy the richness of life. It would not be possible to do this anyway; reality and the interactions within it will always "overflow" beyond the boundaries of any logical analysis. Logic is intended only as a disciplined way of testing trails and points of view. So, let's see what points of view logic can reveal in this situation.

Statement #1 is vague. Depending on the context, *unless* sometimes means *or* and sometimes *if and only if.* If the mother made this statement intending the *or meaning,* then she would be saying essentially, "Either you clean your room or you do not go out tonight." This statement in turn is equivalent to "If you do go out tonight, then you must clean your room." Now, if the son has had a logic course, he will realize that his mother is only specifying a necessary condition for his freedom to go out with his friends. He will realize that statement (1) is saying the same thing as (3), but what he would rather hear is (2). Here's why.

If his mother says, "You do go out tonight with your friends, if you clean your room," she is telling her son, and committing herself to the position, that cleaning his room is **sufficient** for his going out with his friends, that this is "all he has to do." Thus, this statement leaves open the possibility that the son does not have to show proper respect or do anything else deemed appropriate by the mother. For this reason, most experienced mothers would not say this. They instead say or intend something along the lines of (1) or (3). These statements are saying, "If you do go out tonight, then you must clean your room," that it is **necessary** for the son to clean his room, but his going out is not guaranteed by his cleaning the room. He must clean his room to have a chance to go out, but there may be other conditions as well. At this point, if both the mother and the son have had a logic course, they would realize that the appropriate statement to agree on would be the very specific "You do go out tonight if and only if you clean your room." Here's why.

In stating the sufficient condition (2), although there is a clear consequence implied if the son does clean his room—he is allowed to go out—there is no clear consequence if he does not clean his room! From this statement, if he does not clean his room, it would be invalid to infer that he should not be allowed to go out. And obviously, it would also be invalid to conclude that he should be allowed to go out. If he does not clean his room, the situation as to what should happen next is vague. Perhaps the son tells his mother, "But, Mom, there's no time to clean my room, and Billy's father is taking us to the championship football game, and it's the last time I will see Billy because his family is moving." If the mother decides to let him go out with his friends, she is **not** changing her mind, because the only position she has committed herself to is to let her son go out if he cleans his room. She has made no commitment as to what should happen if the son does not clean his room, because saying that "You do go out if you clean your room" is not the same as saying "If you don't clean your room, you do not go out tonight." To summarize this formally:

Valid   If you clean your room, you do go out tonight with your friends.
You clean your room.
Therefore, you do go out tonight with your friends.

Invalid   If you clean your room, you do go out tonight with your friends.
You do not clean your room.
Therefore, you do not go out tonight with your friends.

Invalid   If you clean your room, you do go out tonight with your friends.
You do not clean your room.
Therefore, you do go out tonight with your friends.

Similarly, in stating the necessary condition (3), although there is a clear consequence if the son does not clean his room—he does not go out—as we have seen, there is no clear consequence if he does clean his room. After he has cleaned his room, his mother is not committed to letting him go out and may consistently impose other conditions at that point. Summarizing:

Valid   You do go out tonight with your friends only if you clean your room.
You do not clean your room.
You do not go out tonight with your friends.

Invalid   You do go out tonight with your friends only if you clean your room.
You do clean your room.
Therefore, you do go out tonight with your friends.

Invalid   You do go out tonight with your friends only if you clean your room.
You do clean your room.
Therefore, you do not go out tonight with your friends.

Thus, from the mother's point of view, if she has really had it with her son's messy room and wants the condition of the room being cleaned to be absolute, the sufficient condition statement is not fair—it is too noncommittal as to what happens if the son does not clean the room. (It is also unfair for another reason. The son could break furniture in the living room or dishes in the kitchen in protest, but if he cleaned his room his mother is committed to letting him go out with his friends.) On the other hand, from the son's point of view, the necessary condition is unfair because it is also not strong enough—he could clean his room and still not be allowed to go out if his mother was in a bad mood and at the last minute wanted something else done. Hence the appropriateness of the very specific contractual "You do go out tonight with your friends *if and only if* you clean your room." From this it follows that if the son does not clean his room he cannot go out, and if he does clean his room he can go out.

Valid   You do go out tonight with your friends if and only if you clean your room.
You clean your room.
Therefore, you do go out tonight with your friends.[1]

Valid   You do go out tonight with your friends if and only if you clean your room.
You do not clean your room.
Therefore, you do not go out tonight with your friends.

Now, if you feel a little dizzy after all this, you are beginning to understand why symbolic logic was invented. It has taken me a couple of pages to describe the various logical implications of the mother-son contracts, and you probably had moments when you had to stop and think about what I was saying. Later, with proficiency in symbolic logic, the above logical implications can be presented in a couple of lines and you will be able to understand them in seconds.

But one thing at a time. Let's return to the task of translating ordinary English statements into symbolic logic. What follows is a dictionary of examples of how common sentences in normal English would be translated into symbolic logic. Because one of the first stages of learning is imitation, we can use this dictionary as a model to imitate when translating similar English sentences, even though initially you may not be fully comfortable with what you are doing or why the translations end up the way they do. At this point, notice how each italicized connective gets translated and then (don't think too much yet!) attempt to mimic this process by doing the exercises at the end of the dictionary. For instance, item (11) in the dictionary uses a *not both* phrase and is translated ~(J • K), so we mimic this by translating item (4) in Exercises I as ~(A • G) and (9) in Exercises II as ~(S • D).

Usage Dictionary of Logical Connectives

1. John passed the final exam *and* the course. (F, C)
$$F • C$$

---

[1] Because this argument is valid, can you see why even the stronger "if and only if" contract is still something of a disadvantage for the mother? Being a parent is not easy.

2. Either John passed the final exam *or* he passed the course. (F, C)
$$F \lor C$$

3. *If* John passes the final exam, *then* he will pass the course. (F, C)
$$F \supset C$$

4. John will pass the course *if and only if* he passes the final exam. (C, F)
$$C \equiv F$$

5. John passed the course *but not* the final exam. (C, F)
$$C \bullet {\sim}F$$

6. John did *not* pass the course, *but* he did pass the final exam. (C, F)
$${\sim}C \bullet F$$

7. John passed the course *even though* he did *not* pass the final exam. (C, F)
$$C \bullet {\sim}F$$

8. *Even though* he did *not* pass the final exam, John passed the course. (F, C)
$${\sim}F \bullet C$$

9. John did *not* pass *both* the final exam *and* the course. (F, C)
$${\sim}(F \bullet C)$$

10. John did *not* pass the final *and* he did *not* pass the course. (F, C)
$${\sim}F \bullet {\sim}C$$

11. Johnson *and* Kaneshiro will *not both* be hired. (J, K)
$${\sim}(J \bullet K)$$

12. Johnson *and* Kaneshiro will *both not* be hired. (J, K)
$${\sim}J \bullet {\sim}K$$

13. John either passes the final exam *or* he does *not* pass the course. (F, C)
$$F \lor {\sim}C$$

14. Either John did *not* pass the final exam *or* he did *not* pass the course. (F, C)
$${\sim}F \lor {\sim}C$$

15. John passed *neither* the final exam *nor* the course. (F, C)
$${\sim}(F \lor C) \quad \text{or} \quad {\sim}F \bullet {\sim}C$$

16. John will take the bus to school *unless* his girlfriend drives him in her car. (B, D)
$$B \lor D \quad \text{or} \quad {\sim}D \supset B$$

17. John will pass the course, *if* he passes the final exam. (C, F)
$$F \supset C$$

18. John will pass the course *only if* he passes the final exam. (C, F)
$$C \supset F$$

19. *If only* John passes the final exam, he will pass the course. (F, C)
$$F \supset C$$

20. *Provided that* John passes the final exam, he will pass the course. (F, C)
$$F \supset C$$

21. John will pass the course, *provided that* he passes the final exam. (C, F)
$$F \supset C$$

22. Passing the final exam is a *necessary condition* for passing the course. (F, C)
$$C \supset F$$

23. Passing the final exam is a *sufficient condition* for passing the course. (F, C)
$$F \supset C$$

24. *If* John does *not* pass the final exam, *then* he will *not* pass the course. (F, C)
$${\sim}F \supset {\sim}C$$

25.. *It is not true* that *if* John passes the final exam, he will pass the course. (F, C)

$$\sim(F \supset C)$$

EXERCISE
I
Now use the above examples as models for translating the following abbreviated statements into symbolic notation.

1. A, but not B.
*2. If not A, then B.
3. Z only if not B.
4. Not both A and G.
5. P, if not G.
6. Either not P or not D.
7. Neither R nor H.
8. S unless not P.
*9. Not Z, if and only if Y.
10. A necessary condition for P is not Y.

In the above exercises you should have tried simply to imitate the dictionary. But, as in most learning, we want to be able to obtain a deeper level of understanding. In translating, we want to know why the translations end up the way they do. Why, for instance, in the dictionary, are (18) and (19) translated differently? The implicit claim of this dictionary is that the translations must be this way to capture faithfully the meaning of the original English statements. So, before we continue with more complex translations, let's elaborate on the dictionary examples.

Examples (5) through (8) show that we will use the • symbol to translate any conjunctive expression such as *but* and *even though*. Other English conjunctive expressions translated with a • would be *however, moreover, although, yet,* and *whereas.* The latter is often used with a semicolon (;), as in "If Johnson does not get the promotion, Smith will not be able to finish the Japan project; whereas, if Johnson does get the promotion, Kaneshiro will not be able to finish the Singapore project." $(\sim J \supset \sim S) \cdot (J \supset \sim K)$.

The examples in the dictionary show that we will adopt the following convention when translating the negation sign (~). If there are no parentheses in the translation, then the negation sign will refer to only the capital letter immediately following it. So, $\sim F \cdot C$ would be a correct translation for the statement, "John did not pass the final, but he did pass the course." Whereas, $\sim(F \cdot C)$ would be a translation of the statement, "John did <u>not</u> pass <u>both</u> the final exam and the course." Note that if there is a negation outside parentheses, then the negation is being applied to the *entire statement inside the parentheses.*

In just looking at the translations of (11) and (12), many students will think that these statements must have the same meaning. But the negation in (11) is not referring to each letter individually inside the parentheses; it is negating the entire statement J • K. So, (11) is saying that at least one of the men will not be hired, leaving open the possibility that at least one will; whereas (12) is much stronger stating that neither of the men will be hired. Consider the difference between (9) and (10). Number (10) is very precise in what it is claiming. If (10) is true, then we know that John did not pass the final and he did not pass the course. On the other hand, (9) is less precise. It only claims that John did not pass both the final and the course, leaving open the possibility that he passed one or the other.

The significant difference in the meaning between **not both** and **both not** shows how easy it is for us to misread the implications of language and make logical mistakes. By now you should be thoroughly aware of how important it is to avoid such misreadings. If you are not aware of the potential pitfalls, then it will be easy, as the examples in Chapter 1 show, for someone to take advantage of you. For very good reasons, language is rich and complex. But there are times when tools are needed to surgically trace our way through this richness.

Consider how much is implied in the simple statement, "Alice and Barbara are not both coming to your party." Suppose I am giving a party, and I have invited both Alice and Barbara. My friend, knowing this, comes to me before the party and tells me that unfortunately Alice and Barbara have had a monstrous fight and have vowed never to be seen together again on the face of this earth. He tells me that he is sure that they will *not both* be coming to my party. If his statement is true, what will I be sure of? Will I know that Alice is not coming? No. Will I know that Barbara is not coming? No. Will I know that neither of them are coming? No. Will I know that Alice is coming? No. Will I know that Barbara is coming? No. All that I am sure of is: If Alice is planning to come to the party and Barbara knows this, then Barbara will not come; if Barbara is planning to come to the party and Alice knows this, then Alice will not come; and finally, if Alice thinks Barbara is coming and Barbara thinks Alice is coming, then neither of them will come! All of this is packed into the phrase *not both*. Either one will not come or the other will not come, possibly both will not come.

"Possibly *both* will *not* come"! Suppose in a different set of circumstances my friend tells me that "Alice and Barbara will both not be coming to your party." Although the same little words are used—*both* and *not*—the different word order makes this now a very different statement that implies a very different set of circumstances. Suppose my friend tells me, "I don't know what you did, but both Alice and Barbara don't like you, so I can guarantee that they will *both not* be coming to your party." This statement is much more precise. If it is true, then I will be sure that Alice is not coming to my party and Barbara is not coming to my party. Whereas the *not both* statement implies the possibility of neither coming, the *both not* statement implies that this is guaranteed. Changing the order of the words (both not . . . not both) is a very big deal.

If you find the unsuspected complexity of these examples overwhelming at this point, consider that we can distinguish between a full **understanding** of the dictionary and a practical **mimicking** of the dictionary. Because our first goal is simply to mimic the dictionary, you must remember only this:

$$\textbf{not both} = \sim(\underline{\qquad} \cdot \underline{\qquad})$$
$$\textbf{both not} = \sim\underline{\qquad} \cdot \sim\underline{\qquad}$$

You should eventually be able to understand why the translations are this way. In fact, you likely already understand this, considering that we learn how to use these words at a very young age. But for most of this chapter, all you need do is mimic.

Similar considerations apply in using negations with *or* statements. Note the difference between (14) and (15). These statements are not saying the same thing: $\sim(F \vee C) \neq \sim F \vee \sim C$. Number (15) is much stronger than (14). If (15) is true, then we know that John did not pass the final, and he did not pass the course; whereas, in (14) we only know that he did not pass at least one. When we negate an entire *or* statement, $\sim(\underline{\qquad} \vee \underline{\qquad})$, we are stating that the usual claim of the *or* statement (that at least one thing happened) is not true. Notice that **neither . . . nor** statements have the same translations as **both . . . not** statements. With a little reflection we also see that (14) could be expressed using *not both*, as is (9). So, to summarize and for future reference,

$$\textbf{not both} = \textbf{either not . . . or not . . .} \quad = \sim(\underline{\qquad} \cdot \underline{\qquad}) = \sim\underline{\qquad} \vee \sim\underline{\qquad}$$
$$\textbf{both not} = \textbf{neither . . . nor . . .} \qquad\quad = \sim\underline{\qquad} \cdot \sim\underline{\qquad} = \sim(\underline{\qquad} \vee \underline{\qquad})$$

BUT

$$\textbf{not both} \neq \textbf{both not}$$
$$\sim(\underline{\qquad} \cdot \underline{\qquad}) \neq \sim\underline{\qquad} \cdot \sim\underline{\qquad}$$
$$\textbf{neither . . . nor . . .} \neq \textbf{either not . . . or not . . .}$$
$$\sim(\underline{\qquad} \vee \underline{\qquad}) \neq \sim\underline{\qquad} \vee \sim\underline{\qquad}$$

Although the context of *unless* statements sometimes indicates that an *if and only if* meaning is intended, until clarification is given that this stronger meaning is intended it is best to always start with a minimum *or* interpretation. The easiest procedure to adopt at a mimic stage of translating is to simply replace *unless* with *or* and translate accordingly. Thus, the meaning of (16) can also be captured by "Either John will take the bus *or* his girlfriend will drive him in her car." Note that this *either-or* statement is equivalent to "If his girlfriend does not drive him in her car, John will take the bus to school." Thus, either translation shown in (16) can be used for translating the minimum interpretation of *unless* statements.

Numbers (17), (18), and (19) are very important and are most often confused, both by students doing translations and in general by the average person drawing invalid inferences when these phrases are part of the premises in arguments. Number (17) is an *if . . . then* statement, but it shows that sometimes for emphasis we will state the consequent first and then the antecedent. The essential meaning of (17) would not change if it were restated, "If John passes the final exam, then he will pass the course." Number (18), however, shows that the order of the conditional is reversed when *only* occurs in front of an *if*. Consider this statement:

A person is pregnant only if that person is female.

If we were to follow the procedure adopted in (17) and make whatever follows the *if* an antecedent, we would end up with **F ⊃ P**. But this incorrectly claims that being female is all that is needed to be pregnant; whereas, the intention of the original statement is that being female is a condition for pregnancy only, and that, as we all know after a little sex education, some male sperm is also needed. So, the correct translation of this statement would be **P ⊃ F**.

Number (19) shows that sometimes we use *only* to emphasize the importance of the antecedent. But placed after the *if* in this statement, the *only* does not change the indication that passing the final is the antecedent. The difference between (18) and (19) is very important. If a professor said (18), he or she would be implying that it is absolutely necessary to pass the final exam to pass the course, that if John does not pass the final he will not pass the course. But this leaves open the possibility of other course requirements that John must meet, such as quizzes and other exams. On the other hand, (19) implies that passing the final exam is all that John needs to do to pass the course.

Here is an easy way of implementing these differences from a mimic point of view: Whatever statement (simple or compound) follows an *only if* that is not part of an *if and only if*, translate that statement as a consequent. Otherwise what follows an *if* will be translated as an antecedent. To illustrate, consider the following statements and their translations:

The number of jobs in marine maintenance will increase *only if* the number of boat slips increases. J only if S, J ⊃ S

The number of jobs in marine maintenance will increase *if* the number of boat slips increases. J if S, S ⊃ J

The number of jobs in marine maintenance will increase *if and only if* the number of boat slips increases. J if and only if S, J ≡ S

Note, however, that sometimes some words may intervene between the *only* and the *if* as in "These proceedings *only* can be legally concluded *if* there is an agreement on the financial arrangements." This statement must be read carefully to see that *only if* is intended and hence, (P only if A) P ⊃ A.

One way of avoiding the tricky nature of *only if* and *if* statements is to use the very precise designation of a necessary or sufficient condition (22) and (23). When a professor tells her class at the beginning of the semester that her curriculum is organized in such a way that the course will be passed only if the final exam is passed, she means that passing the final exam is a necessary condition for passing the course, but that it is not necessarily sufficient—that passing the final is an absolute condition, but that there are other course requirements as well. On the other hand, if just before the final exam a student asked the professor what his standing was in the course and the professor responded with "Well, if you pass the final you will pass the course," the professor would be implying that given the student's performance to date passing the final at this point would be sufficient for passing the course.

In terms of the mimic rule for translating necessary and sufficient conditions, **necessary conditions** will always be translated as consequents, and **sufficient conditions** will always be translated as antecedents. As easy as this rule is to remember, take care to identify what is specified as a necessary or sufficient condition in a statement. For instance, if we were to reword (22) to read "A necessary condition for passing the course is passing the final exam," note that although the words *passing the course* occur immediately after the words *necessary condition,* the condition phrase refers to passing the final. Students will often mistranslate this statement as $F \supset C$ simply because the words *passing the course* appear closer to the words *necessary condition.* The intention of the words *for* and *is* must be recognized.

We can now see that another way of understanding *if and only if* is to see that its use specifies a necessary (only if) and sufficient (if) condition. So, the mother's statement to her son, "You go out tonight with your friends if and only if you clean your room" could be translated as either $O \equiv R$ or $(O \supset R) \cdot (R \supset O)$.

Finally, the same care in using negations with parentheses in combination with *and* and *or* statements applies to *if. . .then* statements. Number (24) is a very different statement from (25). It would be a mistake to think that $\sim F \supset \sim C$ equals $\sim(F \supset C)$. As we have seen, a negation outside parentheses cannot simply be pushed inside and applied to the parts inside the parentheses without altering the connective. Number (24) says that not passing the final is sufficient for not passing the course; whereas, (25) denies that passing the final is a sufficient condition for passing the course. Number (25) says that it is possible for John to pass the final and still not pass the course $(F \cdot \sim C) = \sim(F \supset C)$; whereas, (24) says that if he does not pass the final he is guaranteed to fail the course, that it is not possible for him to not pass the final and still pass the course $\sim(\sim F \cdot C) = (\sim F \supset \sim C)$. Number (25) refers to a consequence related to passing the final, and (24) refers to a consequence related to not passing the final.

If this makes you dizzy again, note the mimic way of capturing the meaning of *if . . . then* statements mixed with negations. In (24), in the English statement the negation occurs after the *if.* Thus the negation refers only to the antecedent. In (25), the negation occurs before the *if,* so it refers to the entire *if . . . then* statement. The statement, "*If* we do *not* have a winning team this year, the manager's contract will not be renewed," would be translated as $\sim W \supset \sim R$. Whereas, the statement, "It is *not* true that *if* we have a winning team this year, the manager's contract will be renewed," would be translated as $\sim(W \supset R)$.

EXERCISE
II
Now use the dictionary as a model for translating the following full statements into symbolic notation.

1. The economy will perform poorly again next year, and the Republicans will have a difficult time at the ballot box. (E, R)

2. This man must either be drunk or have a brain tumor. (D, T)

3. If the economy performs poorly again next year, then the Republicans will have a difficult time at the ballot box. (E, R)

*4. If the economy does not perform poorly again next year, then the Republicans will not have a difficult time at the ballot box. (E, R)

5. Neither the economy nor Republican chances at the ballot box will improve next year. (E, R)

6. Children will be promoted to the next grade only if they pass the essential competency test. (G, C)

7. Passing the essential competency test is a necessary condition for promoting children to the next grade. (C, G)

8. If you don't clean your room, then you don't go to the dance tonight. (R, D)

*9. You will not both study tonight and go to the dance. (S, D)

10. The economy and the Republican chances at the ballot box will both not improve next year. (E, R)

Complex Translations, the Use of Parentheses, and Arguments

We now must consider a method for mapping more complex statements. As we proceed, keep in mind that complex wholes are made up of simple parts. To avoid sensory overload, maintain a mental calmness and discipline and you will see the combination of simple parts masquerading as a perplexing jumble of symbols.

In translating complex statements, we use parentheses in much the same way that we use punctuation in ordinary language. Consider the difference between the following two statements, noting the position of the commas in the originals and the use of parentheses in the translations.

1. If the president implements his tax program, then the deficits will continue to increase and the economy will not improve. (T, D, E)
$$T \supset (D \cdot \sim E)$$

2. If the president implements his tax program then the deficits will continue to increase, and the economy will not improve. (T, D, E)
$$(T \supset D) \cdot \sim E$$

These are very different statements. Statement (1) claims that increasing deficits *and* a poor economy will be consequences of the president's tax program. Statement (2) claims that although the deficits will be a consequence of the president's tax program, the economy will perform poorly regardless of what the president does with taxes.[1] Someone might claim (1) who believed that the president's tax program will be bad for the deficits and the economy; whereas (2) might be claimed by someone who believed that the economy's poor performance will result from circumstances other than the president's tax program. To understand the difference in meaning, we see that (1) is an *if . . . then* statement, whereas (2) is an *and* statement. To map this difference, we use parentheses around D • ~E in (1) to show the ⊃ as the major connective, and parentheses around T ⊃ D in (2) to show the • as the major connective.

Thus, the general rule to remember in translating complex statements is to first identify the major connective, then decide what parts need parentheses to set off the major connective. The ***major connective*** is the distinguishing or basic connective of the statement. For instance, in translating,

3. If a student does <u>not</u> achieve at least a 2.0 grade point average <u>or</u> does <u>not</u> complete at least 50 percent of all credits attempted during a semester, <u>then the</u> student will be on probation. (A, C, P)

we identify the basic statement as a conditional and write down the ⊃ symbol first. We know that an *if . . . then* statement has two parts and the ⊃ symbol has two sides. To keep things as simple as possible and to build up your confidence as you translate, next identify the simplest part and translate it, placing the translation on the appropriate side. In this case, the simplest part is the consequent. Using the letter P, we now have ⊃ P. Next, take the more complex part of the statement and translate it separately, identifying the connective of this part and letters for simple statements. In the example, the minor connective is an *or* statement, so we would have ~A ∨ ~C. Now we combine this with ⊃ P to get ~A ∨ ~C ⊃ P. Finally, to show that the ⊃ is the major connective, we put parentheses around ~A ∨ ~C, so that our final translation is (~A ∨ ~C) ⊃ P.

Without the parentheses, the translation ~A ∨ ~C ⊃ P is ambiguous; it could mean

(~A ∨ ~C) ⊃ P   If a student does not achieve at least a 2.0 grade point average or does not complete at least 50 percent of all credits attempted during a semester, then the student will be on probation.

or

~A ∨ (~C ⊃ P)   Either a student does not achieve at least a 2.0 grade point average, or if a student does not complete at least 50 percent of all credits attempted during a semester then the student will be on probation.

Believe it or not, this last statement is equivalent to "If a student *achieves both* (!?) at least a 2.0 grade point average and does not complete at least 50 percent of

---

[1] Grammatically, the meaning of (2) might be better stated as "The economy will not improve, but if the president implements his tax program then deficits will continue to increase."

all credits attempted during a semester, then the student will be on probation."
Clearly this is not the statement intended.[1]

Before we try another set of exercises, some comments are in order regarding
the translation of negations. How would we translate the following statement?

4. It is impossible for the president to be reelected next year, even though his
   foreign policy record is excellent.

As we have seen in our simple propositional language, the choice of a capital
letter to represent a simple statement is relatively arbitrary. However, if you used a
G, and I used an A for the simple statement "A student achieves a 2.0 grade
point average" we would have a hard time evaluating each other's translation. So,
consistency is a constraint: We need to decide on a letter and maintain the use of
that letter throughout a complex statement or argument. We could decide to translate
(4) as I • F. However, if this statement were part of an argumentative exchange in
which the statement "It is possible for the president to be reelected next year"
occurred, we would now have a problem. If we use I to stand for "It is impossible
for the president to be reelected," I can't use P for "It is possible for the president
to be reelected next year," because the proper opposition would not be captured.
A better approach, and the one we will adopt, is to have a capital letter always
represent a non-negated expression. Thus, statements with expressions such as
impossible, uncommon, and *illegal* are best translated as ~P, ~C, and ~L respectively.
A better translation for (4) would be ~P • F.

However, this does not mean that any statement with a negative connotation
should involve a ~ symbol when translated. For instance, in general, the simple
statement "The economy will perform poorly next year" should not be translated
as ~E because we would then have to use ~~E for the statement "The economy
will not perform poorly next year." Such translations are unnecessarily complicated.
However, there is no absolute rule other than the general goal to be as simple as
possible while capturing the essence of any logical relationships. Sometimes it will
be relatively arbitrary which simple statement will have the ~ symbol when translated.
Given an argumentative exchange with simple statements that contained the words
impartial and *biased,* a decision would definitely have to be made on the use of the
~ symbol to capture the logical relationship. But it's a toss-up whether we translate
the simple statement that contained *impartial* as ~B and the simple statement that
contained *biased* as B or the statement with *impartial* as I and the statement with *biased*
as ~I. The important point is that we would need to agree and then stay consistent.[2]

Next, although this will be rare, there are times when the complexity of statements
requires that in addition to parentheses ( ), we must also use brackets [ ], and
sometimes even braces { }. Suppose someone claimed that (1) on page 243 above
is not true. To translate this claim we would need to negate T ⊃ (D • ~E). In doing

---

[1] If you think this statement is farfetched and would not be asserted by anyone, consider how often professors
read from student writing such statements caused by the lack of proper punctuation.
[2] If the argumentative exchange also involved a simple statement with the phrase *not biased,* then it would be best
to translate the simple statement with *impartial* as ~B. Otherwise, if *impartial* were translated as I and *biased* as
~I, then *not biased* would be translated as ~~I, and this is unnecessarily complicated.

so, we would need to put brackets [ ] around the entire statement and place the ~ symbol outside the brackets as follows:

5. ~[T ⊃ (D • ~E)]

In the United States it is not unusual to hear football announcers exchanging very complicated hypothetical playoff scenarios toward the end of the regular football season. They say things such as,

6. "If Chicago wins a playoff spot only if San Francisco and Los Angeles are both eliminated from the playoffs, then Green Bay is eliminated, provided that Chicago beats Philadelphia by more points than Atlanta did." (C, S, L, G, B)

    C = Chicago wins a playoff spot.
    S = San Francisco wins a playoff spot.
    L = Los Angeles wins a playoff spot.
    G = Green Bay wins a playoff spot.
    B = Chicago beats Philadelphia by more points than Atlanta did.

This statement would be translated as

$$B \supset \{ [C \supset (\sim S \bullet \sim L)] \supset \sim G \}$$

No doubt at this stage this seems like a very difficult translation. But we ought to be able to capture what the average football fan can understand. For practice, see if you can identify and write out the compound parts that make up (6).

Finally, although we have concentrated in this chapter on translating statements, because our goal is to analyze arguments, we must learn how to translate arguments. Consider the following:

7. If a star is 13 billion years old, then the universe cannot be 11 billion years old. This is so, because stars make up galaxies, and galaxies make up the universe. Moreover, if stars make up galaxies, and galaxies make up the universe, then it is not possible both for a star to be 13 billion years old and the universe 11 billion years old.

    T = A star is 13 billion years old.
    E = The universe is 11 billion years old.
    S = Stars make up galaxies.
    G = Galaxies make up the universe.

As in our previous argument-structuring exercises, you should first identify the conclusion. The phrase "This is so because" indicates that the first statement is the conclusion. Next, we identify the number of statements other than the conclusion. Usually, each of these statements indicates a premise. Here is a translation, then, of the above argument: (Note that the symbols (/ ∴) are used to designate the conclusion, which is positioned adjacent to the last premise.)

1. S • G
2. (S • G) ⊃ ~(T • E)   /∴ T ⊃ ~E

As noted previously, do not expect to be an expert at translating right away. You would not expect to be an expert at Japanese or Russian after only one introductory chapter on either of these languages. Continue to practice translating a little at a time while we develop additional symbolic techniques of analysis in the following chapters. In addition to the following exercises, to try your hand at translating additional arguments, see Chapter 8, Exercise IV, and Chapters 9 and 10, translations and formal proofs.

EXERCISE III

The following translations are more complex, requiring parentheses ( ) for punctuation. Hint: Translate each part separately, then recombine into a whole using parentheses to show the major logical connective.

1. If not A, then neither B nor C.

*2. X, only if not both P and Z.

3. P and Z, if not D.

4. Either D or not both P and Z.

5. If Alice is going to the party, then Barbara and Carol will also go. (A, B, C)

6. If Johnson gets the job, then neither Smith nor Kaneshiro will be hired. (J, S, K)

*7. Either Alice is not going to the party or Barbara and Carol are both not going. (A, B, C)

8. The economy will not improve next year, but if the president decides to run again he still has his foreign affairs record to sustain his popularity with the voters. (I, R, F)

9. If the labor contract is renewed at current salary levels, then there will either be an illegal strike or a work slowdown. (R, S, W) Note: S = There will be a legal strike.

10. (cartoon) If I find something in the laundry that shouldn't be there, someone is going to be in big trouble unless it is money. (L, T, M) Note: L = There is something in the laundry that should be there.

11. Syria will recognize Israel only if Israel gives back all the Arab lands it captured in the 1967 Mideast war and accepts past U.N. resolutions. (R, G, A)

12. If the basketball team has an exciting team and a winning team this year, then coach Little's contract will be renewed. (E, W, L)

13. North Korea will not allow inspection of its nuclear research plant unless U. S. nuclear involvement in the peninsula is stopped and international teams carry out simultaneous inspection of South Korean weapons sites. (A, S, C)

*14. If the United States and Israel both don't support Gemayal's Christian government in Lebanon, then Jumblatt will not remain in the Syrian camp for long. (U, I, J)

15. It is not true that having an argument with true premises is a sufficient condition for a valid argument. (T, V)

16. From an article commenting on the little-known Japanese nuclear bomb program in the 1940s: "If Japan considers American hands to have been soiled by the atomic bomb then Japanese hands are equally dirty, although most Japanese are unaware." (S, D, A)

17. Political commentary on the 1992 elections: "The 1992 election choices will be serious if and only if political parties debate plans for escaping depression and militarism, and heed the people who have long been deprived of competent public services." (C, D, M, H)

C = The 1992 election choices will be serious.
D = Political parties debate plans for escaping depression.
M = Political parties debate plans for escaping militarism.
H = Political parties heed the people who have long been deprived of competent public services.
Note: [ ] should be used in this translation.

18. The United States has pledged to withdraw all of its landbased nuclear weapons from the Korean peninsula, although it would still be able to strike North Korean targets with nuclear missiles launched from submarines and will still have a substantial naval presence in Japan. (P, S, N)

P = The United States has pledged to withdraw all of its landbased nuclear weapons from the Korean peninsula.
S = The United States would still be able to strike North Korean targets with nuclear missiles launched from submarines.
N = The United States will still have a substantial naval presence in Japan.

19. An explanation for why the Kansas City Chiefs were out of the football playoffs: The Chiefs beat the 49ers and Buffalo, however, the Raiders beat the 49ers and Buffalo both by more points.

A = The Chiefs beat the 49ers.
B = The Chiefs beat Buffalo.

C = The Raiders beat the 49ers by more points.
D = The Raiders beat Buffalo by more points.

*20. Translate the argument on page 229 of this chapter. Number each premise and remember to indicate the conclusion with the symbols / ∴ . Use these capital letters for the simple statements:

G = God exists.   H = Human suffering exists.
A = Heaven exists.   E = Eternal suffering exists.
B = Hell exists.   C = Human suffering contributes to fulfilling God's purpose.

EXERCISE IV

Some of the following translations require a rephrasing into connectives such as *and, or, if . . . then.*

1. An adequate condition for being excused from the final is having a quiz average better than 90 percent. (F, Q)

2. (Snoopy) "The commanding officer only offers me a root beer when there's a dangerous mission to be flown." (R, D)

*3. (Ecology poster) "Without you, it won't get done." (H = you help, J = the job gets done)

4. There is only one possibility for a negotiated settlement: Konner must resign from the negotiation team. (S, K)

5. Sports announcer Dan Derdorf at the Minnesota Metrodome, during a Bears and Minnesota football game. "It's noisy here even when it is quiet." (Q = quiet, ~Q = noisy)

6. U.S. Rep. Charles Rangel, D-N.Y. commenting on the plight of Haitian refugee treatment: "There's no question, if they were not poor, if they were not black, that we would find some compassion to let these people in." (P, B, I)

7. Australian court decision ruling against a man charged with the rape of his wife. The man had argued that sexual intercourse with his wife was a marital right. "If it was ever the common law that by marriage (a woman) gave irrevocable consent to sexual intercourse with her husband, it is no longer the common law." (C)

*8. Putting political issues aside and rescinding some of the adverse tax legislation produced in 1986 and 1989 is the only way we can avoid being economically killed by the Japanese or enslaved by the new European Common Market. Hint: sometimes *or* means *and.* (P, R, K, E)

9. Should the economy improve and consumer confidence return, the Republicans will do better at the ballot box, provided that they get a handle on the embarrassment of the right wing. (I, C, R, H)

10. Rain and humidity are not uncommon in Hawaii. (R = rain is common in Hawaii, H = humidity is common is Hawaii.)

Writing Exercises.

1. Write a short essay explaining why the statement, "An employee gets a day off during the week if and only if the employee works on a Saturday," is fair to both employees and employer. Hint: Think of the different meanings of $O \equiv W$, $O \supset W$, and $W \supset O$. Remember that $O \equiv W$ is equivalent to $(O \supset W) \cdot (W \supset O)$.

2. If W stands for "We have a winning team this year," and R stands for "The manager's contract will be renewed," write out an English equivalent for each of the following. Then explain how each differs in meaning. What is (1) saying? How does what it says differ from (2)? And so on.

   1. $W \supset R$
   2. $\sim W \supset \sim R$
   3. $\sim(W \supset R)$
   4. $\sim(\sim W \supset \sim R)$

3. Write out an explanation for the difference between (1) $F \supset (C \cdot G)$ and (2) $(F \supset C) \cdot G$.   F = "John passes the final exam," C = "John passes the course," and G = "John's GPA is high enough for eligibility for the dean's list." How would (2) best be expressed in English?

4. Explain in writing why the statement, "It is not true that being female is a sufficient condition for being pregnant," $\sim(F \supset P)$, is not equivalent to the statement, "If a person is not female, then that person is not pregnant," $\sim F \supset \sim P$.

5. If T stands for "An argument has true premises," and V stands for "An argument is valid," write out English equivalents for the following:

   1. $\sim(T \supset V)$
   2. $\sim T \supset \sim V$
   3. $V \supset T$
   4. $\sim(V \supset T)$

   Based on what you have learned from Chapter 1 concerning the concept of validity, which statements are true and which are false? Explain.

ANSWERS TO STARRED EXERCISES:

I. 2. ~A ⊃ B

9. ~Z ≡ Y

II. 4. ~E ⊃ ~R

9. ~(S • D)

III. 2. X ⊃ ~(P • Z)

7. ~A ∨ (~B • ~C)

14. (~U • ~I) ⊃ ~J

20. 1. ~G ⊃ ~(A ∨ B)
    2. G ⊃ (H ⊃ C)
    3. (H • E) ⊃ ~C
    4. B ⊃ (H • E)   /∴ ~B

IV. 3. ~H ⊃ ~J or J ⊃ H

8. (~K • ~E) ⊃ (P • R)

Chapter 8

BIT-BRAINS, LOGICAL CONNECTIVES, AND TRUTH TABLES

Introduction

In the previous chapter we learned symbolic translations for the five key logical connectives *not* (~), *and* (•), *or* (∨), *if . . . then* (⊃), and *if and only if* (≡). In this chapter we will learn our first symbolic method for judging arguments to be valid or invalid. To accomplish this, we must be able to devise a way of capturing symbolically the essential meaning of these words and phrases. As we have seen, logicians have discovered that a key element in deciding whether an argument is valid or invalid is the argument's structure or form. In this chapter we will see that it is the particular configuration of statements, component statements, and logical connectives that determine an argument's structure. Our goal will be to work from the smallest commonsense parts of the way we use our language to analyze more difficult reasoning trails. Thus, we now need to provide rules of usage for these key logical connectives and some method of representing or picturing these rules in our symbolic language. This task is more technologically and philosophically interesting than it may first seem.

By the time we are adults we take much for granted. So much so that there is not near the amount of mystery, apparent magic, excitement, and sense of continual discovery that existed for us when we were children. For instance, we take for granted that we know what the words *table, chalk, eraser,* and *father* refer to. But consider the possible initial difficulty you had with these words when you were a child. Consider the situation of a mother teaching her child what the word *chalk* means. Suppose she holds a piece of chalk in her hands and then, pointing to it, says "chalk" several times. Initially, how does the child know what the mother is referring to? Viewed from the fresh and imaginative perspective of a child, there are a virtually infinite number of *things* the child could identify as the denotative meaning of the strange new word *chalk*. The child might notice that the mother is holding the piece of chalk at a particular angle, that it is being held against a particular background, that the mother is holding her hand in a certain way, that she is gesturing in a strange way; that this new object is being held this way in a particular room, at a particular time, on a particular day; that when she makes this gesture no one else is in the room, or someone else is in the room, or the family dog is in the room, or that there are or are not chairs in the room, or that she has a particular expression on her face. From the open perspective of a child not yet locked into a "correct" view of things, any of these circumstances could mean *chalk*. Furthermore, because the mother may utter the word *chalk* several times when gesturing emphatically with the piece of chalk in her hand, the child may also think that any one of these circumstances may be the referent for the repetitive phrase "chalk . . . chalk"! Eventually, through surprisingly little trial and error and use of the word *chalk* in different contexts, we understand what is actually a very abstract word. Consider, too, the greater difficulty of learning words such as *and* and *or*. At least pieces of chalk are physical objects. With words such as *and* and *or* parents have nothing to point to when trying to teach a child their different meanings. There are no objects or little creatures running around in the world called *and* and *or* that we could trip over.[1] Eventually we understand the difference between saying, "Lisa, go outside to see if the cat *and* the dog are there waiting for dinner," and "Lisa, go outside to see if the cat *or* the dog is there waiting for dinner." What do we eventually understand when we learn to use these logical connectives correctly?

Symbolic Pictures of Logical Connectives: *And, Or,* and *Not*

Let's begin with *not* (~) and statements we call **negations**. If someone were to say "Alice is going to the party" (A), and I later found this statement to be true (T), then I would know the statement "Alice is not going to the party" (~A) to be false (F). On the other hand, if I found out that the statement "Alice is not going to the party" is true, then I would know the statement "Alice is going to the party" to be false. This is how we use *not,* and this will be our rule for *not:* If a statement (A) is true, then the negation of that statement (~A) is false, and if a statement (~A) is true, then (A) is false. This is how we picture this rule symbolically.

A	~A
T	F
F	T

---

[1] Except on the children's educational TV show *Sesame Street,* where children are eased into the abstractness of these words by playful creatures called *and* and *or.* Here they are shown that the creature called *and* has a different function with words than the creature called *or.*

When we use the word *not* correctly this is what we mean: ***the negation of a true statement is false (~T = F), and the negation of a false statement is true (~F = T).*** Because this rule of usage applies to *any* statement using a negation— statements about AIDS, presidential elections, the environment, or the passing of a final exam, not just a statement about Alice going to a party—logicians introduce the technique of using a ***variable*** to summarize the rule. The variables used in logic (p, q, r), play a similar role to the variables (x, y, z) used in algebra. Here is the rule for negation where the small **p** stands for ***any statement whatsoever:***

p	~p
T	F
F	T

Next, we have the connective *and* (•). Statements using this connective you will recall are called ***conjunctions***. If someone said "Alice and Barbara are going to the party" (A • B) and I found that it is true that Alice is going and true that Barbara is going, then I would know this statement to be true. But if I found that although Alice is going, Barbara is not, then I would know the above statement to be false, because the statement claims both are going, and in this case only Alice is going. I would also know the above statement to be false in the case Alice is not going, even though Barbara is, and in the case in which neither Alice nor Barbara are going. This, then, will be the rule for *and*: ***A conjunction (A • B) is true when all the components are true, and false otherwise.*** The following is how we picture this rule symbolically:

A	B	A • B
T	T	T
T	F	F
F	T	F
F	F	F

Furthermore, because this table result would be the same for the statement "Nguyen passed the final exam, and he also passed the course" (F • C), we picture the conjunction rule for any statement whatsoever using variables as follows:

p	q	p • q
T	T	T
T	F	F
F	T	F
F	F	F

To conclude this section, we have the logical connective *or* (∨) and statements called ***disjunctions***. As you recall from the previous chapter, providing a symbolic picture of this connective is complicated by the fact that we find two different usages of this connective in our language. First, there is the **inclusive** sense of *or,* such as when someone says "Alice or Barbara is going to the party," and they mean "Well, I'm not sure. They might both be going, but I think at least either Alice is going or Barbara is going." This sense of *or* means **one or the other, possibly both.** Secondly, there is an **exclusive** sense of *or,* such as the case where a menu at a restaurant states that you may have soup or salad and the intention is clear that you may have **one or the other but not both.** We will choose to picture only the inclusive sense for two reasons. First of all, any statement that has the exclusive sense intended can be pictured using the inclusive symbol plus other symbols such as *not* (~) and *and* (•). For instance, if a menu stated that each customer may have pea soup or salad we could picture this in the following way: (P ∨ S) • ~(P • S), that is, pea soup or salad, but not both pea soup and salad. Second, there is no reason to complicate our translation process by adding another symbol. So, for the sake of convenience we will use the symbol (∨) to stand for the inclusive sense of *or.*

So, if someone said (intending the inclusive sense) "Alice or Barbara is going to the party" (A ∨ B), and I discovered it true that Alice is going and true that Barbara is going, then I would know the above statement (A ∨ B) to be true. The statement claimed that one or the other, possibly both, are going, and both are going. Similarly, if only one of them is going, the statement would still be true. The only time the disjunction would be false would be when neither are going. This, then, will be our rule for disjunction. *An or statement (A ∨ B) is true when either one or both of the simple statement parts are true, and false only when both parts are false.* Accordingly, this is how we picture this rule symbolically, along with its representation using variables:

A	B	A ∨ B		p	q	p ∨ q
T	T	T		T	T	T
T	F	T		T	F	T
F	T	T		F	T	T
F	F	F		F	F	F

We can now use all three of our rules presented thus far to make one symbolic picture or table.

p	q	~p	~q	p • q	p ∨ q
T	T	F	F	T	T
T	F	F	T	F	T
F	T	T	F	F	T
F	F	T	T	F	F

This symbolic picture shows what we eventually understand when we learn to use these abstract words correctly. We know how a computer represents and processes these concepts. Exactly how our brains or minds represent these concepts is a matter of much debate in philosophy. From a practical point of view, the value of this table is that in using it we can mechanically discover the truth value of a statement if we know the truth value of its component parts. We can compute complicated statements by analyzing their simple parts. For instance, if we had the following statement, "We will hire either Johnson or Kaneshiro, but not both," and we discovered it is true that Johnson will be hired and false that Kaneshiro will be hired, then we can mechanically conclude that the statement is true in the following way:

Translation:	$(J \lor K) \cdot \sim(J \cdot K)$	$J = T$
Substitution:	$(T \lor F) \cdot \sim(T \cdot F)$	$K = F$
Result:	$(T \lor F) \cdot \sim(T \cdot F)$	
	$\quad T \quad \cdot \quad \sim(F)$	
	$\quad T \quad \cdot \quad T$	
	$\quad\quad T$	

We could, of course, have known the truth value result of this statement without symbolic logic. Because the above statement is the expression of an exclusive *or* and we were told that Johnson will be hired but Kaneshiro will not, we know that the meaning of the exclusive disjunction is fulfilled (one or the other but not both) and the statement is true. However, the practical value of mapping our common sense in this way is not only to understand the process of analysis, but to be able to compute much more complicated expressions without getting lost. This mechanical method is useful for such complicated examples as the following: "It is neither the case that the Chiefs and the Raiders will both not be participating in the playoffs nor the case that the Patriots will be participating, even though either the Chiefs will not be participating or the Raiders will not be participating." If we discovered that the Chiefs will be participating, the Raiders will not be participating, and the Patriots will be participating, then we can calculate that the above statement is false as follows:

Translation:		$\sim[(\sim C \cdot \sim R) \lor P] \cdot (\sim C \lor \sim R)$	$C = T$
Substitution:		$\sim[(\sim T \cdot \sim F) \lor T] \cdot (\sim T \lor \sim F)$	$R = F$
Result:	(1)	$\sim[(\sim T \cdot \sim F) \lor T] \cdot (\sim T \lor \sim F)$	$P = T$
	(2)	$\sim[(F \cdot T) \lor T] \cdot (F \lor T)$	
	(3)	$\sim[\quad F \quad \lor T] \cdot (\quad T \quad)$	
	(4)	$\sim[\quad\quad T \quad] \cdot (\quad T \quad)$	
	(5)	$\quad\quad F \quad \cdot \quad T$	
	(6)	$\quad\quad F$	

More simply put, this statement says that "The Chiefs or (exclusive) the Raiders will be participating in the playoffs, but the Patriots will definitely not be participating."[1] Because we were told that it is true that the Patriots are participating, we know this statement is false because it says that the Patriots are definitely not going to be in the playoffs. But would you have been able to figure this out using your common sense?

Before we try some exercises, let's focus on some of the mechanical features of this truth value calculation. Notice that step (1) involves simply substituting the

---

[1] Later, we will learn techniques for simplifying such statements.

indicated truth values for C, R, and P. Next, a series of simple calculations are made using the table on page 256, following a principle of working from the "inside (of parentheses) out." So, if you were a computer rather than a human being, you would first scan line (1) and convert all ~Ts to Fs and all ~Fs to Ts, because this is what your program (the table on page 256) tells you to do when you process a ~T or a ~F. This procedure gives us line (2). Next, the computer would analyze each piece of this line that involves a single connective, go to its program, and then transfer a result. For instance, the (F • T) piece in line (2) is converted to F in line (3), because the third row of the table on page 256 is the F–T possibility and under the • column the computer would see that the result is F. Likewise, the (F ∨ T) piece on line (2) would be converted to T, because the computer would see that the F–T case (third row again) produces a T under the ∨ column for this row. Next, the computer would convert the F ∨ T piece in line (3) to T. Notice that nothing happens to the first negation until line (5). Not until the F ∨ T piece in line (3) is converted, can the negation be applied. Not until we have a single T within the brackets in line (4) can the negation be applied, that is, ~[T] = F. Finally, the F • T in (5) is converted to F, the final result and the truth value for this statement, given the input of C = T, R = F, and P = T.

It is important to understand and get a feel for how apparent complexity can be broken down into simple "bits." Essentially, computers are bit-brains; they may work very fast but each step in their thinking is no more than answering "yes" or "no" to a question. Should electricity be allowed to go through a channel or not? This process is identical to what we are doing here: Does the bit we are faced with convert to a T or an F? In order to find the following exercises simple, you must think like a computer, staying calm, focused, and bit-oriented. Although we derived the table on page 256 from our common sense, we can forget for now where this table came from and just follow it like a computer or a Komodo dragon. In other words, don't think! Your entire intellectual effort amounts to no more than looking at this table and staring at the correct row and column combination—to see that T • F = F, that F ∨ F = F, that ~F = T, and so on.

EXERCISE
I

If A, B, and C are true statements, and X, Y, and Z are false statements, which of the following are true and which are false? Use another sheet of paper and follow the format used above: Translation, Substitution, and Result.

1. (C ∨ Z) • (Y ∨ B)

2. (A • B) ∨ (X • Y)

3. ~(B ∨ X) • ~(Y ∨ Z)

4. ~(C ∨ B) ∨ ~(~X • Y)

5. ~B ∨ C

6. ~(B ∨ X)

7. ~X ∨ A

*8. ~(X ∨ Y)

9. ~[(~B ∨ A) ∨ (~A ∨ B)]

10. ~[(~Y ∨ Z) ∨ (~Z ∨ Y)]

11. ~[(~C ∨ Y) ∨ (~Y ∨ C)]

*12. ~[(~X ∨ A) ∨ (~A ∨ X)]

13. ~[A ∨ (B ∨ C)] ∨ [(A ∨ B) ∨ C]

14. ~[X ∨ (Y ∨ Z)] ∨ [(X ∨ Y) ∨ Z]

15. [A • (B ∨ C)] • ~[(A • B) ∨ (A • C)]

*16. ~[X • (~A ∨ Z)] ∨ [(X • ~A) ∨ (X • Z)]

17. ~{[(~A ∨ B) • (~B ∨ A)] • ~[(A • B) ∨ (~A • ~B)]}

18. ~{[(~C ∨ Z) • (~Z ∨ C)] • ~[(C • Z) ∨ (~C • ~Z)]}

19. [A ∨ (B • C)] • ~[(A • B) ∨ (A • C)]

20. [B ∨ (~X • ~A)] • ~[(B ∨ ~X) • (B ∨ ~A)]

Logical Connectives Continued: *If... then* and *If and only if*

Having now come to an understanding of the connectives (~), (•), and (∨), we need to provide a symbolic picture for the usage of (⊃) and (≡). We begin with (⊃), which is called a **conditional**.

Similar to disjunction, providing a symbolic picture of conditional statements is complicated by the fact that we find many different usages of the *if . . . then* phrase in our language. We have seen that necessary and sufficient conditions are translated as conditionals, as well as *only ifs* and straightforward *if . . . thens*. So, let's see if we can come to an intuitive understanding of the core meaning of a conditional by thinking about the following sentence: "If Friday is a nice day, then I am going to the beach (you will not have to come to class)." (N ⊃ B) Suppose I made this unexpected statement on Wednesday. Suppose that, in spite of how unorthodox just taking off on a regular class day would be for a college professor, on Friday it is a nice day and I go to the beach. We would agree that my statement is now true. I said that if Friday was a nice day I would go to the beach, and it was a nice day and I did go to the beach. Now, suppose on Friday the weather is beautiful, but instead of going to the beach I come to class and give a surprise quiz! It is now obvious that my statement is false and that any attempt on my part to claim the statement had become "inoperative" would be met with the justifiable contention that my statement is just plain false.

To summarize what we have agreed upon so far, we see that when the **antecedent** (that part of the statement that follows *if* and translated on the left side of the (⊃) symbol) is true and the **consequent** (that part of the statement that follows *then* and translated on the right side of the (⊃) symbol) is true, then a conditional statement is true. When the antecedent is true, but the consequent is false, then the conditional statement is definitely false.

There are two remaining possible combinations to cover. Suppose now that it is false that the weather is nice on Friday and also false that I go to the beach. I had claimed that if the weather was nice, then I would go to the beach. The weather is not nice and I do not go to the beach. Clearly, my statement is not false and I am not lying as in the previous case. My action is not inconsistent with what I claimed I would do, so we conclude the statement is true. When we think about what a conditional statement claims, it is not surprising to see that a conditional will be true when both the antecedent and the consequent are false.

Finally, we come to the most puzzling case. Suppose it is false that the weather is nice, but I go to the beach anyway. I said that *if* the weather was *nice*, then I would go to the beach. I did not say what I would do if the weather was bad. By going to the beach anyway I did not lie, nor have I acted in any way inconsistent with my statement. It is invalid to infer from my going to the beach that the weather is nice. In using a conditional statement, we are thinking hypothetically, not categorically. A conditional statement does not assert that its antecedent is true, but only that **if the antecedent is true, the consequent is true also.** Furthermore, a conditional statement does not assert that its consequent is true, but only that **its consequent is true, if its antecedent is true.** Hence, we cannot say my statement is false, if the antecedent is not true. So, we conclude that my statement is true. This then will be our full rule for (⊃) in propositional logic: **A conditional statement is false only when the antecedent is true and the consequent is false, and true otherwise.** Here is how we will picture this rule symbolically:

N	B	N ⊃ B		p	q	p ⊃ q
T	T	T		T	T	T
T	F	F		T	F	F
F	T	T		F	T	T
F	F	T		F	F	T

The T that we put for the third possibility is admittedly a "shaky" true. It reveals that we are capturing only a partial meaning common to the use of conditional statements. If we mistakenly thought that this interpretation represented the absolute meaning of all conditional statements, we would be forced to declare as true such statements as "If I sneeze within the next five minutes, then the whole world will disappear"—provided I don't sneeze within the next five minutes! If I don't sneeze within the next five minutes, then the antecedent is false and, based on the above interpretation, the statement is true. Worse, the statement "If I don't sneeze within the next five minutes, then the world will disappear" would also be true—provided I do sneeze within the next five minutes. Worse yet, it follows that the world will disappear within the next five minutes whether I sneeze or not.

The argument   1. S ⊃ D   is valid!

             2. ~S ⊃ D   /∴   D

This result shows that not only are there symbolizations and interpretations of meaning that are parts of more advanced logics, but that there will always be some "holes" in any attempt to completely capture the world with logic.[1] As has been emphasized repeatedly in this book, logic is simply a set of practical tools that can be used when appropriate to provide clarity in a complex situation, to follow a disciplined trail that will help us test our beliefs, and to organize our thoughts for decision making. Logic is a set of *human* tools, and the fact that these tools are useful in many situations should not make us so overconfident that we believe that we are somehow capturing the essence of life or of reality when we are logical. Some modern philosophers believe that past philosophical mistakes were made due to this overconfidence. In Western culture, many historical figures believed that in practicing logic and mathematics they were reading the mind of God. This in turn led to the belief that if one was logical or rational, one was automatically morally superior and had a divine sanction to "educate" and control people of a different culture.

Some twenty-first-century philosophers and logicians accept the admonishment of the philosopher and scientist Pierre Duhem, that "Pure logic is not the only rule of our judgments." For Duhem, human beings are also capable of judgments of "good sense" that are necessarily less rigorous and more vague than the sharp knife of deductive logic. As we saw in Chapter 3, such causal implication claims (claims regarding causal connections) as to the relationship between sneezing and the state of the world or between cigarette smoking and lung cancer must be appraised using the less precise standards of inductive inference. Applying such standards to the above examples, we can say with the relative confidence of good inductive sense that such claims regarding a connection between sneezing and the state of the world are false. All our background knowledge, all our scientific theories about the physical world, and all our historical experience lead us to believe that there is no connection between sneezing and the world disappearing. As with the connection between cigarette smoking and lung cancer we could be wrong—it may be that the world works in such a way that there is no connection between sneezing and the world disappearing until ten minutes from now!—but based on the evidence it is very unlikely that we are wrong. Thus, the F that we put for *if . . . then* statements regarding sneezing and the state of the world may not be as categorical and as certain as the F we put for "If Friday is a nice day, then I am going to the beach" when the antecedent is true and the consequent is false, but our commitment to the false nature of these statements is equally rational. There is a joke that goes something like this: "Just because you're paranoid doesn't mean they AREN'T after you." True, but the possibility that someone is after me does not mean that it would be rational for me to believe that they ARE after me. As ɪ tool, the rules of propositional deductive logic have a particular range of applicability. But just as it would be stupid to use a carpenter's hammer during a medical operation or a surgeon's scalpel to cut a piece of wood, so it would be inappropriate to use the rules of propositional logic for all situations.

---

[1] This result is often called the paradox of **material implication**. Chapter 12 discusses some additional paradoxes and some novel approaches for defining conditional statements.

Finally, to define *if and only if* (≡), a **biconditional**, we could follow up on the discussion from Chapter 7 on the statement "You go out tonight if and only if you clean your room." We could use our commonsense intuition, but because few of us use this phrase much in daily life, a better way is to show the power of what we have accomplished thus far. Because a statement such as "You go out tonight if and only if you clean your room" is actually a combination of two conditionals— "You go out tonight *only if* you clean your room" (O ⊃ R) *and* "*If* you clean your room, *then* you go out" (R ⊃ O)—we can figure out what a biconditional means by analyzing it in terms of what we have already learned about conditionals and conjunctions. We simply have to put together our definitions of (⊃) and (•) to figure out the statement (O ⊃ R) • (R ⊃ O), which is an equivalent way of saying O ≡ R. We can do this by making a table as follows.

O	R	O ⊃ R	R ⊃ O	(O ⊃ R) • (R ⊃ O)
T	T	T	T	T
T	F	F	T	F
F	T	T	F	F
F	F	T	T	T

To summarize: We defined O ⊃ R and R ⊃ O according to the (⊃) rule and then put them together by the (•) rule. This then will be our rule for (≡): *A biconditional statement (O ≡ R) is true only when both components have the same truth value and false otherwise.*

At this point, a little commonsense reflection shows why this is so. When a mother tells her son that he will go out if and only if he cleans his room, the mother's statement would be true in only two cases: when the son cleans his room and the mother fulfills her promise by letting the son go out (row 1 above), and when the son does not clean his room and suffers the consequence of not going out (row 4 above). In the other two cases—when the son does not clean his room and the mother lets him go out anyway (row 2 above), and when the mother does not let her son go out even after he has cleaned his room (row (3) above)—the mother's statement would be false. Further reflection on this case reveals how understanding the difference between *only if* and *if and only if,* and the related difference between a necessary condition, a sufficient condition, and a necessary and sufficient condition is crucial in many communication and negotiation situations, not only in family relationships, but in public and international relations as well. As we saw in Chapter 7, if a mother were to say to her son "You go out tonight only if you clean your room" (O ⊃ R), a son trained in logic would negotiate for a better condition! He would know that his mother only specified a necessary condition for going out, and that his mother would not be lying if after he cleaned his room, his mother then asked him to also do the dishes. The son would want to negotiate for an *if and only if* rather than just an *only if.* In this way, it would be sufficient for him to clean his room and, assuming his mother is consistent, he would go out. Similarly, when Iran held U.S. hostages in the late 1970s, statements from Iranian officials such as "We will release the hostages only if you release Iranian assets held in U.S. banks" (H ⊃ A) would not have been acceptable to United States officials. Like the son's case, the United States would want to negotiate an *if and only if.*

All of this may be a little too much for you to digest at this point, and we will be going over the logical implications like this imbedded in our language much more thoroughly later. But everything that was said in the last paragraph can be summarized in a simple symbolic picture. Like the old adage that a picture is worth a thousand words, symbolic logic attempts to summarize complicated linguistic relationships, just as mathematics attempts to summarize complicated numerical relationships. So, to make a long story short, here is our symbolic picture (our final rule) for a biconditional, a picture that summarizes 362 words, the number of words in the previous paragraph.

p	q	p ≡ q
T	T	T
T	F	F
F	T	F
F	F	T

We can now summarize all our rules again in one table using only variables, reflecting the important point that these rules apply to any statements whatsoever.

Final Truth Table for Logical Connectives:

p	q	~p	~q	p • q	p ∨ q	p ⊃ q	p ≡ q
T	T	F	F	T	T	T	T
T	F	F	T	F	T	F	F
F	T	T	F	F	T	T	F
F	F	T	T	F	F	T	T

EXERCISE II   Repeat the procedure as in the previous exercise: A, B, C = true; X, Y, Z = false.

1. A ⊃ (B ⊃ C)
2. A ⊃ (B ⊃ Z)
3. A ⊃ (Y ⊃ C)
4. A ⊃ (Y ⊃ Z)
5. X ⊃ (B ⊃ C)
6. X ⊃ (B ⊃ Z)

7. X ⊃ (Y ⊃ C)

8. X ⊃ (Y ⊃ Z)

9. (A ⊃ B) ⊃ Z

10. (X ⊃ Y) ⊃ Z

*11. [(X ⊃ Y) ⊃ B] ⊃ Z

12. [(B ⊃ Z) ⊃ B] ⊃ Z

13. [(X ⊃ A) ⊃ X] ⊃ X

14. [X ⊃ (Y ∨ Z)] ⊃ [(X ⊃ Y) ⊃ Z]

15. {[A ⊃ (B ⊃ C)] ≡ ~X} ⊃ {X ⊃ [(A • B) ⊃ C]}

*16. [(A ⊃ Z) • (Z ⊃ A)] ⊃ ~[(A • Z) ∨ (~A ∨ ~Z)]

17. {[X ⊃ (Y ⊃ Z)] ⊃ [(X • Y) ⊃ Z]} ≡ [(X ⊃ A) ⊃ (B ⊃ Y)]

18. (B ⊃ A) ≡ (~A ⊃ ~B)

19. ~(A • B) ≡ (~A ∨ ~B)

20. [A ⊃ (B ⊃ C)] ≡ [(A • B) ⊃ C]

Shortcuts and Human Learning

Before we continue, it is worth reflecting on the differing effects doing the above exercises would have on a simple computer and the average human being. If you did all forty of the above exercises, you probably noticed some shortcuts after awhile. For instance, a close look at the truth table result for (•) shows that any time a part of a conjunction is false the entire statement is false. So, if we had the computation

$$\mathbf{F} • \{[~X ⊃ (~Y ⊃ Z)] ⊃ [(X • Y) ⊃ ~Z]\}$$

we instantly know that the result is false without having to do any computations for the connectives inside the braces on the righthand side. Similarly, because a disjunction is true when any part of the statement is true, we would know instantly that

$$\mathbf{T} ∨ \{[~X ⊃ (Y ⊃ ~Z)] ⊃ [(X • ~Y) ⊃ Z]\}$$

is true without having to do any other computations.

For conditional statements there are two shortcuts. Because the only time a conditional is false is when the antecedent is true and the consequent is false, whenever the antecedent is false or the consequent is true, a conditional is true. At a glance we can see that

$$\mathbf{F} ⊃ \{[~X ⊃ (Y ⊃ Z)] ⊃ [(X • Y) ⊃ ~Z]\}$$

and

$$\{[X \supset (\sim Y \supset \sim Z)] \supset [(X \cdot Y) \supset Z]\} \supset \mathbf{T}$$

are both true.

You may not have noticed these shortcuts, but you would have if you were forced to do hundreds of exercises like the above. Your mind would naturally get bored being a Komodo-dragon-like computer that must do every single computation, step by step. After awhile you would be able to just glance at a problem like 15 in Exercise II and see that the statement is true. Because the antecedent X within the major consequent is false, you would be able to quickly see that

15. $\{[A \supset (B \supset C)] \equiv \sim X\} \supset \{\mathbf{F} \supset [(A \cdot B) \supset C]\}$
    $\{[A \supset (B \supset C)] \equiv \sim X\} \supset \quad \mathbf{T}$
    $\qquad\qquad\qquad \mathbf{T}$

A simple computer programmed to follow our truth table on page 263 would not learn these shortcuts. One of the goals of present-day computer science and efforts in what is called "artificial intelligence," or AI, is to see if more sophisticated and powerful computers can be constructed that can "learn" higher-order connections on their own without human beings programming them to specifically see the shortcuts. The fact that a simple computer that thinks in terms of step-by-step bits cannot learn connections on its own may mean that the human brain is not just a bigger or more complex bit brain, as was once thought by computer scientists. It may mean that the human brain works in terms of networking systemic connections and patterns.

Truth Tables, Validity, and Logical Pictures

We are now ready to learn our first symbolic method for analyzing arguments to be valid or invalid. To accomplish this task, you need to know two things: (1) You need to be proficient in the use of the symbolic picture/table of our commonsense understanding of the logical connectives on page 263; and, (2) most important, you need to remember what a valid argument is—you must understand the concept of deductive validity covered in Chapter 1. Many students become quite good using the table on page 263, but many also do not know what they are doing because they are still struggling with all the ramifications of the concepts of validity and invalidity. Luckily, the **truth table method** that we will be learning in this section will offer not only a technique for analyzing arguments, but a pictorial way of reviewing and reinforcing the concepts of validity and invalidity. The one-picture-is-worth-a-thousand-words adage will again be applicable. With a few simple truth tables we will be able to summarize all the basic points made in Chapter 1.

Let's review. Recall that the technical definition of a valid deductive argument given in Chapter 1 was **an argument that does not allow for the possibility of true premises and a false conclusion**. The key word in this definition for us now is *possibility*. Recall how difficult it was in some of the Chapter 1 exercises to see if it was possible, given the truth of the premises, not to be locked into the truth of the conclusion. A truth table eliminates this uncertainty by painting a picture of an argument's total logical possibilities, showing us directly whether a given argument allows for true premises and a false conclusion or not.

To illustrate this, let's start with a simple argument that we know is valid, an argument in which our common sense is sufficient for seeing that we are locked into the conclusion. Suppose a history teacher told John just before the final exam that he was guaranteed to pass the course (C), if he passed the final exam (P).[1] Suppose John then passed the final exam. Taken as premises, if this information were true, we would be guaranteed that John passed the course. We would have a valid argument as follows:

1. $P \supset C$
2. P $/\therefore$ C

A truth table proof that this is a valid argument would look like this:

TABLE T1	**P**	**C**	**P $\supset$ C**	**P**	**C**
	T	T	T	T	T
	T	F	F	T	F
	F	T	T	F	T
	F	F	T	F	F

Constructing this table involves a number of simple steps.

1. Decide on the possible logical combinations of true and false, given the number of simple statements. Because we have two simple statements (P, C), we have four possible combinations. The two columns on the left side show the combinations. P could be true and C could be true (row 1). P could be true and C could be false (row 2). P could be false and C could be true (row 3). And P and C could both be false (row 4).
2. Make a column for each premise and the conclusion.
3. Figure out the truth value for each row in each column by using the given possibilities decided in 1 and the table on page 263. (For instance, row 2 under the premise P $\supset$ C is false because P is true in that row and C is false, and our table on page 263 dictates that when the antecedent of a conditional is true and the consequent is false, the conditional is false.)
4. After completion of the table, look across each row under the premises and conclusion and see if *any row* has *all true premises* and a *false conclusion*. If not, the argument is valid. If there is at least one case of all true premises and a false conclusion, then the argument is invalid.

---

[1] We will use a capital P to translate "John passed the final exam" so as not to confuse our translation with the F that stands for false.

In looking at the table we have constructed for the above argument, we see that no row contains a case of all true premises and a false conclusion. This shows that this argument will not allow (given any possibility) for true premises and a false conclusion. Thus, it is a valid argument. However, it is important to understand and remember that valid arguments will allow for false premises (rows 2, 3, and 4) and false conclusions (rows 2 and 4). Also note that if the conclusion of a valid argument is false, at least one of the premises is false (rows 2 and 4). Remember that judging an argument to be valid is not to judge the specific content of an argument, but only the reasoning trail. So, if John did not pass the course (rows 2 and 4), we would know only that either his teacher did not tell him the truth (row 2) or John did not pass the final (row 4). But this possible content outcome would not change the validity of the reasoning.

Understand that row 1 by itself (all true premises and a true conclusion) does *not* show validity. As we will see shortly, and as was emphasized in Chapter 1, invalid arguments can also have true premises and a true conclusion. What shows the argument above to be valid is that inspection of *all* the rows reveals no case with all true premises and a false conclusion. The first row contains the ideal case of a sound argument, but only because we know the argument is valid. Validity must be determined before soundness can be determined.

To reinforce these points, let's look at a truth table of an invalid argument. Suppose our history teacher made the same statement to the effect that if John passed the final exam, he would pass the course. Suppose John's friend finds out that he did not pass the final exam, and infers from this that John must not have passed the course. This would be an invalid inference, because the history teacher did not say that passing the final exam was the only way that John could pass the course. The history teacher said that if he passed the final exam, he would be guaranteed of passing the course—which left open the possibility that there might be other ways of passing the course. He might have failed the final exam, but obtained a sufficiently high score on the final that, combined with his other grades (other exams and quizzes), he just made a passing score for the course. Or, an extra credit option might have allowed him to get enough points to pass the course. Neither of these possibilities are inconsistent with the history teacher's statement to John. The history teacher was telling John what would guarantee passing the course, not telling him the only way to pass the course.

To show that this argument is invalid and to reinforce the conceptional difference with a symbolic picture between validity and invalidity, here is a truth table for this invalid inference.

TABLE T2 1. $P \supset C$
2. $\sim P$ $\quad / \therefore \quad \sim C$

P	**C**	**P ⊃ C**	**~P**	**~C**	
T	T	T	F	F	
T	F	F	F	T	
F	T	T	T	F	****
F	F	T	T	T	

Notice the third row. Unlike our first argument, this argument does allow for a possible case of all true premises and a false conclusion. It also allows for all true premises and a true conclusion (row 4). This table underscores pictorially the important point made about invalid inferences in Chapter 1. Valid inferences are preferred over invalid ones not because valid arguments always have true conclusions or because invalid arguments always have false conclusions, but because when we have all true premises and a valid argument we are guaranteed a true conclusion. Whereas even if we have all true premises in an invalid argument, we are guaranteed nothing, the conclusion could be true or false. We want our effort in finding reliable premises and thinking about what follows from them to count. If our premises are true and the reasoning valid, then we are guaranteed of learning something in the conclusion. Even if we find the conclusion of a valid argument to be false, we still learn something important—we learn that at least one of our premises is false. Remember, we test our beliefs with valid reasoning. With invalid reasoning we learn nothing; we do not test our premises when it is discovered that the conclusion is false.

The above cases are models for constructing a truth table for any argument. However, a truth table would not have been needed in either case—our common sense would have been sufficient for judging these arguments. A truth table serves a more useful purpose for arguments such as the following:

> If we pay our medical bills and buy the new car, then we will not have enough money for basic necessities. We have to pay our medical bills. So, if we don't buy the car, then we will have enough money for basic necessities. (M, C, B)

Perhaps you can tell that something does not seem quite right with the conclusion of this argument. But are you sure it is an invalid argument? To be sure, let's do a truth table. First, the translation:

1. $(M \cdot C) \supset \sim B$
2. M $\qquad$ $/ \therefore$ $\sim C \supset B$

Note that in following the steps for truth table construction listed above, the first step of determining all of the logically possible combinations given M, C, and B will now be more complicated. There are now three simple statements that could be true or false, so our table will have to be longer and involve more rows down. We could use our common sense to do this: We could start with M, C, and B all true, then M and C true, and B false, and so on until we ended with M, C, and B all false. However, you should be able to see that in very complicated arguments in which four, five, or six or more simple statements could be true or false, figuring out the possible combinations would be a nightmare. It would be difficult to keep track of which combinations we already had and which ones we still required to cover all the logical combinations.

It is in cases such as this that mathematics comes to our rescue. As noted earlier, mathematics has been built up through the centuries to make life easier for us. In our situation, mathematicians have discovered a little equation that will enable us to mechanically and effortlessly figure out the logical combinations given any number of simple statements: $2^y = n$. Like all mathematical equations, it looks intimidating until you are told what it means. To use it, you need only be able to multiply by two. The **y** stands for the number of simple statements, and the **n** stands for the number of rows down needed to cover all the logical combinations of true and false, given **y** (the number of simple statements). In our case, we have three simple statements. Plugging this into the equation, we have $2^3 = 8$. $(2^3 = 2 \times 2 \times 2 = 8)$ Think how much time this equation would save if we had a truth table with six simple statements! Using the equation we would know instantly that the truth table would need sixty-four lines down to cover all the logical possibilities. It would probably take at least several hours to figure this out with our common sense, and then more time to make sure that we did not make a mistake in repeating a line of possibilities or omitting one.

However, this equation does not tell us how to arrange the various Ts and Fs; it only tells us with certainty how many rows down we need to cover all the possibilities. But arranging the Ts and Fs also is simple. We make the first column (the first simple statement), in our case M, half Ts and half Fs. So, if we have eight lines down, the first four rows will be true and the second four rows will be false. Then we take the next column, in our case C, and make it half (in a sense) of what M was. So, because M was four Ts and four Fs, C will be two Ts and two Fs all the way down. The next column, B, will then be half of C's distribution, one T and one F all the way down. Here is what this will look like symbolically.

M	C	B
T	T	T
T	T	F
T	F	T
T	F	F
F	T	T
F	T	F
F	F	T
F	F	F

Three simple statements = 8 rows.
M = 4 Ts and 4 Fs.
C = 2 Ts and 2 Fs.
B = 1 T and 1 F.

Note the elegance of this process (mathematicians and logicians love elegant processes). Without thinking very much, we have covered all the logical combinations of true and false for three simple statements. The first row has all true and the last row has all false, and between these two rows every other combination is covered. This combination of simple equation and assignment of truth values will work for any number of simple statements. If we had four simple statements, we would need sixteen rows. The first simple statement would be assigned eight Ts and eight Fs, the second simple statement would be assigned four Ts and four Fs all the way down, the third simple statement would be assigned two Ts and two Fs all the way down, and finally, the last simple statement would be assigned alternating Ts and Fs all the way down. As in the case of three simple statements, we would have all Ts in the first row and all Fs in the last row, and every other combination in between. (Take a break here and try this. Make up four letters and set up a table of logical combinations of true and false to verify what I have just said.)

We can now finish steps 2–4 listed on page 266 for constructing a complete truth table. It would look like the following.

TABLE T3

M	C	B	(M • C) ⊃ ~B	M	~C ⊃ B	
T	T	T	F	T	T	
T	T	F	T	T	T	
T	F	T	T	T	T	
T	F	F	T	T	F	****
F	T	T	T	F	T	
F	T	F	T	F	T	
F	F	T	T	F	T	
F	F	F	T	F	F	

Notice that this argument allows for all true premises and a false conclusion in the fourth row, verifying the suspicion that the conclusion seemed strange. What should have been concluded from the above premises is that a new car should not be purchased, if enough money is to be had for basic necessities (B ⊃ ~C), or that buying a new car will guarantee not having enough money for basic necessities (C ⊃ ~B). To say that buying the new car will guarantee not having enough money for basic necessities is not the same as saying that not buying the new car will guarantee having enough money for basic necessities. To show that this is so, and to illustrate a more piecemeal method for constructing more complicated truth tables, let's show that concluding B ⊃ ~C would give us a valid argument as follows.

TABLE T4

M	C	B	(M • C)	⊃ (p1)	~B	M (p2)	B	⊃ (c)	~C
T	T	T	T	**F**	F	**T**	T	**F**	F
T	T	F	T	**T**	T	**T**	F	**T**	F
T	F	T	F	**T**	F	**T**	T	**T**	T
T	F	F	F	**T**	T	**T**	F	**T**	T
F	T	T	F	**T**	F	**F**	T	**F**	F
F	T	F	F	**T**	T	**F**	F	**T**	F
F	F	T	F	**T**	F	**F**	T	**T**	T
F	F	F	F	**T**	T	**F**	F	**T**	T

First note that in constructing this table, the first premise and the conclusion are broken down into parts and each part is given a column. The parts are then combined using the major connective to arrive at the final truth value. For instance, in premise 1 in the first row, the antecedent M • C is determined to be true and the consequent ~B is determined to be false. These results are then combined by (⊃) to get false. In row 1 of the conclusion, B is true and ~C is false. This is combined by (⊃) to determine false. The remaining rows are determined in a similar fashion. The value of this more piecemeal approach is to do everything step by step. The disadvantage is that when finished there are so many columns of true and false combinations, that it would be easy to get our eyes crossed and not look at the right results to determine if the argument allows for any case with all true premises and a false conclusion. To remedy this, the key results are in boldface print and a **p1** (for premise 1), **p2** (for premise 2), and **c** (for the conclusion) are placed above the key column results.

Upon inspecting just these columns, we see that this argument does not allow for any row with all true premises and a false conclusion. In the only rows that have a false conclusion (rows 1 and 5), at least one of the premises is false in these rows. Hence, the argument is valid.

Now it is time for you to construct your own truth tables. Use the four examples above as models to do the following exercises.

EXERCISE III

Construct a truth table for each of the following arguments, proving that each is either valid or invalid. If the argument is invalid, circle the key row that shows all true premises and a false conclusion. Do not circle any row if the argument shows valid. Do you know why?

1. A ⊃ B
   B   / ∴   A

2. A ⊃ B
   C ⊃ B   / ∴   A ⊃ C

*3. (A ∨ ~B) ⊃ ~C
    A ≡ B   / ∴ ~C

4. X ⊃ (~Y ⊃ Z)
   Y ∨ X
   ~Y   / ∴ Z • ~Y

5. (G ⊃ X) • (Z ⊃ A)
   ~X ∨ ~A   / ∴   G ≡ ~Z

EXERCISE IV

Translate each of the following arguments into symbolic notation. After checking with your instructor for the correct translations, construct truth tables and prove validity or invalidity.

1. This man is either intoxicated or has diabetes. We know he does have diabetes. So, he must not be intoxicated. (I, D; to provide the correct translation you must decide whether the *or* in the first premise is inclusive or exclusive.)

2. My wife's phone at work is busy. John's phone is also busy. If they are talking to each other, then both phones are busy. Therefore, my wife and John must be conversing over the phone. (W, J, T)

3. If God exists, then He is all-powerful. If He exists and is all-powerful, then the universe is elegantly ordered. Because the universe is elegantly ordered, this shows that God exists. (G, P, U)

4. Only if Jesus was more than a man are Christians right that he made significant statements. Jesus rose from the dead, and this is sufficient to know that he was more than a man. This proves

that Christians are right that Jesus made significant statements. (M, S, R)

5. If I don't qualify for the job, then you don't qualify; whereas, if you don't qualify, then I don't qualify. If both of us don't qualify, then Renee will get the job. So, if either of us does not qualify, Renee will get the job. (I = I qualify; Y = you qualify; R = Renee will get the job)

*6. An old Volkswagen ad featured a photo of a street of identical houses. In front of each home is parked a red and white Volkswagen station wagon. Under the photo is this message:

If the world looked like this, and you wanted to buy a car that sticks out a little, you wouldn't buy a Volkswagen station wagon. But in case you haven't noticed, the world does not look like this. So, if you want a car that sticks out a little, you know just what to do.

Identify the premises and conclusion of this argument and then use W = The world looks like this; S = You want a car that sticks out a little; V = You will buy a Volkswagen station wagon. (What do you think the phrase "you know just what to do" means?)

7. A basketball player named Johnson is suspended and fined for not showing the proper respect to his coach. His union asks for a ruling and an arbitration board rules that although the fine was justified, the suspension was not because Johnson was having "emotional problems" due to a marital separation. The coach responded that the ruling was inconsistent; that if Johnson was responsible for his action, then the fine and the suspension should be justified, and if he was not responsible, then neither should be justified. Here is a formalization of the coach's reasoning:

The view of the arbitration board that the fine was justified but the suspension was not is mistaken, for the following reasons: Either Johnson was or was not responsible. If he was, then the fine and the suspension were both justified. If he was not responsible, then neither the fine nor the suspension was justified. (R, F, S)

8. Presidential blues or invalid reasoning?

If Clinton comes completely clean on the Whitewater investigation, then he must fire his White House counsel. If he evades the press on this issue, the Republicans will continue to gain political points against his presidency. If he comes completely clean on the Whitewater investigation, then he will not evade the press on this issue. So, if Clinton fires his White House counsel, the Republicans will not continue to gain political points against his presidency. (C, F, E, P)

C = Clinton comes completely clean on the Whitewater investigation.

F = Clinton fires his White House counsel.
E = Clinton evades the press on the Whitewater issue.
P = The Republicans will continue to gain political points against his presidency.

9. Often philosophers note that the concept of a reliable belief and the success of science are tied to a belief about how the natural world behaves, that there are uniformities in nature, that certain relationships are repeated and can be recognized if we pay attention. This argument discusses these connections.

It is not true both that human beings can make predictions about the future and not achieve reliable beliefs about how the natural world works. This is so, because if there are uniformities in nature, then human beings can achieve reliable beliefs about how the natural world works. Moreover, human beings can make predictions about the future only if there are uniformities of nature. (P, R, U)

P = Human beings can make predictions about the future.
R = Human beings can achieve reliable beliefs about how the natural world works.
U = There are uniformities in nature.

10. For scientific creationists, one of the grave implications of Darwin's theory of evolution is that there is no guarantee that human beings are special in the universe. They are very concerned about this, because they presuppose that it is not possible to have objective moral values and standards unless human beings are special. If this presupposition is true, then it would follow that if Darwin's theory is true it would not be possible to have objective moral values. But is this presupposition true? Analyze the following argument.

If scientific creationism is true, then human beings are guaranteed to be special creatures in the universe and there is a source for objective values. If Darwin's theory of evolution is true, then human beings are not guaranteed to be special creatures in the universe. Because scientific creationism and Darwin's theory of evolution cannot both be true, it follows that there is no source for objective values unless human beings are special creatures in the universe. (C, S, O, D)

C = Scientific creationism is true.
S = Human beings are guaranteed to be special creatures in the universe.
O = There is a source for objective values.
D = Darwin's theory of evolution is true.

Argument Forms and Variables

Throughout this book we have emphasized the importance of concepts. Many concepts may be abstract, but our ability to abstract is responsible for our success on this planet as a species and also forms a large part of what it means to say that human beings possess "intelligence." When you understand a concept, you have a net, so to speak, to throw over, control, and understand countless details. When a child understands the concept of a chair, he or she understands how to use any of a number of objects that we call *chair,* no matter how they differ in individual details. By grasping concepts we are able to see the forest and not get lost within the trees. The most important part of your academic life, no matter what your major, should consist of learning concepts and their interconnections, not the details of dates, names, and events. The latter are easily forgotten but just as easily looked up in books or computer data bases. Concepts save time and give you a lot of power. With insight into concepts apparent complexity is seen to be simplicity in disguise.

As we have seen, logicians take this abstracting ability of ours very seriously. They slow our thinking down and identify patterns of thought. In the chapters on informal fallacies we saw that once we identified the argument structures or forms of fallacies we had a pattern that could be filled in by any number of details and issues. In the remaining part of this chapter, and especially in the chapters that follow, we will be taking this form-recognition process one step further by expanding considerably the applicability of the notion of a ***variable***—our use of **p**s and **q**s. Consider the following argument:

> **EXAMPLE A1**   If we pay our medical bills and buy the new car, then we will not have enough money for basic necessities this month. We will pay our medical bills this month, but we are also going to buy the new car. So, we will not have enough money for basic necessities this month. (M, C, B)

The translation of this argument would be:

1. $(M \cdot C) \supset \sim B$
2. $M \cdot C$ $/\therefore$ $\sim B$

This argument is obviously valid but if we had to do a truth table, we would have to construct a time-consuming table of eight rows. There is a much faster way. We instantly recognize the validity of this argument without doing a truth table because it has a form or pattern of reasoning that is as common as the concept of a chair. Recall that whenever we assert a conditional statement, we are claiming that if the antecedent is true, then the consequent must be true. So, if we assert a conditional statement as part of an argument and then find out that the antecedent is true, we are locked into asserting that the consequent is true as a conclusion. The above argument has the same form as:

> **EXAMPLE A2**   If John passes the final, he will pass the course. John passed the final. So, John will pass the course. (P, C)

1. P ⊃ C
2. P   /∴ C

Just as we used the logical variables **p** and **q** to capture the essence of our use of logical connectives, just as we used these variables in our table on page 263 to represent any statements whatsoever, so we can use variables to capture argument forms. What both of the above arguments have in common is two premises, the first being a conditional, the second repeating the antecedent of the conditional, and the conclusion repeating the consequent of the conditional. We can picture this commonality with the argument form:

1. p ⊃ q
2. p   /∴   q

We know intuitively that this is a valid argument, and a simple truth table of this argument form corroborates our intuition:

p	q	p ⊃ q	p	q	
T	T	T	T	T	Valid, no row with true
T	F	F	T	F	premises and a false conclusion.
F	T	T	F	T	
F	F	T	F	F	

Thus, we do not need to construct a time-consuming truth table of Example A1, but only its much simpler argument form.[1] In general, if a table is needed to judge the validity or invalidity of a complex argument, we can always construct a table of an argument's form. For instance, if we had the complex translated argument—

EXAMPLE A3   1. {[(A ⊃ B) ≡ (C ∨ ~B)] • ~D} ⊃ (~E ∨ ~F)
2. {[(A ⊃ B) ≡ (C ∨ ~B)] • ~D}   /∴   (~E ∨ ~F)

a truth table that ignores the argument form would require sixty-four lines! But if we look carefully we see that this argument is just another example of the same form identified above and so would require only four lines. Apparent complexity is often simplicity in disguise. Here are some more examples of arguments and their much simpler argument forms.

EXAMPLE A4   1. (P • C) ⊃ D
2. ~D   /∴   ~(P • C)

1. p ⊃ q
2. ~q   /∴   ~p

---

[1] Actually, as we will see in the next chapter, it is not necessary to construct a table at all because this form is already known to be valid. So, any argument that fits this form is automatically known to be valid.

EXAMPLE A5
1. (H ∨ ~T) ⊃ (S ⊃ P)
2. ~D ⊃ (S ⊃ P)   /∴   (H ∨ ~T) ⊃ ~D

1. p ⊃ q
2. r ⊃ q   /∴   p ⊃ r

EXAMPLE A6
1. (A ∨ B) ⊃ (~C • ~D)
2. (~C • ~D) ⊃ (H ∨ ~T) /∴   (A ∨ B) ⊃ (H ∨ ~T)

1. p ⊃ q
2. q ⊃ r / p ⊃ r

A truth table based on the argument form of Example A4 would require only four rows and show this argument to be valid. A truth table of Example A5 would require only eight rows and show this argument to be invalid. A truth table of Example A6 would also require only eight rows and show this argument to be valid.

EXERCISE V   Identify the argument form for each of the following arguments.

1. P ⊃ (S ∨ ~T)
   ~D ⊃ (S ∨ ~T)   /∴   P ⊃ ~D

2. (P • T) ⊃ (S ∨ ~Y)
   P • T   /∴   S ∨ ~Y

3. (X ∨ Y) ⊃ H
   ~H   /∴   ~(X ∨ Y)

*4. ~(S • G) ⊃ (T ⊃ Y)
   (T ⊃ Y) ⊃ ~P   /∴   ~(S • G) ⊃ ~P

5. (A • B) ∨ (Y ⊃ ~T)
   ~(A • B)   /∴   Y ⊃ ~T

6. [(A • B) ⊃ ~X] • [G ⊃ (S ∨ T)]
   (A • B) ∨ G   /∴   ~X ∨ (S ∨ T)

*7. (N ⊃ T) • (S ⊃ Y)
   /∴   N ⊃ T

8. C ≡ (B • D)
   /∴   [C ⊃ (B • D)] • [(B • D) ⊃ C]

9. (A • B) ⊃ (H ∨ Y)
   ~(A • B)   /∴   ~(H ∨ Y)

10. (~A ∨ ~B) ⊃ (S ≡ H)
    S ≡ H   /∴   ~A ∨ ~B

Brief Truth Tables

As we have seen, a concept that plays a major role in logic is validity. Thus, we would expect a considerable amount of saved effort in constructing truth tables by a thorough understanding of this concept. Recall that the technical definition of validity is an argument that does not allow for any possibility of true premises and a false conclusion. Most arguments involve conclusions that are **contingent** statements. That is, they are statements that are true or false depending on the state of affairs in the world and thus their truth table will show a mix of true and false results given the different rows of possibilities. We can take advantage of this by realizing that to judge an argument by the truth table method we need only to check the rows in which the conclusion is false. If we are always looking for cases of true premises and a false conclusion, then cases where the conclusion is true cannot show invalidity.

Consider the following truth table where the result of the conclusion is constructed first.

					p1		p2	c	
A	**B**	**C**	**(A • B)**	**⊃**	**~C**	**~A**	**B ⊃ C**	
T	T	T					T	
T	T	F				F	F	
T	F	T					T	
T	F	F					T	
F	T	T					T	
F	T	F	F	T	T	T	F	****
F	F	T					T	
F	F	F					T	

TABLE T5

The conclusion is a **contingent statement**; its truth table result shows a mix of true and false results that depend on what might be the case in the world. Only in the second row and the sixth row (B = T and C = F) will the conclusion be false. Thus, these are the only possible rows that could show the defective situation of having all true premises and a false conclusion. In working backward starting with the second row we quickly see that the second premise is false. We need go no further regarding this row, because we are looking for a situation with *all* true premises and a false conclusion. Although we have a false conclusion in this row, we can't have all true premises because the second premise is false. However, working backward from the conclusion in the sixth row we find that all the premises in this row are true. Hence, the argument is invalid. If we were to have done this row first, this would have been the only row necessary to construct! Because once we find a row with all true premises and a false conclusion, we know that the entire argument is weak; the argument does not guarantee a true conclusion whenever the premises are true.

To reinforce this shortcut insight, consider the truth table result for 7, Exercise IV. Your table could have looked like this.

TABLE T6

R F S	p1 R ∨ ~R	p2 R ⊃ (F • S)	p3 ~R ⊃ ~ (F ∨ S)	c ~ (F • ~S)
T T T				T T F F
T T F		T F F	F T	F T T T
T F T				T F F F
T F F				T F F T
F T T				T T F F
F T F			T F F(T)	F T T T
F F T				T F F F
F F F				T F F T

Only the second and sixth rows show a false conclusion. Because these are the only rows we need to check, working backward in the second row we find that although the third premise is true (the antecedent ~R is false, so the conditional is automatically true), the second premise is false. Thus, we need go no further on this row; we have discovered that this row cannot show all true premises and a false conclusion. Checking the sixth row, we find that the third premise is false, so we need work no further on this row. Because these are the only rows that had a chance of showing all true premises and a false conclusion, and we discovered that the premises cannot be all true when the conclusion is false, we have demonstrated that this argument is valid: It does not allow for any possibility of all true premises and a false conclusion. Recognizing the full implications of the concept of validity as applied to truth tables saves us a considerable amount of unnecessary busywork. It is unnecessary to work out the details of the remaining rows.

Incidently, your truth table result for the first premise (R ∨ ~R) should have shown that this premise is not a contingent statement because it is never false; there is no row where this premise is false. Statements that are always true, regardless of the possibilities in the world, are called **tautologies**. Such statements provide ultimate security, but at a price: They don't say anything significant about the world and any premise or combination of premises imply them. If I were to tell you that the color of my car is black or it is not black, I would be telling you the truth no matter what the color of my car, but I would not be conveying any information to you about the color of my car. Furthermore, tautologies are useless as conclusions, because they make any argument valid, regardless of the premises. The above statement about my car would follow (be the conclusion of a valid argument) given the premise, "The moon is made of green cheese." It would also follow validly from the premise, "The moon is not made of green cheese." Because arguments are invalid if and only if they allow for a case of all true premises and a false conclusion, an argument that never has a case of a false conclusion is valid by default.

A similar problem occurs with statements that are called **contradictions**. Contradictions are statements that are always false regardless of the situation in the world. If the first premise of the above argument was ~(R ∨ ~R), its truth table result would show F in every row. Given such a premise, an argument is also valid by default, regardless of the nature of the other premises or the conclusion. This is so because with a premise that is always false, it is impossible to have a case with all true premises—a necessary condition for invalidity. So, just as any conceivable premise or set of premises imply a tautology, contradictions (as premises) imply any conceivable conclusion. With a contradiction as a premise, the statements "The moon is made of green cheese" and "The moon is not made of green cheese" could both be conclusions of valid arguments. Although Eastern philosophers have generally believed that contradictions are valuable in understanding the true nature of reality, Western philosophers have generally believed that contradictions should be avoided like the plague, not only because we wish to make true statements but also because they "prove" everything to be true.[1]

So, arguments that involve contradictory premises, such as A and ~A, would be valid by default, but useless. It should be apparent from this why logicians are so concerned with consistency. If our beliefs are inconsistent, they cannot all be true. In addition, they imply not only what we want them to imply—other beliefs that we have—but also the beliefs we think are not true. If I believed that the fetus is a human being and should be granted all the rights of citizenship, and I also believed that the fetus is not a human being and thus should not be granted the rights of citizenship, then my contradictory beliefs imply the conclusions of both positions on abortion (pro-choice and pro-life), as well as that the moon is made of green cheese! Contradictions don't get us anywhere in the belief reliability game, precisely because they get us everywhere, so to speak.

Thus, from a logical point of view a large part of personal growth involves becoming more rational by recognizing that each of us has many inconsistent beliefs, and then making some attempt to resolve the inconsistency. A large part of mature reflective thinking, of philosophical and logical analysis, involves looking for and resolving hidden contradictions. From a scientific point of view, our goal should be to have noncontradictory contingent statements as beliefs and then investigate the world empirically to see which ones are true. Simplified, life and the goal of finding reliable beliefs can be seen as a combination of **conceptual** and **empirical problems**. If our concepts are not consistent, we have a conceptual problem. For instance, in the mid-1990s the Hubble telescope provided evidence that the universe was younger than the stars in some star clusters. Our best theory of how the universe began seemed conceptually inconsistent, because we had thought that all stars evolved after the universe began. On the other hand, if our beliefs are not consistent with what experience continues to tell us, we have an empirical problem. If I believe that the world should end at 12:00 a.m., January 1, 2000, then I have an empirical problem if I am still experiencing the world going about its normal business on January 2, 2000. Recognizing and solving problems is an inevitable part of life. And logical analysis, both for recognizing hidden contradictions and for testing beliefs against experience, plays a major role.

---

[1] *Prove* here means of course in terms of validity, not soundness. Arguments with a contradiction as a premise can never be sound. For what Eastern philosophers have to say about contradictions, see Chapter 12 on fuzzy logic.

To return now to truth-table analysis, by carrying out the validity shortcut insight further, we see that we don't even need to construct a table of possibilities at all! If an argument can only show invalidity when the conclusion is false, why not "make" the conclusion false and then see if the argument we are checking can consistently have true premises. If it can, we know the argument is invalid; if it cannot, we know the argument is valid. As an illustration of this, consider the following argument and brief proof of invalidity.

TABLE T7	**T**	1. A ⊃ B				A = T
		T T				B = T
						C = F
		2. C ⊃ B	/∴ A ⊃ C			**Invalid**
	T	F T	T F		
			F		

We have shown this argument to be invalid, because we can make the conclusion false and then consistently make the premises all true. Here are the steps in the construction of brief truth tables.

1. Start with the conclusion. *Make the conclusion false.* (Because the above argument has a conditional for its conclusion, we assign a T to the antecedent and an F to the consequent.)
2. *Substitute the assigned truth values consistently.* Whatever truth assignments were made in the conclusion must also be made in the premises. (We are committed to A being T, so A must also be T in the first premise. C is F in the conclusion, so C must be F in the second premise.)
3. *Try to make all the premises true.* Attempt to assign uncommitted values in such a way to make the premises true. (Because C is F in the second premise, we are free to put either T or F for B to make this premise true. We choose T because we are not free to put an F for B in the first premise. B must be T in the first premise, and we are free to choose T because C is F in the second premise.)

When a result of true premises and a false conclusion is created, we have essentially isolated the row in a long table that would show the argument to be invalid. However, for valid arguments we will not be successful in making the premises all true when we make the conclusion false, because valid arguments will not allow for this; if the conclusion is false, at least one of the premises will be false. See what happens when we try to show the following argument to be invalid.

TABLE T8	**T**	1. A ⊃ B				A = T
		T T				B = T
						C = F
		2. B ⊃ C	/∴ A ⊃ C			**Not invalid**
	F	T F	T F		
			F		

OR

F 1. A ⊃ B   A = T
     T   F   B = T
                       C = F

2. B ⊃ C   /∴ A ⊃ C   **Not invalid**
 T   F   F   T   F
             F

In attempting to make the first premise true, we must assign T to B. But this makes the second premise false. If we started with the second premise and made B = F, this would make the second premise true, but it would then make the first premise false. In all valid arguments, there will be a similar flip-flop in the premises. In making one premise true, another premise will be made false. Each attempt by itself does not show the argument to be valid, because we must check all possibilities. Each attempt only shows that we were not successful in creating the case of all true premises and a false conclusion. However, because B must be either T or F, in combining both cases we have covered all the possibilities and have shown that this argument must have at least one premise false, if the conclusion is false. Hence, it is valid.

Before trying some exercises, it should be noted that as nice as this brief method is, there are limitations. The brief truth table method works best when there is only one way to make the conclusion false. If an argument has a conclusion in which there are several ways to make the conclusion false, it would be a hasty conclusion to assume validity, if in checking for true premises given only one way of making the conclusion false, not all the premises could be made true. Because an assessment can be made only after all the possibilities have been checked, each assignment that would make the conclusion false must be checked. It is possible that although one assignment of truth values that makes the conclusion false will not result in all true premises, the next assignment will.

So, the brief truth table method works best with single statements or their negations (A, ~A), or conditional or disjunctive statements as conclusions, because only one way exists of making such statements false. Because biconditional statements can be made false in two ways, and conjunctions can be made false in three ways, the brief truth table method is more cumbersome to apply given conclusions with these connectives. In these cases, it is often more convenient and less confusing to do a long table.

EXERCISE   Show the following arguments to be valid or invalid using the brief
VI   truth table method.

1. (A ∨ B) ⊃ ~C
   C   /∴ A

  2. A ⊃ B
     C ⊃ B
     ~D ⊃ B   /∴   A ⊃ ~D

*3. (A ≡ B) ⊃ C
     B ⊃ A
     ~C   /∴   ~A ∨ B

  4. A ⊃ (B ∨ D)
     A
     D ⊃ (S • ~Y)   /∴   ~B ⊃ ~Y

  5. (D • C) ⊃ (H ⊃ ~Y)
     P ⊃ (D • C)
     Y
     (P • H) ∨ (D • C)   /∴   ~(D • C)
     Hint: Put this argument into its argument form first.

ANSWERS TO STARRED ITEMS:

I.   8. True   ~(X ∨ Y)
                         ~(F ∨ F)
                         ~(F)
                             T

       12. False   ~[(~X ∨ A) ∨ (~A ∨ X)]
                      ~[(~F ∨ T) ∨ (~T ∨ F)]
                      ~[(T ∨ T) ∨ (F ∨ F)]
                      ~[T   ∨   F   ]
                      ~[T   ]
                             F

       16. True   ~[X • (~A ∨ Z)] ∨ [(X • ~A) ∨ (X • Z)]
                      ~[F • (~T ∨ F)] ∨ [(F • ~T) ∨ (F • F)]
                      ~[F • (F ∨ F)] ∨ [(F • F) ∨ (F • F)]
                      ~[F •   F   ] ∨ [F   ∨   F]
                      ~[F   ] ∨   F
                         T   ∨   F
                               T

II.   11. False   [(X ⊃ Y) ⊃ B] ⊃ Z
                      [(F ⊃ F) ⊃ T] ⊃ F
                      [T   ⊃ T] ⊃ F
                      [T   ] ⊃ F
                             F

       16. True   [(A ⊃ Z) • (Z ⊃ A)] ⊃ ~[(A • Z) ∨ (~A ∨ ~Z)]
                      [(T ⊃ F) • (F ⊃ T)] ⊃ ~[(T • F) ∨ (~T ∨ ~F)]
                      [(T ⊃ F) • (F ⊃ T)] ⊃ ~[(T • F) ∨ (F ∨ T)]
                      [F   •   T   ] ⊃ ~[F ∨   T   ]
                            F   ⊃ ~[T   ]
                            F   ⊃   F
                                 T

III. 3. Valid

			p1					p2	c
A	B	C	(A	∨	~B)	⊃	~C	A≡B	~C
T	T	T	T	T	F	F	F	T	F
T	T	F	T	T	F	T	T	T	T
T	F	T	T	T	T	F	F	F	F
T	F	F	T	T	T	T	T	F	T
F	T	T	F	F	F	T	F	F	F
F	T	F	F	F	F	T	T	F	T
F	F	T	F	T	T	F	F	T	F
F	F	F	F	T	T	T	T	T	T

IV. 6. Invalid

1. (W • S) ⊃ ~V
2. ~W   /∴ S ⊃ V

			p1			p2	c
W	S	V	(W • S)	⊃	~V	~W	S ⊃ V
T	T	T	T	F	F	F	T
T	T	F	T	T	T	F	F
T	F	T	F	T	F	F	T
T	F	F	F	T	T	F	T
F	T	T	F	T	F	T	T
F	T	F	F	T	T	T	F	★★★★
F	F	T	F	T	F	T	T
F	F	F	F	T	T	T	T

V.

4. p ⊃ q
   q ⊃ r /∴ p ⊃ r

   Why is it unnecessary to map ~(S • G) and ~P as ~p and ~r respectively?

7. p • q
   /∴ p

   It is possible for arguments to have only one premise.

VI.

3. **T**   (A ≡ B) ⊃ C   A=T
           T F   F   B=F
              F   F   C=F

   T   B ⊃ A   **Invalid**
           F T

   T   ~C   /∴ ~A ∨ B
           ~F   ~T ∨ F
                          F ∨ F
                           F

Chapter 9

SYMBOLIC TRAILS AND FORMAL PROOFS OF VALIDITY

Introduction

Throughout this book we have used the metaphor of a reasoning trail. The cultural roots for our use of logic and mathematics can be traced back to the ancient Greeks. The ancient Greeks believed that our reasoning ability gave us a special mystical power to "see" or detect unseen realities. They thought we could start with what we immediately experience and then follow a trail using logic and mathematics to transport our minds places unaccessible to our immediate experience. Thus, the blind man was able to see, in a sense, that he had on a white hat, just as Eratosthenes was able to "see" the size and shape of the Earth even though he was visually limited to a small piece of our large Earth. What gave the Greeks and much of our past Western culture confidence that we indeed had this power was the metaphysical belief in a resonance between our thinking and reality. Reality was thought to have particular trails or laws, and when we think correctly it was thought that we are mirroring those trails or laws.[1]

Most modern philosophers no longer accept this metaphysical view, but see our logic and mathematics instead as human constructions that we impose on reality, as practical tools that we use to successfully interface or work with reality. Although we may no longer possess the same confidence in our intellectual specialness—the confidence that somehow God has given us a head start by supplying us with the same thoughts with which He has constructed reality—the results of this initial confidence are with us today as never before. For good or ill, we live in a scientific-

---

[1] For a summary and critique of this past cultural perspective, see Richard Rorty's, *Philosophy and the Mirror of Nature* (Princeton, New Jersey: Princeton University Press, 1979).

technological culture, where people daily sit down in front of desks and follow or analyze reasoning trails symbolically. Most often this is now in the form of computer programs or with the application of computer programs, but using computers or not, the process is the same: Using assumptions based on accepted knowledge, telecommunication specialists attempt to design the most efficient phone lines connecting cities from Los Angeles to New York; the engineer wants to see or discover prior to actual construction how a bridge will look and function; using the laws of nature, the physicist wants to see the course of a space craft and the navigational corrections necessary to keep it on course when it encounters gravitational influences; the chemist wants to see the properties and use of a new combination of known elements, and so on.

In this chapter and the next you will be matching these processes of following a symbolic reasoning trail. You will be learning how to create your own symbolic reasoning trails, and you will be experiencing the adventure of symbolic problem solving: the trial and error, the back-to-the-drawing-board frustration and play, the tension of a problem not solved, and hopefully, at least some of the time, the feeling of completion when the trail ends successfully. Most people come to enjoy this in spite of their initial fear of mathematics and any kind of symbolic reasoning. There is something in our nature that makes us enjoy problem solving, that makes us enjoy being a detective, being in the hunt for a solution to a puzzle, seeking something just out of our grasp. At night you will even dream of solutions to some of the symbolic problems we will be working on. Don't worry about this when it happens; your brain loves this stuff.[1]

Truth tables are mechanical; the process of symbolic reasoning that we will be learning in this chapter is not. This process will involve creativity, discipline, and perseverance, and for this reason individual personality factors will emerge as one is tested by the adversity of not knowing in any mechanical fashion the right path to take. For all of this to happen, we will need to take some time and learn this new process in several steps. For an overview of where we will be going, consider the following example.

Most of us when we solve problems or analyze must slow down our thinking and examine possible trails piece by piece. Occasionally we meet individuals who have something like photographic minds and the special ability to keep everything focused, seeing where every step leads instantly. When Ronald Reagan was president of the United States I once went to a lecture by a government expert on the implications of what was called then the Iran-Contra scandal. Ronald Reagan had been elected in 1980 in part by convincing the voters that he would be a stronger leader than Jimmy Carter. During the last stages of the Carter presidency, U. S. embassy staff were being held hostage in Iran and all the U. S. military power and influence seemed impotent in stopping this injustice and embarrassment. Further, this weakness seemed to coincide alarmingly with the United States' increasing economic weakness in the world order. Carter was portrayed by Republicans in 1980 as a "wimp"; Reagan was the "Duke," the hero in a John Wayne movie, the

---

[1] It will not be permanent! Within a few weeks after your logic course ends the dreams will cease.

Marlboro man riding off into the sunset after dealing with all the bad guys. With Reagan, the United States would not let terrorists push us around anymore.

So, it was very embarrassing for Reagan, during his second term in office, to admit that he had negotiated with terrorists. The Reagan administration had promised never to negotiate with terrorists because this would legitimize and encourage their illegal and immoral actions. However, at that time Reagan admitted to the American people that he had allowed the sale of military equipment to Iran in exchange for their influence in getting U. S. hostages in Lebanon released.[1] More serious than just politically embarrassing was the revelation that the proceeds from the sale of arms to Iran were used to support the Nicaraguan Contras. The sale of arms to Iran may have been embarrassing or stupid, but supporting the Contras militarily was strictly illegal—Congress had passed a law prohibiting the U.S. government from supporting the Contras other than with humanitarian aid. Reagan claimed never to have known about the Contra diversion; if he did he would have been impeached, because as a government of law, not of men, even the president cannot violate the law without sanction. In fact, it is one of the president's principal duties to make sure all the laws of the land are upheld.

For about an hour I listened to the government expert analyze the various political and legal ramifications of the Iran-Contra scandal. He talked about constitutional issues and precedents, the various administration officials involved and their responsibilities, the U. S. foreign policy, the evidence whether Reagan knew or not about the Contra diversion of funds, and the implications of impeachment. At the end of the talk there was a question-and-answer discussion session with the audience. A man stood up and announced very confidently that the most immediate implication of what the speaker had said was that Ed Meese, the attorney general, should resign from office. The speaker seemed a little stunned. He had spoken for over an hour, painstakingly analyzing detail after detail, conveying years of experience and reflection on government matters, and the questioner had the audacity to state that everything boiled down to one simple implication. But the speaker was intrigued; there was something in the implication that seemed appealing. Like a nibble at the end of a fishing line, something needed to be pulled out of the muddy water. So, he asked the questioner to elaborate, and the man responded with a quick summary that went something like this:

> Well, it is clear that Reagan lied about the Iran deal. He has admitted this to the American people and asked for their forgiveness, explaining how important it was for him to win the release of U. S. hostages. But if he lied about the Iran deal, then, as you have explained, he also lied about the Contra deal or he should have known about the Contra deal. The buck stops with the president and he is responsible for guaranteeing laws are not broken, especially by his own staff. Now, Reagan claims not to have known about the Contra deal. Let's assume this is true. Well, since the chief of staff is responsible for the flow of information to the president (actually one

---

[1] During this time, Iran and Iraq were engaged in a bloody and futile war of human attrition, and we were also supporting Iraq militarily.

of the most powerful persons in government, and not elected), he should assume responsibility for the president not knowing about the Contra diversion and resign. McFarlane, who has obviously been the scapegoat, should be exonerated. And finally, since the attorney general (Ed Meese) is responsible for ensuring that all laws are followed, the necessary conditions of responsibility that apply to the chief of staff also apply to Meese. So, Meese should resign immediately.

As the man's rapid fire logic cascaded about the room, eyebrows were raised, and the speaker seemed to be getting embarrassed. When he finished, there was a hushed silence. What would the speaker say? The man's reasoning seemed to flow; especially as he had put it together so quickly, meshing the thoughts together one by one like bricks in a sturdy cemented wall. His conclusion seemed like a novel idea that was hiding behind the complexity of a million facts, now pulled out for all to see. But was his insight correct? The speaker gave an answer the specifics of which I don't remember. It was obvious that he did not know what to say, and he basically gave the type of answer that changed the subject and evaded the issue.

In the immediate flow of experience, it is often hard to hold on to all the details, to isolate what is important and arrange what is important as premises for a reasoning trail. So, the questions and discussion jumped around, moving on to other topics raised by the speaker, and the subject of Meese resigning did not come up again that night. But I suspect that when the speaker had more time to think about what the questioner had said, he began to talk with his friends about the necessity of Meese resigning. Let's slow down our thinking also and analyze the man's logic step by step. First, a more formal presentation of his argument.

1. If Reagan lied about the Iran deal, then he either lied about the Contra deal or he should have known about the Contra deal.
2. It is clear he did lie about the Iran deal.
3. But (let's assume) he did not lie about the Contra deal.
4. If he should have known about the Contra deal, then his chief of staff should not continue in office and McFarlane should be exonerated.
5. Also, Meese should continue in office only if Reagan's chief of staff continues in office.

Therefore, Meese should not continue in office.

Next, a translation of the argument.

1. $I \supset (C \lor K)$
2. I
3. $\sim C$
4. $K \supset (\sim S \cdot E)$
5. $M \supset S \qquad / \therefore \quad \sim M$

Now, if we combine some of these premises in isolation, we can derive some "mini" conclusions by just using our common sense. The first premise states that if I is true, then $C \lor K$ is true. Because the second premise states that I is true, we can conclude that $C \lor K$ is true. So,

STEP 1 $I \supset (C \vee K)$
                    I /∴ $C \vee K$

But the third premise says that C is not true. If we have concluded that $C \vee K$ is true, but now know that C cannot be true, then we can conclude that K is true. So,

STEP 2 $C \vee K$
                    $\sim C$ /∴ K

But the fourth premise says that if K is true, then S is *not* true and E is true. Because we now know that K follows from the previous premises, we can conclude that S is *not* true and E is true as follows,

STEP 3 $K \supset (\sim S \cdot E)$
                    K /∴ $\sim S \cdot E$

Well, if we know that S is *not* true and E is true, then we know that S is *not* true. If we know two things, then we surely know one thing. So,

STEP 4 $\sim S \cdot E$
                    /∴ $\sim S$

Finally, the last premise states that M is true only if S is true. So, because we have discovered that S is *not* true, M is *not* true.

STEP 5 $M \supset S$
                    $\sim S$ /∴ $\sim M$

Reflect now on what we have done. We have shown that if the above premises (1–5) are true, then ~M is true. In other words, we have shown this argument to be valid by creating a chain of reasoning in which each mini-step is valid, such that starting with the premises, we created a number of steps until we arrived at the conclusion ~M. To have proven this argument to be valid using the truth table method would have required a complicated table with sixty-four lines! We have better things to do with our time.

Constructing Formal Proofs of Validity

What you will be learning to do in this chapter is to formalize the method of proof of validity that we just used. We will be learning a rigorous way of presenting our common sense. We will be creating reasoning trails, such that complicated arguments will be proved to be valid by creating a chain of **elementary valid arguments**. Here is an example of what this rigorous method of presentation will look like applied to the above argument.

1. I ⊃ (C ∨ K)
2. I
3. ~C
4. K ⊃ (~S • E)
5. M ⊃ S   /∴   ~M
6. C ∨ K   (1) (2) MP
7. K   (6) (3) DS
8. ~S • E   (4) (7) MP
9. ~S   (8) Simp.
10. ~M   (5) (9) MT

This rigorous method of presentation is called a *formal proof.* Formal proofs are simply objective methods of presenting reasoning trails, such that anyone who learns the method of presentation can check the steps against their own common sense. Lines 6 through 10 show our chain of reasoning. The numbers adjacent to each line show the premises used to infer each line as a conclusion, as a link in the chain of reasoning. To derive line 6 as a valid conclusion we used premises 1 and 2 (Step 1 above). We then used line 6 with line 3 to derive line 7 (Step 2 above). Then putting line 7 together with line 4 we derived line 8 (Step 3). From line 8 we knew line 9 (Step 4), and finally from line 9 with line 5 we concluded line 10 (Step 5).

The capital letters next to the lines of justification refer to names of elementary valid arguments. We saw in Chapters 4 and 5 that fallacies have been named. Similarly, logicians have studied our common sense carefully and named many of our elementary common sense inferences. Notice that lines 6 and 8 have the same capital letters (MP) adjacent to the lines of justification. Although these lines involve different content (different letters), if you look carefully the *form* or pattern of reasoning is the same.

I ⊃ (C ∨ K)   K ⊃ (~S • E)
I   /∴   C ∨ K   K   /∴   ~S • E
                p ⊃ q
                p   /∴   q

Modus ponens is the fancy name that logicians give to this form of reasoning. This Latin name, which means to be in the mode of affirmation (of the antecedent), shows that this form of reasoning has been recognized to be valid for centuries. We know intuitively that it is an elementary valid argument, and we saw in the last chapter that a truth table also shows this pattern to be valid (p. 275). Remember the timesaving virtue of form recognition. Complicated arguments such as Example A3 in Chapter 8 (p. 275) can be seen to be valid at a glance. Once we know that the form of an argument is valid, we know an infinite number of arguments to be valid. We know that any argument that fits the form is valid, just as once a child learns what a chair is, he or she can apply this concept to a multitude of different things that all have in common the fact that they are chairs.

Step 1: Recognizing Forms: Copi's Nine Rules of Inference

We are now ready for the first step in constructing formal proofs. Lines 7, 9, and 10 of the formal proof above have the justifications DS, Simp, and MT. These abbreviations stand for disjunctive syllogism, simplification, and modus tollens. Along with modus ponens, we must examine and learn the forms of these elementary common sense inferences as well as five other rules. We will call these rules *Copi's Nine Rules of Inference* after Irving Copi, the first logician to systematize these rules in textbook form for previous generations of logic students for constructing formal proofs.[1] What follows is a presentation of each rule with three examples of application. Your task is to continue the form-recognition process that we began at the end of Chapter 8. Examine each of the three examples presented for each rule and make sure you see how each fits the argument form presented with variables. If you do not see how any example fits the argument form of a rule, mark it and ask your instructor to explain the fit.

Modus Ponens (MP)

For the sake of completeness and to be able to compare the valid form of modus ponens with the form of a very common fallacy, let's review this rule one more time.

MODUS PONENS (MP)

Egs.   1. $A \supset B$ 2. $R \supset J$ 3. $\sim(I \equiv R) \supset \sim P$
$A \quad / \therefore \quad B$ $R \quad / \therefore \quad J$ $\sim(I \equiv R) \quad / \therefore \quad \sim P$

Argument Form: $p \supset q$
$p \quad / \therefore \quad q$

To help you recognize an application of modus ponens in the future, reflect on the essence of this rule. All three examples above have two premises, one of the premises has ($\supset$) as a major connective, the other premise matches exactly the antecedent of the first premise, and the conclusion matches the consequent of the first premise exactly. Note that it does not matter what the antecedent or consequent are, they can be simple, complex, or involve negations as in 3. All that matters is that they match in this way.

Here are three more arguments. Do they match the pattern of modus ponens?

---
[1] See Copi's classic *Introduction to Logic*, 8th ed. (New York: Macmillan Publishing Co., 1990).

1. A ⊃ B   2. (S • R) ⊃ T   3. ~H ⊃ (P • D)
   B   /∴   A   T   /∴   (S • R)   (P • D)   /∴   ~H

You should have answered "no" in all three cases. The second premise does not match the antecedent of the first premise; instead it matches or "affirms" the consequent. Furthermore, the conclusion matches the antecedent rather than the consequent as in modus ponens. The argument form that fits all three examples is

p ⊃ q   **Invalid**
q   /∴   p

A truth table of this argument form will show it to be invalid. This form represents a very common mistaken inference. It is called the **Fallacy of Affirming the Consequent (FAC)** and should never be used in a formal proof. Its persuasiveness is no doubt caused by its close resemblance to modus ponens. For instance, the argument

1. If John passes the final exam, he will pass the course.
2. John passed the course.
   Therefore, John passed the final.

might sound good and seem to flow, but as we have seen, the first premise does not specify that passing the final is the only way John can pass the course. The first premise specifies what would be sufficient for John to pass the course, but not what is necessary. Thus the conclusion could be false even if the premises are true. Compare this argument with

1. John will pass the course only if he passes the final.
2. John passed the course.
   Therefore, John passed the final.

This is an example of modus ponens. We are locked into the conclusion, because if the first premise is true that passing the final is a necessary condition for passing the course, then because John passed the course, he must have passed the final.

Modus Tollens (MT)

Step 5 in the formal proof on the Iran-Contra scandal above was an example of modus tollens (Latin for being in the mode of denying the antecedent in the conclusion). Here are some more examples followed by the argument form:

MODUS TOLLENS (MT)

Egs. 1. $A \supset P$ 2. $\sim G \supset \sim(A \lor B)$ 3. $S \supset L$
          $\sim P$ /∴ $\sim A$ $\sim\sim(A \lor B)$ /∴ $\sim\sim G$ $\sim L$ /∴ $\sim S$

Argument Form: $p \supset q$
                          $\sim q$ /∴ $\sim p$

Reflect on the essence of this rule. Modus tollens has two premises, one of which has ($\supset$) as a major connective; the other premise negates whatever the consequent is of the ($\supset$) premise, and the conclusion is always a negation of the antecedent of the ($\supset$) premise. Note that although complex, 2 stays true to the rule. The consequent of the ($\supset$) premise is $\sim(A \lor B)$, so, to be an example of modus tollens the second premise must be $\sim\sim(A \lor B)$. Also, the antecedent of the ($\supset$) premise is $\sim G$, so the conclusion must be a negation of this or $\sim\sim G$.[1]

Examine the following arguments. Do they fit the form of modus tollens?

1. $S \supset L$ 2. $(A \lor B) \supset H$ 3. $X \supset \sim(Y \supset A)$
    $\sim S$ /∴ $\sim L$ $\sim(A \lor B)$ /∴ $\sim H$ $\sim X$ /∴ $\sim\sim(Y \supset A)$

You should have answered "no" in all three cases. Rather than negate the consequent of the first premise as in modus tollens, the second premise in all three cases negates or "denies" the antecedent of the first premise. Also, rather than concluding with the negation of the antecedent of the first premise, these arguments conclude with the negation of the consequent. The argument form for these arguments is

$$p \supset q \qquad \textbf{Invalid}$$
$$\sim p \quad /∴ \ \sim q$$

and is called the ***Fallacy of Denying the Antecedent (FDA)***. No doubt the persuasiveness of this form of reasoning is due to its closeness to modus tollens. But this argument form is invalid and should never be used in a formal proof. Consider the difference between

1. If John passes the final, he will pass the course.
2. John did not pass the final.
    Therefore, John did not pass the course.
    (fallacy of denying the antecedent—invalid)

---

[1] Although our common sense tells us that G is opposite of $\sim G$, technically G is not a negation of $\sim G$. The negation of a negation is a double negation ($\sim\sim G$). So, if G were the conclusion in 2, the argument would be valid but not yet an example of modus tollens.

and

1. If John passes the final, he will pass the course.
2. John did not pass the course.
   Therefore, John did not pass the final.
   (modus tollens—valid)

Because the first premise in both examples specifies only one guaranteed condition for passing the course—passing the final—we cannot say for sure that John did not pass the course if he fails to fulfill this one condition. There may be other ways to pass the course, even if the first premise is true. However, if we find that John did not pass the course, then we can conclude that John did not fulfill the guaranteed condition of passing the final.

Let's summarize what we have learned about arguments involving conditionals. *In the valid arguments MP and MT the premises involve a relationship whereby the antecedent is affirmed or the consequent is denied; whereas in the invalid arguments FAC and FDA the consequent is affirmed or the antecedent is denied.*

Disjunctive Syllogism (DS)

Step 2 in the above formal proof was an example of disjunctive syllogism. Here are some more examples and the argument form:

DISJUNCTIVE SYLLOGISM (DS)

Egs. 1. $C \vee K$ 2. $\sim G \vee \sim(A \vee B)$ 3. $(I \cdot P) \vee R$
           $\sim C \; / \therefore \; K$ $\sim\sim G \; / \therefore \; \sim(A \vee B)$ $\sim(I \cdot P) \; / \therefore \; R$

Argument Form: $p \vee q$
                       $\sim p \; / \therefore \; q$

Note that the essence of disjunctive syllogism is that one premise must be an *or* statement, (a disjunction) and the other premise must negate the left side of the *or* statement. Then the conclusion exactly matches the right side of the disjunction. Note 2. If the left side of the disjunction premise is $\sim G$, then to stay true to the rule the second premise must be $\sim\sim G$.

An argument that has a premise that negates the right side of the disjunction and concludes the left side of the disjunction would be valid, but would not yet be an example of disjunctive syllogism.[1] Disjunctive syllogism is valid for both inclusive and exclusive disjunctions. However, arguments of the form

---

[1] We say "not yet an example" because we will be combining our rules of inference with other rules to turn the example just given into a disjunctive syllogism.

p or q
p   /∴ ~q

are invalid for inclusive disjunctions, but valid for exclusive disjunctions. Consider the following.

1. This man must either be drunk or have diabetes.
2. He is drunk.
   Therefore, he must not have diabetes.
   (Inclusive *or*, so invalid; the man could also have diabetes.)
1. We will hire either Aweau or Kaneshiro for the new electronics position.
2. We will hire Aweau.
   So, we will not hire Kaneshiro.
   (Exclusive *or*, so valid; there is only one new position.)

This second example is valid because its translation would be

(A ∨ K) • ~(A • K)
A   /∴ ~K

The important point for now is that disjunctive arguments that have a premise that negates the right side of the disjunction, or ones that have a premise that matches the left side of a disjunction are not examples of disjunctive syllogism. To be an example of disjunctive syllogism, a premise must negate the left side of a disjunction, and the conclusion must be an exact match of the right side of the disjunction.

Hypothetical Syllogism (HS)

Although not used in our formal proof, a common form of valid reasoning is to "chain" *if . . . then* statements together, as in the following:

1. If the cold war is over, then less tax money can be spent on defense.
2. If less tax money can be spent on defense, then a tax reduction is possible or more money is available for rebuilding our infrastructure and supporting education.
   So, because the cold war is over, a tax reduction is possible or more money is available for rebuilding our infrastructure and supporting education.

The reasoning form for such *if . . . then* chaining is called hypothetical syllogism. Here are some more examples and the argument form. (The third example is a translation of the above argument.)

HYPOTHETICAL SYLLOGISM (HS)

Egs.

1. $A \supset B$
   $B \supset C$
   $\therefore A \supset C$

2. $\sim(B \lor K) \supset K$
   $K \supset H$
   $\therefore \sim(B \lor K) \supset H$

3. $C \supset L$
   $L \supset [T \lor (I \cdot E)]$
   $\therefore C \supset [T \lor (I \cdot E)]$

Argument Form:

$p \supset q$
$q \supset r \quad / \therefore \quad p \supset r$

Note the essence of hypothetical syllogism. The major connective of both premises and the conclusion is ($\supset$). The consequent of one premise matches the antecedent of the other premise. The conclusion then links the antecedent of one premise with the consequent of the other premise.

Two common invalid inferences that are often confused with hypothetical syllogism are:

$p \supset q$ **Invalid**   and $p \supset q$ **Invalid**
$r \supset q \quad / \therefore \quad p \supset r$ $q \supset r \quad / \therefore \quad r \supset p$

Neither should ever be used in a formal proof.[1]

Constructive Dilemma (CD)

A slightly more complicated form of valid reasoning than the rules presented thus far involves combining conditional statements with a conjunction and a disjunction. Here is an example:

1. If we continue to accept biological-weapons research grants for the university, then there will be student protest; whereas, if we do not accept these grants, there will be serious financial difficulties.
2. Either we accept biological weapons research grants or we do not.
   Therefore, either there will be student protest or there will be serious financial difficulties.

This form of valid reasoning is called constructive dilemma. Here are translated examples followed by the argument form. (The second example is a translation of the above argument.)

---

[1] Number 12, Exercise III, of Chapter 1 is an example of the second invalid form. It is much easier to see that this argument is invalid on the basis of recognizing its form.

CONSTRUCTIVE DILEMMA (CD)

Egs. 1. $(A \supset B) \cdot (C \supset D)$ 2. $(B \supset P) \cdot (\sim B \supset F)$
          $A \vee C$ $/\therefore$ $B \vee D$ $B \vee \sim B$ $/\therefore$ $P \vee F$

        3. $[(A \cdot B) \supset \sim C] \cdot [\sim D \supset (X \vee Y)]$
           $(A \cdot B) \vee \sim D$ $/\therefore$ $\sim C \vee (X \vee Y)$

Argument Form: $(p \supset q) \cdot (r \supset s)$
                  $p \vee r$ $/\therefore$ $q \vee s$

Although more complicated than any of our previous rules, with a little concentration you should be able to see the essence of constructive dilemma. To form a constructive dilemma, we must have two premises, one of which is a conjunction of two conditionals, the other of which must be a disjunction of the two antecedents of the conditionals. Then the conclusion must be a disjunction of the two consequents of the conditionals. The third example is a good test of your form recognition ability. Notice that the major connective of the first premise is •, and it connects two $\supset$ statements. The antecedents of these two conditionals are $(A \cdot B)$ and $\sim D$, and these two statements are matched exactly and connected by a $\vee$ statement in the second premise. The consequents of the two conditional statements are $\sim C$ and $(X \vee Y)$, and these are matched exactly and connected by a $\vee$ statement in the conclusion. Remember that the p, q, r, and s are *variables* and can stand for anything whatsoever. It does not matter whether they stand for something simple or complex; all that matters is that the content of an argument matches up to fit the above form.

Conjunction (Conj.)

To construct formal proofs, logicians have discovered that even our most trivial and obvious commonsense inferences must be identified. It is a trivial, simple inference to say that if we know two things separately, then we know two things together. If we know that Galileo was a religious man, and we also know that Kepler was a religious man, then we know that Galileo and Kepler were both religious men. In essence, if we know two separate premises are true, then we know the conjunction of those premises is true. This rule of inference is called conjunction. Here are some translated examples followed by the argument form:

CONJUNCTION (CONJ.)

Egs. 1. A 2. $K \cdot G$ 3. $G \supset L$
        B $T \cdot B$ $\sim G \supset U$
        $/\therefore$ $A \cdot B$ $/\therefore$ $(K \cdot G) \cdot (T \cdot B)$ $/\therefore$ $(G \supset L) \cdot (\sim G \supset U)$

Argument Form: p
                   q $/\therefore$ $p \cdot q$

Notice the simple essence of this rule. No matter what the premises are, the conclusion is a simple combination of these premises by the • connective. All this rule says is that you can combine premises by a conjunction.

We will now look at some rules that are also needed for the completeness of constructing formal proofs, but which have only one premise.

Absorption (Abs.)

Similar to the conjunction rule is the following obvious inference: If I know that passing the final is sufficient for John passing the course, then I know that if John passed the final, he passed both the final and the course. This commonsense inference is called absorption. Here are some examples and the argument form, the first of which is a translation of the example just given:

ABSORPTION (ABS.)

Egs.   1. $F \supset C$
$\quad /\therefore \ F \supset (F \cdot C)$

2. $J \supset \sim(S \cdot K)$
$\quad /\therefore \ J \supset [J \cdot \sim(S \cdot K)]$

3. $W \supset \sim W$
$\quad /\therefore \ W \supset (W \cdot \sim W)$

Argument Form: $p \supset q$
$\quad /\therefore \quad p \supset (p \cdot q)$

Reflect on the essence of the absorption rule. It must have only one premise. Both the premise and the conclusion must have $\supset$ as a major connective, and the consequent of the conclusion must combine the antecedent and the consequent of the premise by the • connective.

Simplification (Simp.)

Step 4 in the above formal proof involved the simple commonsense inference: If we know two things, then we know one thing. Because we had concluded that the chief of staff should resign and McFarlane should be exonerated, we can of course conclude that the chief of staff should resign. If we know that John passed the final and the course, then we know that John passed the final. This elementary inference is called simplification. Here are some examples and the argument form:

SIMPLIFICATION (SIMP.)

Egs. 1. ~S • M   2. (L ⊃ D) • (~D ⊃ L)   3. (F ∨ G) • ~E
          /∴ ~S   /∴   L ⊃ D   /∴   F ∨ G

Argument Form: **p • q**
                      /∴ **p**

Please note carefully the essence of simplification. As elementary as this rule is, it is the rule most often abused by beginning logic students. This rule has only one premise, *the major connective of the premise must be a conjunction*, and the conclusion must be the first part of the conjunction. The following examples are common misapplications of simplification by beginning logic students.

  1. ~(F • C)   2. X ⊃ (Y • Z)   3. A • B
     /∴   ~F   /∴   X ⊃ Y   /∴ B

Example 1 is an invalid argument. If we knew that John did not pass both the final and the course, we could not be sure that it was the final that he did not pass. The negation is the major connective of the premise in example 1, and simplification must have a conjunction as a major connective in the premise. Example 2 is valid, but the major connective in the premise is a conditional rather than a conjunction. We will learn a way in the next chapter to prove example 2 valid using a combination of rules. Example 3 is valid and has a conjunction as a major connective in the premise, but the conclusion does not match the first part of the conjunctive premise. We also will learn a way to prove this obviously valid argument by using simplification in combination with another rule.

Examples 1 and 2 reveal a rule about all the rules of inference: *The rules of inference apply to whole lines only.* Simplification cannot be applied if only a part of a premise is a conjunction; the major connective must be a conjunction. Similarly, given the premise A ⊃ (B ∨ C), and the premise ~B, we cannot conclude C by the disjunctive syllogism rule. The B ∨ C is part of a line and must be a whole line by itself for the rule to apply.

Addition

Our last rule of inference is called addition. It is an elementary argument that is valid because of the commitment we have already made to the meaning of the inclusive *or*. Recall that if one part of a disjunction is true, then the entire statement is true. So, if a conclusion has ∨ as a major connective, then the conclusion would have to be true if the first part of the disjunction is true. It would be impossible for a premise to be true, which is the first part of a disjunctive conclusion, and that disjunctive conclusion false. Here are some examples and the argument form.

ADDITION

Egs. 1. A   2. ~X   3. A ≡ ~B
              /∴   A ∨ B   /∴   ~X ∨ (C ⊃ D)   /∴   (A ≡ ~B) ∨ ~D

Argument Form: **p**
                                    /∴   **p ∨ q**

Note the essence of the addition rule. There is always a single premise, and that premise can be anything. The major connective of the conclusion is always a disjunction, and the premise is the first part of the disjunction.[1] What is added to the premise by a disjunction can be anything. Do not confuse the "addition" name of this rule with a conjunction. Concluding p • q from just p is invalid. If we knew John passed the course, we would know that he passed the course *or* the final. But we would not know that he passed the course *and* the final. We would also know that he passed the final *or* the moon is made of green cheese! This is not as counterintuitive as it may seem, because we clearly would not know that he passed the final *and* the moon is made of green cheese.

This concludes the introduction to the nine rules of inference. We will now test your pattern recognition ability with some name-the-rule exercises. Attempting to apply each rule is no different than walking into a room and recognizing the difference between a chair and a table. You must notice that a particular object fits the pattern we call *chair* before you sit down. This is a very complex neurological activity that we take for granted. We have learned in the past few decades that getting a computer to recognize a chair is not easy. Some fairly complex computer programs do fine until the chair is turned upside down. Yet a normal human being has little trouble in walking into a room, seeing a strange chair turned upside down, turning it right side up, and then sitting on it.

In some of the following exercises you may have a similar experience to that of the confused computer. Just as you have learned to be flexible in applying the pattern of a chair to strange instances of chairs that are not right side up, so you will need to be flexible in applying the nine rules. For instance, the following are examples of modus ponens, modus tollens, and hypothetical syllogism respectively.

  1. A   2. ~(A ∨ B)   3. ~Z ⊃ A
     A ⊃ ~C /∴ ~C   X ⊃ (A ∨ B)/∴ ~X   B ⊃ ~Z /∴ B ⊃ A
       MP   MT   HS

These examples show that nothing is absolute about the order of the premises of those rules that have two premises. Just as you learned in grammar school that

---

[1] Obviously, an argument would be valid that had the second part of the disjunction of the conclusion as a premise (q). However, because we are interested in identifying one elementary argument at a time, we will not classify this as addition. We will have a way of proving the argument form q /∴ p ∨ q to be valid later.

nothing was absolute about the addition order of a rule such as 3 + 2 = 5. You eventually (you have probably forgotten the original confusion and trauma) learned that both

$$\begin{array}{r} 3 \\ +2 \\ \hline 5 \end{array} \qquad\qquad \begin{array}{r} 2 \\ +3 \\ \hline 5 \end{array}$$

were applications of the rule. So even though the premises are not presented in the same order as the rule was originally presented, all three examples above fit the essential features of their respective rule. As in all examples of modus ponens, 1 has a conditional premise, another premise that matches the antecedent of the conditional premise, and then a conclusion that matches the consequent of the conditional premise. As in all examples of modus tollens, 2 has a conditional premise, another premise that negates the consequent of the conditional premise, and then a conclusion that is the negation of the antecedent of the conditional premise. As in all examples of hypothetical syllogism, 3 has two conditional premises and a conditional conclusion, the consequent of one premise matches the antecedent of the other premise (~Z), and the antecedent of one of the conditional premises (B) is linked with the consequent of the other conditional premise (A) in the conclusion.

In other words, the presentation of these rules using variables could have been

p	~q	q ⊃ r
p ⊃ q /∴ q	p ⊃ q /∴ ~p	p ⊃ q /∴ p ⊃ r
MP	MT	HS

This flexibility in application applies to all rules that have two premises. But there is no similar flexibility for rules with only one premise. For instance, it would not be an example of the addition rule to have A ∨ B as a premise and conclude A from this. If we knew that either Lisa or Jorge is coming to the party, we would not know for sure that it was Lisa who was coming. The nine rules of inference "move" in only one direction.

Suggestion: Before doing the following exercises, you should write on a separate sheet of paper each of the translated examples, the *form* of each rule in variables, and a representation of the form in the following way as demonstrated with modus ponens.

MODUS PONENS (MP)

Egs.	1. A ⊃ B	2. R ⊃ J	3. ~(I ≡ R) ⊃ ~P
	A /∴ B	R /∴ J	~(I ≡ R) /∴ ~P

$$p \supset q$$
$$p \quad /\!\therefore \quad q$$

○ ⊃ □
○   /∴   □

STEP 1 EXERCISES   For each of the following arguments state, if applicable, the argument form that justifies the argument as a valid argument. If the argument is not an example of one of the nine rules of inference, indicate this with an X.

1. (S ∨ E) • (A ∨ G)
   /∴ S ∨ E

2. (A • B) ⊃ C
   /∴ (A • B) ⊃ [(A • B) • C]

*3. B ⊃ C
   /∴ (B ⊃ C) ∨ (B ⊃ C)

4. G ⊃ H
   ~H /∴   ~G

5. ~(A • C) ⊃ (D ∨ E)
   ~(A • C)   /∴   D ∨ E

6. ~(D • K) • (L ∨ Y)
   /∴   ~(D • K)

7. A ⊃ (B • C)
   /∴ A ⊃ B

8. (S • T) ∨ [(U • V) ∨ (U • W)]
   ~(S • T) /∴   (U • V) ∨ (U • W)

9. O ⊃ P
   ~O /∴   ~P

10. (T • U) ∨ (C • B)
    T • U /∴   C • B

11. [N ⊃ (O ∨ P)] • [Q ⊃ (O • R)]
    N ∨ Q /∴   (O ∨ P) ∨ (O • R)

*12. (X ∨ Y) ⊃ ~(Z • A)
    ~~(Z • A) /∴   ~(X ∨ Y)

13. F ⊃ (G • D)
    (G • D) ⊃ ~X
    /∴ F ⊃ ~X

14. ~(A ⊃ I) ⊃ (A ≡ I)
    (I ≡ H) ⊃ ~(A ⊃ I)
    /∴ (I ≡ H) ⊃ (A ≡ I)

15. (C ⊃ K) ∨ (P ⊃ L)
    L ⊃ M
    /∴ [(C ⊃ K) ∨ (P ⊃ L)] • (L ⊃ M)

*16. ~A ∨ (D • B)
    ~A /∴   D • B

17. A ⊃ B
    /∴ (A ⊃ B) ∨ C

18. [(A • B) ∨ C] ⊃ D
    (A • B) ∨ C   /∴   D

19. (A ∨ B) ⊃ C
    C /∴   A ∨ B

20. [E ⊃ (F ⊃ G)] ∨ (C ∨ D)
    ~[E ⊃ (F ⊃ G)]   /∴   C ∨ D

21. C ⊃ (D ∨ G)
    (D ∨ C) ⊃ (H • T)
    /∴   C ⊃ (H • T)

22. (X ⊃ T) • (H ⊃ P)
    X ∨ H   /∴   T ∨ P

*23. P ∨ T
    (P ⊃ ~X) • (T ⊃ ~Y)
    /∴ ~X ∨ ~Y

24. (X • P) • (T ⊃ P)
    /∴ X • P

25. H ⊃ T
    (H ⊃ T) ⊃ ~X
    /∴ ~X

26. ~(N • P)
    T ⊃ (N • P)
    /∴ ~T

*27. (A • B)
    (X • P) ⊃ (A • B)
    /∴ X • P

28. (H • T) ∨ (S ⊃ P)
    ~(H • T)
    /∴ S ⊃ P

29. A ∨ ~(C ≡ D)
    [A ∨ ~(C ≡ D)] ⊃ ~~Z
    /∴ ~~Z

30. ~~(I • Z)
    ~(I • Z) ∨ [H ≡ ~(P • V)]
    /∴ H ≡ ~(P • V)

Strategies For Pattern Recognition

Before we continue with the next step in learning formal proofs, we should stop and reflect on some strategies that you may have unconsciously used in identifying the answers to Step 1.

From one point of view, three of the rules of inference conclude with a match of a part of a premise. Modus ponens, disjunctive syllogism, and simplification all have part of one of the premises as a conclusion. From another point of view, many of the rules of inference always have the same major connective in the conclusion. Modus tollens will always have a negation as the major connective of the conclusion. Absorption and hypothetical syllogism will always have a conditional as a conclusion. Addition and constructive dilemma will always have a disjunction as a conclusion. And the conjunction rule will always have a conjunction as the major connective of its conclusion.

Because a large part of pattern recognition is simply the ability to "stare" at the right features at the right time, we can use the insights just mentioned to develop strategies of staring. When confronted with a complex problem, try the following.

> **Strategy 1** See if the conclusion is a part of a premise, and then try to match the problem with either modus ponens, disjunctive syllogism, or simplification.

or

> **Strategy 2** Focus on the connective of the conclusion, and then try to match with an appropriate rule. If the major connective of the conclusion is a negation, try to match with modus tollens; if the major connective of the conclusion is a conditional, try to match with either absorption or hypothetical syllogism. And so on.

Recall the super-messy argument (A3) from Chapter 8.

1. {[(A ⊃ B) ≡ (C ∨ ~B)] • ~D} ⊃ (~E ∨ ~F)
2. {[(A ⊃ B) ≡ (C ∨ ~B)] • ~D}   /∴ (~E ∨ ~F)

There is so much that we could look at in this argument. It is easy to suffer sensory overload and be overwhelmed by all the detail. Many students panic when looking at a problem with this much detail. Where do we start? There is so much detail and nine possible rules to look at for a possible match. However, anyone should at least be able to stay calm enough to notice that this problem has two premises, so it could not possibly be an example of any of our rules with only one premise. This simple insight immediately cuts down on the amount of staring we need to do. We don't have to look at simplification, absorption, or addition.

Next we use the above strategies. In spite of this problem's complexity, we see that the first strategy applies—the conclusion ~E ∨ ~F is a part of one of the premises. This cuts down on our staring even more; now we will focus only on modus ponens and disjunctive syllogism. (Remember that simplification was already eliminated because it has only one premise.) Next, we notice that the major connective of the premise with ~E ∨ ~F in it is a conditional. At this point we should be staring at modus ponens. It is important to reflect that at this point our strategy has enabled us to follow a trail by eliminating possibilities and arrive at a *hypothesis* that this problem may be an example of modus ponens. The rest of the parts still have to fit. The second premise must match the antecedent. If it did not, then we would have to try something else.

Let's see how the second strategy would be applied. Number 14 was one of the hardest problems in the above exercises.

14. ~(A ⊃ I) ⊃ (A ≡ I)
   (I ≡ H) ⊃ ~(A ⊃ I)   / ∴   (I ≡ H) ⊃ (A ≡ I)

Here is how to stay disciplined and calm applying the strategies. First, we notice that there are two premises. This immediately eliminates simplification, absorption, and addition. Next, we notice that the conclusion is not a self-contained part of any premise. (The antecedent of the conclusion, (I ≡ H), and the consequent of the conclusion, (A ≡ I), are themselves parts of premises, but the entire conclusion, (I ≡ H) ⊃ (A ≡ I), is not a part of any premise.) This eliminates the first strategy and the rules modus ponens, disjunctive syllogism, and simplification from consideration. Using the second strategy, we focus on the major connective of the conclusion. Because the major connective is a conditional, we now have a hypothesis that the answer might be hypothetical syllogism. (Absorption also has a conditional as a major connective for its conclusion, but this rule was already eliminated as a possibility because it has only one premise.) We now check the rest of the problem for a pattern match with hypothetical syllogism. Is the major connective of both premises a conditional? Yes. Does one of the consequents of the premise conditionals match one of the antecedents. Yes, ~(A ⊃ I) is a consequent of the second premise and an antecedent of the first premise. Is an antecedent of one of the premises linked with a consequent of the other premise in the conclusion? Yes, (I ≡ H), the antecedent of the second premise, is linked with (A ≡ I), the consequent of the first premise. All of the parts fit the pattern, hence, this is an example of HS.

Notice that these strategies are only rules of thumb; each one by itself will not always work. Combined, however, they provide a way of focusing and disciplining our attention when faced with confusing details. As noted previously in this book, without some logical strategies our minds are like a radio whose channel selector is moving back and forth chaotically, never quite focusing on a particular channel long enough to receive anything more than confused noise. In Step 2 of learning how to construct formal proofs we will be increasing the amount of "noise" that you will need to filter into coherent channels of pattern recognition, and the above strategies will be very helpful.

Step 2: Justifying Reasoning Trails with the Rules of Inference

Much of learning involves an initial step of mimicking. Painters often learn to imitate the styles of previous masters before they develop their own style. Musicians study and play the works of the great composers before they compose their own music. In this section, you will study formal proofs that have already been completed. The reasoning trail will be presented, but the premises and the rules used to justify each line will be omitted. Your job will be that of a detective—to reconstruct the thought processes in terms of premises and rules used to create the trail of reasoning. Let's use our original formal proof as an example.

1. $I \supset (C \lor K)$
2. I
3. $\sim C$
4. $K \supset (\sim S \cdot E)$
5. $M \supset S$ $/\therefore \quad \sim M$
6. $C \lor K$
7. K
8. $\sim S \cdot E$
9. $\sim S$
10. $\sim M$

Because our job is to provide a justification for the reasoning trail (lines 6–10), we start with line 6. We know that line 6 is a conclusion from some line or combination of lines of the premises 1–5, but how do we find this justification when there is so much detail to consider? We use the strategies discovered in the last section.

Starting with the first strategy, we ask if line 6 is a part of one of the premises above. Right away we have a possible connection—line 6 is part of premise 1. We then consider the rules modus ponens, disjunctive syllogism, and simplification. Because line 6 is a consequent of premise 1, we should consider modus ponens. Of the three rules, only modus ponens concludes a consequent of a premise. But modus ponens requires two premises, and the second premise must be a match of the antecedent. Because the antecedent is I, we must have an I by itself as a premise to complete the connection. Bingo. Premise 2 is I, so we have discovered that premises 1 and 2 fit the pattern of modus ponens and justify the conclusion $C \lor K$.

So, we write this down as the justification for line 6:

1. I ⊃ (C ∨ K)
2. I
3. ~C
4. K ⊃ (~S • E)
5. M ⊃ S   /∴   ~M
6. C ∨ K   (1) (2) MP
7. K
8. ~S • E
9. ~S
10. ~M

The next line to justify is line 7. Using the first strategy, we see that K occurs as a part of lines 1, 4, and 6. In such situations you must know the rules well. You must know by thoroughly practicing Step 1 exercises that no rule of inference would pull out a K from a premise such as I ⊃ (C ∨ K). No rule allows us to conclude part of a consequent. Similarly, no valid rule concludes the antecedent of a conditional as in line 4. (The fallacy of affirming the consequent concludes with the antecedent of a conditional, but this form should never be used in a formal proof of validity.) Thus, only line 6 remains as a possibility for applying strategy 1. Focusing on line 6 then, because the connective is a disjunction and the K is in the right location, disjunctive syllogism is the best hypothesis. But disjunctive syllogism uses two premises, and to confirm our hypothesis we must find a ~C. Premise 3 is ~C, so we have discovered that line 7 is justified by lines 6 and 3 and the rule of disjunctive syllogism. Our proof now looks like this:

1. I ⊃ (C ∨ K)
2. I
3. ~C
4. K ⊃ (~S • E)
5. M ⊃ S   /∴   ~M
6. C ∨ K   (1) (2) MP
7. K   (6) (3) DS[1]
8. ~S • E
9. ~S
10. ~M

Next, line 8. Again we see that strategy 1 looks promising. The ~S • E of line 8 is a part of line 4. It is also in the right location within a conditional statement to be a modus ponens, which means that to complete the match we will need to find the antecedent K as a line by itself. The line we just completed is K, so we have a match in lines 4 and 7 and the rule modus ponens as a justification for line 8. Our proof will now look like this:

---

[1] It does not matter what order the lines are referred to; we could have also written (3)(6) DS.

1. I ⊃ (C ∨ K)
2. I
3. ~C
4. K ⊃ (~S • E)
5. M ⊃ S   / ∴   ~M
6. C ∨ K   (1) (2)   MP
7. K   (6) (3)   DS
8. ~S • E   (4) (7)   MP
9. ~S
10. ~M

Perhaps you have noticed another strategy that we could combine with strategies 1 and 2. Because we are justifying a chain of reasoning, there is a high probability that the line we just completed will be used to get the next line. Line 6 was used to get line 7, and line 7 was used to get line 8. Like all our strategies, this strategy is only a rule of thumb—it will not always work—but as another technique for focusing our attention systematically it will be very useful. Let's try it now. Our next line is 9. Checking line 8, we see that strategy 1 works again. The ~S is a part of line 8, and because this line is a conjunction and the ~S is in the appropriate location, the simplification rule applies. Our proof will now look like this:

1. I ⊃ (C ∨ K)
2. I
3. ~C
4. K ⊃ (~S • E)
5. M ⊃ S   / ∴   ~M
6. C ∨ K   (1) (2)   MP
7. K   (6) (3)   DS
8. ~S • E   (4) (7)   MP
9. ~S   (8) Simp.
10. ~M

Line 10 completes our proof, but this time strategy 1 fails. The ~M does not occur as a part of any premise.[1] So we must now shift our mental gears to strategy 2 and focus on the connective of the line we want to justify. The connective in line 10 is a negation. The only rule that always has a negation as the major connective of the conclusion is modus tollens. Applying this strategy requires a thorough understanding of the rules. In this case we must be able to reconstruct what the premise would be for a conclusion ~M by the rule modus tollens. Because modus tollens concludes the negation of the antecedent of a conditional, we need a premise M ⊃ ___ and then the negation of the consequent. Line 5 has an M as an antecedent of a conditional, and behold, the line we just completed, ~S, negates the consequent. So, we have discovered that line 10 is justified by lines 5 and 9 and the rule of modus tollens. Our complete proof will now look like this:

---

[1] Remember that the ~M adjacent to line 5 is not a premise. It is the conclusion as indicated by (/ ∴) and only shows us the end point of our proof. So, it is never used as a line of justification in a proof.

1. I ⊃ (C ∨ K)
2. I
3. ~C
4. K ⊃ (~S • E)
5. M ⊃ S   / ∴   ~M
6. C ∨ K   (1) (2)   MP
7. K   (6) (3)   DS
8. ~S • E   (4) (7)   MP
9. ~S   (8) Simp.
10. ~M   (5) (9)   MT

A word of warning before you try some exercises. You must be flexible in applying the strategies. When a strategy does not work, you must give it up. In working on line 7 above we noticed that the K is a part of line 4. But no matter how hard we try and how long we stare, line 4 cannot be made to work with any of our rules. So, strategy 1 failed for this line, and we had to stop staring at line 4 and move on. Like giving up a belief that does not work, this is often hard to do. This is why a playful attitude of trial and error using a hypothesis-and-test method is most appropriate for finding justifications. The discipline you are learning is like being able to turn to a particular radio station, focus clearly on it for awhile, but then be ready to move to another station, whether you are comfortable with the music or not. Your attitude must be experimental.

Similar remarks apply to strategy 2. Suppose we had the following proof and we were working on line 7:

1. P ⊃ Y
2. (X • Y) ⊃ ~H
3. D ⊃ [P ⊃ (X • Y)]
4. D
5. ~H ⊃ X   / ∴   (P ⊃ X) • (P ⊃ Y)
6. P ⊃ (X • Y)   (3) (4) MP
7. P ⊃ ~H
8. P ⊃ X
9. (P ⊃ X) • (P ⊃ Y)

Line 7 is not part of any premise, so we shift to strategy 2. Because line 7 is a conditional, either absorption or hypothetical syllogism must apply. Absorption cannot apply because line 7 does not have a consequent that contains a conjunction. So, focusing on hypothetical syllogism we try to reconstruct the premises that would give us P ⊃ ~H as a conclusion. We could use some scratch paper and set up our staring as follows:

$$P \supset \underline{\quad\quad}$$
$$\underline{\quad\quad} \supset \sim H \quad / \therefore \quad P \supset \sim H$$

We then notice that line 1 is P ⊃ Y, a possibility for the first premise of the hypothetical syllogism. But to match we would need a second premise of Y ⊃ ~H, and there is no such line anywhere above line 7. What gives? Strategy 1 did not work and now strategy 2 also fails. In the Step 2 exercises that follow, there will be no "bogus" steps. A correct justification will exist for each line. We must be flexible enough to try strategy 2 again. Because premise 1 (P ⊃ Y) did not work, we must try another premise that is a conditional and has P as an antecedent. Premise 6 is a possibility, so we set up a possible scenario for hypothetical syllogism using this premise as follows:

1. P ⊃ (X • Y)
2. _____ ⊃ ~H   /∴   P ⊃ ~H

Then, looking for a match to complete the hypothetical syllogism we see that the second premise, (X • Y) ⊃ ~H, completes the match. Keep these flexibility points in mind as you work on the following exercises.

STEP 2 EXERCISES   State the justification for each line that is not a premise of the following arguments.

1.
1. Z • A
2. (Z ∨ B) ⊃ C   /∴   Z • C
3. Z
4. Z ∨ B
5. C
6. Z • C

2.
1. K ⊃ (B ∨ I)
2. K
3. ~B
4. I ⊃ (~T • N)
5. N ⊃ T   /∴   ~N
6. B ∨ I
7. I
8. ~T • N
9. ~T
10. ~N

*3.
1. (D ∨ G) • (H ∨ I)
2. (D ⊃ H) • (G ⊃ I)
3. ~H   /∴   I
4. D ∨ G
5. H ∨ I
6. I

4.
1. H ⊃ I
2. I ⊃ J
3. K ⊃ L
4. H ∨ K   /∴   J ∨ L
5. H ⊃ J
6. (H ⊃ J) • (K ⊃ L)
7. J ∨ L

5.
1. ~R ⊃ S
2. ~T ⊃ (U ⊃ V)
3. T ∨ (~R ∨ U)
4. ~T   /∴   S ∨ V
5. U ⊃ V

6.
1. O ⊃ P
2. (O • P) ⊃ Q
3. ~(O • Q)   /∴   ~O
4. O ⊃ (O • P)
5. O ⊃ Q

6. (~R ⊃ S) • (U ⊃ V)
7. ~R ∨ U
8. S ∨ V

7.
1. X ⊃ Y
2. (X ⊃ Z) ⊃ (A ∨ Y)
3. (X • Y) ⊃ Z
4. ~A   /∴   Y
5. X ⊃ (X • Y)
6. X ⊃ Z
7. A ∨ Y
8. Y

9.
1. G ⊃ ~H
2. ~G ⊃ (I ⊃ ~H)
3. (~J ∨ ~I) ⊃ ~~H
4. ~J   /∴   ~I
5. ~J ∨ ~I
6. ~~H
7. ~G
8. I ⊃ ~H
9. ~I

6. O ⊃ (O • Q)
7. ~O

*8.
1. (~A ∨ B) ⊃ D
2. (D ∨ B) ⊃ [~A ⊃ (C≡E)]
3. ~A • D   /∴   C ≡ E
4. ~A
5. ~A ∨ B
6. D
7. D ∨ B
8. ~A ⊃ (C≡E)
9. C≡E

10.
1. A ⊃ B
2. ~(A • B)
3. A ∨ (~~L • ~~K)
4. P ⊃ ~L   /∴   ~P ∨ ~Y
5. A ⊃ (A • B)
6. ~A
7. ~~L • ~~K
8. ~~L
9. ~P
10. ~P ∨ ~Y

Step 3: On Your Own, Constructing Formal Proofs with the Rules of Inference

The purpose of Step 1 and 2 exercises was to build up your pattern recognition ability to the point that you can construct your own formal proofs. In the next set of exercises you will face problems like the following.

1. A ⊃ B
2. B ⊃ C
3. ~C   /∴   ~A

There will be nothing but empty space under the last premise. Your task will be to put down valid steps, to create your own chain of reasoning until you arrive at the conclusion. Like life, you must create your own trail into an uncertain future. Also like life, not all of the trails created will lead to the desired conclusion, and even of those that do, some will be better (more elegant) than others. Consider two solutions to the problem of fixing the flat tire of a car that is parked on a steep hill. The car cannot simply be jacked up, because the steepness of the hill and gravity will cause the car to slip off the jack. Person one sees a large sturdy tree not too far from the parked car, and so borrows some rope to tie the car securely to the tree to keep it from sliding toward the bottom of the hill. But none of the pieces of rope are long enough to reach the tree from the parked car. (Assume also that the pieces tied together do not reach. So, this person gets the bright idea of putting a shopping cart between the car and the tree, such that one piece of rope can reach from the car to the shopping cart and another from the shopping cart to the tree. For further stability, this person gets some heavy bricks and puts them in the shopping cart.

Consider person two passing by and watching person one struggling with the exertion of moving the heavy bricks and trying to make the ropes as tight as possible so the car will not roll off the jack. Person two sees another car parked a few feet downhill of the car with a flat tire and simply backs this car up against the car with the flat tire, leaving a very small space between the bumpers, and then securely fastens the emergency brake and turns the wheels into the curb. The solution of person one may have worked, but surely the solution of person two is more elegant. It is simpler, faster, requires less effort, and less can go wrong.

From one point of view, the goal of life is not only to solve our problems, but to do so elegantly, like person two. Many people "bump" into the world like person one. They survive, but the solutions they have for their problems lack grace and simplicity, and these solutions are often a little scatterbrained, involving lots of wasted energy. In constructing formal proofs for the first time, many logic students will be like person one. They will blunder forward and perhaps arrive at the desired conclusion. But their proofs will be a unnecessarily complicated.

However, as an initial learning strategy, blundering forward is exactly what most students need to do. At least it is better than just staring aimlessly, putting down no steps at all, or putting down steps that are not valid. Applied to formal proofs, a strategy of **blundering forward** means this: In looking at the above argument do not worry about the conclusion; focus on the premises and ask yourself if you recognize any premise or combination of premises that would fit a rule of inference, such that the premise or combination of premises would entitle you to create a conclusion, a beginning line in a proof. If you see a fit that entitles you to create a conclusion, put it down as a step and don't worry yet if you are on the right trail that will lead to the final conclusion. Your strategy is that if you put down enough steps, you will eventually blunder across the conclusion.

Following this strategy, looking at the above premises your pattern recognition ability should by now allow you to see that premises 1 and 2 fit a pattern of hypothetical syllogism.[1] We would be entitled to create the step $A \supset C$. Because we are blundering forward—just trying to create any steps we see—we can put this step down as follows:

1. $A \supset B$
2. $B \supset C$
3. $\sim C$ $/ \therefore \sim A$
4. $A \supset C$ (1) (2) HS

We may or may not be on the right track. We don't worry about it yet. Taking a look at premises 2 and 3, we notice another pattern—modus tollens. Should we

---

[1] If you do not see that premises 1 and 2 fit the pattern of hypothetical syllogism, you should re-do Step 1 exercises. If you can't do Step 1 quickly, do it again, otherwise you will be wasting your time trying to do Step 3.

do it? Why not? We are just trying to get down steps at this point. Premises 2 and 3 entitle us to create the step ~B. So, our proof could now look like this:

1. A ⊃ B
2. B ⊃ C
3. ~C   /∴   ~A
4. A ⊃ C   (1) (2) HS
5. ~B   (2) (3) MT

See anything else to do? Remember that lines 4 and 5 can now be used also to create more steps. Most people can see several ways to create the conclusion at this point. Sometimes, however, even the most obvious connections right in front of us are hard to see if we are a little nervous or confused. I once had a mature female student who was returning to college after rearing a family. She was very nervous and intimidated by the fact that she was competing with much younger students and much younger minds. (Because she was nervous and intimidated, she studied and did very well.) I noticed that every time she did a formal proof, she would take the first two premises and apply the rule of conjunction, no matter what the conclusion or level of difficulty for the proof. She would then finish the proof in a most elegant way, and her conjunction step was seldom needed. So I finally asked her one day why she did proofs this way. She answered, "To relax! I know that I can always take any two premises and combine them by conjunction. I just need to get started and get rid of some of that blank space, then I'm okay." Well, she was right; there is nothing wrong with her conjunction step from the strict point of view of valid applications of the rules. A conjunction can be done at any time. So, another valid step that could be done would be to take premises 1 and 2 and create a conjunction. Remember that from the point of view of the strategy of blundering forward all we are trying to do is create valid steps. Our proof would now look like this.

1. A ⊃ B
2. B ⊃ C
3. ~C   /∴   ~A
4. A ⊃ C   (1) (2) HS
5. ~B   (2) (3) MT
6. (A ⊃ B) • (B ⊃ C)   (1) (2) Conj.

We could continue to blunder like this for many more steps, but at this point it should be apparent that we don't need to. Either the combination of lines 1 and 5 or lines 4 and 3 by modus tollens would produce ~A. Here is how our finished proof might appear:

1. A ⊃ B
2. B ⊃ C
3. ~C   /∴   ~A
4. A ⊃ C   (1) (2) HS
5. ~B   (2) (3) MT
6. (A ⊃ B) • (B ⊃ C)   (1) (2) Conj.
7. ~A   (4) (3) MT

Although there is nothing wrong with the above proof from the strict point of view of the rules of inference (each step is a valid application of one of the rules), overall it is "ugly." Like the solution of person 1 above, there are unnecessary extra steps. But we are just learning to feel our way in creating symbolic reasoning trails, so the goal is to put down a trail and worry about its elegance later. Gradually, as your recognition ability increases you will learn other strategies for making proofs more elegant.

We can show you one now. Once you are very comfortable with the rules, you can try the strategy of **working backward**. Unlike the blundering forward method where the conclusion is ignored at first, the working backward method requires that we focus on the conclusion and ask a hypothetical question: Given the premises, which premise is most likely to be involved in the last step of a proof for the desired conclusion, and what other step would need to be created that, combined with this premise, would give us the conclusion? Our conclusion is ~A, so premise 1 looks promising. Because premise 1 is A ⊃ B, we know that IF we find a ~B, then we would be entitled to conclude a ~A. We don't have a ~B yet. So, we ask the same question again but now directed at finding a ~B. Given our premises, what premise might be involved in creating a ~B, and what other step would we need to make the conclusion of a ~B valid? Premise 2 looks promising. Because premise 2 is B ⊃ C, we know that IF we find a ~C we can create a ~B again by the rule of modus tollens. We are done. We do not need to create a ~C, because this is premise 3. This is the goal of working backward: To work backward by asking "what if" questions until the trail ends in a premise that we already have. Here is a symbolic picture of our reasoning.

Want	~A	
Have	A ⊃ B	
Need	~B	MT
Want	~B	
Have	B ⊃ C	
Need	~C	MT
Want	~C	Already have ~C

This process can be mapped out on scratch paper. To reconstruct this backward reasoning into the forward reasoning of a formal proof, follow the advice of the children's TV show *Sesame Street*, where children are taught that if you are lost make everything that was last, first, and you will find your way home. What we wanted last was a ~C, but we already have this step in premise 3. What we wanted next to last was a ~B, so this will be the first step to put down in our proof. We needed ~B so we could create what we wanted first, a ~A. Our next, and in this case last, step in our proof will be ~A. Here is how this proof will look placed alongside our original proof using the method of working backward.

Working Backward

1. A ⊃ B
2. B ⊃ C
3. ~C   /∴   ~A
4. ~B   (2) (3) MT
5. ~A   (1) (4) MT

Blundering Forward

1. A ⊃ B
2. B ⊃ C
3. ~C   /∴   ~A
4. A ⊃ C   (1) (2) HS
5. ~B   (2) (3) MT
6. (A ⊃ B) • (B ⊃ C)   (1) (2) Conj.
7. ~A   (4) (3) MT

The working backward proof is clearly more elegant. It is shorter and there are no extra steps. But the blundering forward proof is not incorrect. Each step in this proof is a valid application of one of the rules of inference. As in life, note that one can be logical but a little crazy. The mere use of logic and technology does not guarantee elegance and wise applications. Obviously, the ultimate goal is to have elegant proofs and reasoning trails. But ideal goals are not always achievable. There are 40 million lines of computer code in the computer programs that run the U.S. space shuttle. Undoubtedly from a divine point of view there are more elegant program trails that could accomplish the same tasks. And each year, as the shuttles fly into space and we gain more experience, the hardworking men and women who run the space shuttle program will discover more elegant and efficient reasoning trails to get the job done, but perfect programs will probably remain elusive.

From a pragmatic point of view you should not worry about if your proof is elegant, especially in the beginning. The important thing is to get started on a trail and eventually reach the conclusion with all valid steps. Every step must be a valid application of one of the rules of inference, or the proof will be worthless. A single line of invalid computer syntax or logic will cause the space shuttle's computers to produce dangerous output. The following is a typical example of student error in producing a formal proof.

1. A ⊃ B
2. B ⊃ C
3. ~C   /∴   ~A
4. A ⊃ (A • B)   (1) Abs.
5. A   (4) Simp. **X**
6. (A ⊃ B) • (B ⊃ C)   (1)(2) Conj.
7. B   (1)(5) MP
8. A   (1)(7) MP **X**
9. C   (2)(7) MP
10. C ∨ ~A   (9) Add.
11. ~A   (10)(3) DS

There are some nice, creative steps in this proof—especially the series of steps 9 through 11. Unfortunately, the proof is worthless because steps 5 and 8 are not correct applications of simplification and modus ponens respectively. Step 4 cannot yield A by simplification, because it has a conditional as a major connective, and

simplification must have a conjunction. Step 8 is the fallacy of affirming the consequent, not modus ponens.

In the exercises that follow, you should produce as many steps as you possibly
can, but you should check your steps periodically to see if each is a correct application
of a rule of inference. Your basic goal should be to produce a valid reasoning trail
and avoid any Xs. Then, if you are successful in deriving the conclusion, inspect
your proof for superfluous steps to see if you could rewrite your proof in a more
elegant way. There may be more than one elegant way to do a proof. As in life,
there may be many ways to accomplish a goal. As an illustration of this point, note
that the first problem in the following exercises is the same problem we have been
working on. See if you can create an elegant, two-line proof that is different than
the working backward proof above.

STEP 3 EXERCISES: Construct a formal proof of validity for each of
the following arguments. (For number 1, produce a two-line proof that is different
than the working backward example above.)

1.
1. A ⊃ B
2. B ⊃ C
3. ~C /∴ ~A

2.
1. A ⊃ B
2. A ∨ C
3. ~B /∴ C

3.
1. D ⊃ B
2. F ∨ ~B
3. ~F /∴ ~D

*4.
1. G ⊃ H
2. I ⊃ D
3. G ∨ I /∴ H ∨ D

5.
1. ~D
2. ~D ⊃ ~B
3. C ⊃ B /∴ ~C

6.
1. J ⊃ (K • L)
2. S ∨ J
3. ~S
4. K ⊃ (S ∨ T)/∴ T

*7.
1. S ⊃ L
2. (H • T) ⊃ (P • X)
3. L ⊃ (H • T)
4. ~(P • X)
5. ~S ⊃ (A ⊃ B) /∴ A ⊃ (A • B)

8.
1. A ⊃ B
2. C ⊃ D
3. A ∨ C
4. ~B
5. D ⊃ (S ⊃ B) /∴ ~S

9.
1. (D ≡ ~H) ⊃ (P ∨ ~T)
2. ~~X
3. D ≡ ~H
4. P ⊃ ~X
5. ~T ⊃ Y
6. (~X ∨ Y) ⊃ G /∴ G • ~P

10.[1]
1. X ⊃ (Y ⊃ H)
2. X
3. H ⊃ P
4. Y
5. ~(H • P) /∴ Q

---

[1] **Hint**: A contradiction lurks in the premises of this problem. The proof cannot be solved unless you find the
contradiction. See the discussion of contradictions in Chapter 8 (pp. 278–279) for a further hint on what to do
when you find the contradiction.

TRANSLATIONS AND FORMAL PROOFS:

Translate the following arguments into symbolic notation. Check your answers with your instructor, then provide formal proofs for each translation.

1. If we buy the new car, then we will not have enough money for basic necessities provided we also pay for car insurance. If we buy the new car, we'll have to have car insurance. We are buying the new car. So, we will not have enough money for basic necessities. (C, B, I)

2. If all cars had air bags, car insurance premiums would go down for the following reasons: If all cars had air bags, hundreds of thousands of crippling injuries would be eliminated each year. The elimination of thousands of crippling injuries each year is a sufficient condition for the lowering of medical insurance premiums. The lowering of medical insurance premiums will cause car insurance premiums to go down.

   A = All cars had air bags.
   C = Car insurance premiums would go down.
   I = Hundreds of thousands of crippling injuries would be eliminated.
   M = Medical insurance premiums will be lowered.

3. Johnson must have contracted AIDS in prison, or our theory on AIDS needs revision. This is so because the incubation period for AIDS is approximately ten years. Johnson has been free for only two years since completing his twenty-year sentence. Now, if the incubation period for AIDS is approximately ten years, then provided Johnson has been free for only two years since completing his twenty-year sentence, he must have contracted AIDS in prison.

   I = The incubation period for AIDS is approximately ten years.
   F = Johnson has been free for only two years since completing his twenty-year sentence.
   P = Johnson must have contracted AIDS in prison.
   R = Our theory on AIDS needs revision.

*4. Necessary conditions for solving our country's drug problem involve not only reducing the supply of drugs, but also having effective drug treatment programs and an effective antidrug-education program. If we continue to follow the Bush administration's program, which to date has involved spending more than sixty billion dollars primarily on reducing the supply of drugs, we will not address all the necessary conditions. We are following the Bush administration's program. Hence, either we don't solve our country's drug problem or we don't continue to follow the Bush administration's program.

   S = We solve our country's drug problem.
   R = We reduce (or attempt to reduce) the supply of drugs.
   T = We have effective drug treatment programs.
   E = We have an effective antidrug-education program.
   B = We follow (or continue to follow) the Bush administration's program.

5. If Vietnam falls to communism, then Cambodia falls. If Vietnam and Cambodia

fall, then Laos falls. If Vietnam, Cambodia, and Laos fall, then Thailand falls. If all of the above fall, then all of Southeast Asia falls. Therefore, if Vietnam falls, all of Southeast Asia falls. (V,C,L,T,A) (**Note:** The formal proof of this argument is challenging.)

6. If either taxes are raised again or oil prices rise, then the economy will not continue to improve. Either the economy continues to improve, or the Republicans will have a potentially decisive political issue in November. If the Republicans will have a potentially decisive political issue in November and the Democrats do not have a counterissue of concern to the American people, then control of the White House will change. We know taxes are being raised again. We also know that the Democrats do not have a counterissue of concern to the American people. Accordingly, either control of the White House will change or the Republicans will not have a potentially decisive political issue in November.

   T = Taxes are raised again.
   O = Oil prices rise.
   E = The economy will continue to improve.
   R = The Republicans will have a potentially decisive political issue in November.
   D = The Democrats do have a counterissue of concern to the American people.
   W = Control of the White House will change.

7. Passing algebra with a C grade or better is a necessary condition for Cisa to be eligible to take calculus. Moreover, passing calculus is a necessary condition for Cisa to take engineering physics. Cisa has not passed algebra with a C grade or better. Because Cisa obviously can't pass calculus, if she is not eligible to take calculus, it follows that Cisa cannot take engineering physics.

   A = Cisa passes algebra with a C grade or better.
   E = Cisa is eligible to take calculus.
   C = Cisa passes calculus.
   P = Cisa takes engineering physics.

8. If Spock is emotional, then he is like Doctor McCoy. On the other hand, if Spock is logical, then he has a predominantly Vulcan personality. Either Spock is emotional or he is logical.[1] Everyone knows that Spock is not like Doctor McCoy. So, Spock has a predominantly Vulcan personality.

   E = Spock is emotional.
   M = Spock is like Doctor McCoy.
   L = Spock is logical.
   V = Spock has a predominantly Vulcan personality.

9. It is not true that having all true premises is a necessary condition for being able to have a valid argument. If it is not true that having all true premises is

---

[1] The intention of this *or* statement is exclusive. According to this book, this premise is a questionable dilemma.

a necessary condition for being able to have a valid argument, then one is able to have a valid argument even though that argument has false premises. Now, although having all true premises is a necessary condition for having a sound argument, it is not true that having all true premises is a sufficient condition for a valid or sound argument. If having all true premises is a necessary condition for having a sound argument, then it is not possible to have both a sound argument and one with false premises. Hence, although one is able to have a valid argument and have false premises, one cannot have both a sound argument and one with false premises.

T = One has an argument with all true premises.
V = One is able to have a valid argument.
S = One has a sound argument.

Translate "One has an argument with false premises" as "One does not have an argument with all true premises."

10. The student predicament of our times:
If I take the forty-hour-per-week job and take three classes per semester, then I will let the quality of my family life suffer. If I am to continue to progress at a sustained pace toward my degree and support my family, then I need to take the forty-hour-per-week job and take three classes per semester. If I don't both continue to progress at a standard pace toward my degree and support my family, then I will do neither. If I can neither continue to progress at a sustained pace toward my degree nor support my family, then I will not be happy. I will not let the quality of my family life suffer. However, if I don't continue to progress at a sustained pace toward my degree, I will not be eligible for a scholarship. So, either I am not going to be happy and not eligible for a scholarship, or I will need to win the state lottery or readjust my priorities.

F = I take the forty-hour-per-week job.
C = I take three classes per semester.
Q = I will let the quality of my family life suffer.
P = I continue to progress at a standard pace toward my degree.
S = I support my family.
H = I will be happy.
E = I will be eligible for a scholarship.
L = I will need to win the state lottery.
R = I will need to readjust my priorities.

Note: To allow for a formal proof using the nine rules, translate neither P nor S using the ~P • ~S version.

ANSWERS TO STARRED EXERCISES

STEP 1   3. B ⊃ C
/∴ (B ⊃ C) ∨ (B ⊃ C)
Add   Remember that **q** as a variable can stand for *any* simple or compound statement.

12. $(X \vee Y) \supset \sim(Z \cdot A)$
$\sim\sim(Z \cdot A)$ $/\therefore$ $\sim(X \vee Y)$
MT

16. $\sim A \vee (D \cdot B)$
$\sim A$ $/\therefore$ $D \cdot B$
X   This is a very common student mistake. Note that the second premise should be $\sim\sim A$ to stay true to the rule of DS.

23. $P \vee T$
$(P \supset \sim X) \cdot (T \supset \sim Y)$
$/\therefore \sim X \vee \sim Y$
CD

27. $(A \cdot B)$
$(X \cdot P) \supset (A \cdot B)$
$/\therefore X \cdot P$
X   Fallacy of Affirming the Consequent

STEP 2   3.
1. $(D \vee G) \cdot (H \vee I)$
2. $(D \supset H) \cdot (G \supset I)$
3. $\sim H$ $/\therefore$ I
4. $D \vee G$ (1) Simp
5. $H \vee I$ (2) (4) CD
6. I (3) (5) DS

8.
1. $(\sim A \vee B) \supset D$
2. $(D \vee B) \supset [\sim A \supset (C \equiv E)]$
3. $\sim A \cdot D$ $/\therefore$ $C \equiv E$
4. $\sim A$ (3) Simp.
5. $\sim A \vee B$ (4) Add
6. D (1) (5) MP
7. $D \vee B$ (6) Add
8. $\sim A \supset (C \equiv E)$ (7) (2) MP
9. $C \equiv E$ (4) (8) MP

STEP 3   4.
1. $G \supset H$
2. $I \supset D$
3. $G \vee I$ $/\therefore$ $H \vee D$
4. $(G \supset H) \cdot (I \supset D)$ (1) (2) Conj.
5. $H \vee D$ (4) (3) CD

Note that the conjunction ploy works in this proof! I often get a proof from students that concludes line 5 right away as line 4, justified by (1)(2)(3) CD. This is impossible, because CD does not have 3 premises (none of our rules do). But this is an interesting mistake, because it shows that the student who is making this mistake is beginning to see more than one step at a time. We will often have quick insights like this, but then we must "unpack" them—that is, the wholes that we are intuitively seeing must be broken down into parts. Supposedly, Mozart was able to see his entire fortieth symphony in a split second of inspiration. It then took him months to write it out, detail by detail. This symphony requires about a half hour to play!

7.
1. S ⊃ L
2. (H • T) ⊃ (P • X)
3. L ⊃ (H • T)
4. ~(P • X)
5. ~S ⊃ (A ⊃ B)   /∴   A ⊃ (A • B)
6. S ⊃ (H • T)   (1) (3) HS
7. ~(H • T)   (2) (4) MT
8. ~S   (6) (7) MT
9. A ⊃ B   (5) (8) MP
10. A ⊃ (A • B)   (9) Abs.

Note: This proof can be done several different ways with the same number of steps.

TRANSLATIONS AND FORMAL PROOFS

4.
1. S ⊃ [R • (T • E)]
2. B ⊃ ~[R • (T • E)]
3. B   /∴ ~S ∨ ~B
4. ~[R • (T • E)]   (2) (3) MP
5. ~S   (1) (4) MT
6. ~S ∨ ~B   (5) Add

Chapter 10

SYMBOLIC TRAILS AND FORMAL PROOFS OF VALIDITY, PART 2

Introduction

The previous chapter showed many frustrating signs that something was wrong with our formal proof method that relied on only nine elementary rules of validity. Simple, intuitive valid arguments could not be shown to be valid. For instance, the following intuitively valid arguments cannot be shown to be valid using only the nine rules.

Somalia and Iraq are both foreign policy risks.
Therefore, Iraq is a foreign policy risk.

$$S \bullet I$$
$$/\therefore \quad I$$

Either Clinton or Bush was president of the United States in 1993.
Bush was not president in 1993.
So, Clinton was president of the United States in 1993.

$$(C \lor B) \bullet \sim(C \bullet B)$$
$$\sim B \quad /\therefore \quad C$$

If the computer networking system works, then Johnson and Kaneshiro will both be connected to the home office.
Therefore, if the networking system works, Johnson will be connected to the home office.

$$N \supset (J \bullet K)$$
$$/\therefore \quad N \supset J$$

Either the Start II treaty is ratified, or this landmark treaty will not be worth the paper it is written on.

Therefore, if the Start II treaty is not ratified, this landmark treaty will not be worth the paper it is written on.

$$R \vee {\sim}W$$
$$/ \therefore \quad {\sim}R \supset {\sim}W$$

If the light is on, then the light switch must be on.
So, if the light switch in not on, then the light is not on.

$$L \supset S$$
$$/ \therefore \quad {\sim}S \supset {\sim}L$$

Thus, the nine elementary rules of validity covered in the previous chapter must be only part of a complete system for constructing formal proofs of validity. A *complete* system must allow us to make a formal proof for any valid argument that can be translated into propositional logic form, the level of argumentation consisting of simple sentences and the five logical connectives. So, to complete our system we require ten more elementary arguments. The first nine rules we will call **rules of inference**, the second ten are called **rules of replacement**. Although these rules involve the same level of abstraction, they differ somewhat in application; we will have much more flexibility in applying them. For instance, a rule of inference such as modus ponens can be applied to only whole lines and the direction of inference is only in one direction. The premises cannot be derived from the conclusion. On the other hand, a rule of replacement can be applied to a whole line *or* part of a line, and the direction of inference can go in two directions.

For instance, **De Morgan's theorem** is an example of a rule of replacement: ${\sim}(p \cdot q) \equiv ({\sim}p \vee {\sim}q)$. This rule directs that any statement or part of a statement that has the form of ${\sim}(p \cdot q)$ can be replaced with $({\sim}p \vee {\sim}q)$, and any statement or part of a statement that has the form of $({\sim}p \vee {\sim}q)$ can be replaced with ${\sim}(p \cdot q)$. Whereas a truth table of modus ponens shows this argument form to be valid, a truth table shows De Morgan's theorem to be a *logical equivalence*, the right side of the $\equiv$ symbol has the same truth table result as the left side.

p	q	~	(p	•	q)	≡	(~p	∨	~q)
T	T	**F**	T	T	T	**T**	F	**F**	F
T	F	**T**	T	F	F	**T**	F	**T**	T
F	T	**T**	F	F	T	**T**	T	**T**	F
F	F	**T**	F	F	F	**T**	T	**T**	T

Note that when ${\sim}(p \cdot q)$ is false (first row), ${\sim}p \vee {\sim}q$ is also false. When ${\sim}(p \cdot q)$ is true (rows 2–4), ${\sim}p \vee {\sim}q$ is also true. Because a biconditional ($\equiv$) is true when the truth values of both sides match, the truth value result of De Morgan's theorem is always true. *A logical equivalence is a biconditional that is always true.* All of the new rules of replacement will be logical equivalences, which justifies their use in manipulating statements and in symbolic reasoning trails.

At a less formal level of justification, it should help you to understand that the new rules are, like the nine rules of inference, just formalizations of our common sense. We have already been using De Morgan's theorem when we determined that saying that, "Alice and Barbara are not both coming to the party" ${\sim}(A \cdot B)$, is the same as saying that, "Either Alice is not coming to the party or Barbara is not coming to the party." (~A ∨ ~B) Regardless of how threatening they may look at first, keep in mind that the new rules originate from our common sense, and that we are still following the same procedure of breaking our common sense down into very simple rules so that we can use them to create chains of reasoning and analyze complex arguments.

At a practical level of proof construction, these new rules give us considerable flexibility. Because a rule of replacement can be applied to whole and parts of statements, and the inference direction can be in either direction of the ≡ symbol, each rule of replacement possesses four different possible applications. For instance, here are four different ways that the De Morgan rule could be applied. Because ~(p • q) ≡ (~p ∨ ~q), we could have:

1. ~(A • B) /∴ ~A ∨ ~B   3. ~(A • B) ⊃ C /∴ (~A ∨ ~B) ⊃ C
2. ~A ∨ ~B /∴ ~(A • B)   4. (~A ∨ ~B) ⊃ C /∴ ~(A • B) ⊃ C

Notice that in 1 and 2 the rule is applied to whole lines, but operates in either direction; whereas in 3 and 4 the rule is applied to a part of a line, and also operates in either direction. Because De Morgan's theorem is a logical equivalence, we are entitled to "replace" any statement or part of a statement that fits the form of one side of the biconditional with its equivalent on the other side. This will give us a great deal of flexibility in deriving proofs. However, a major restriction that must never be violated is that *if a rule of replacement is applied to part of a line, the whole line must be concluded*. Notice that in 3 and 4, the conclusion included the consequent C. The conclusion must always be equivalent to the premise when applying a rule of replacement. If the consequent were dropped, then the conclusions in both examples would not be equivalent to the premises.

On the following page is a complete list of all nineteen rules that we will use to construct formal proofs. This system is one of many possible complete systems. Most students find this list overwhelming and quite abstract at first glance, and it takes some time to adjust to applying the new rules just as it did with the first nine. However, keep in mind that each of the new rules has a commonsense equivalent and that the process of pattern recognition is the same as with the first nine rules. We will also gradually become more comfortable using these rules by going through the same three steps (Steps 4–6) as with the nine rules. Most important for your success at this stage is that if you have been successful with the nine rules (if you have not, go back to a previous step in Chapter 9), then you understand the symbolic game: regardless of how esoteric, abstract, and complicated a symbolic statement or transformation may seem, it is just a compilation of simple parts or a chain of simple, commonsense steps. It is simply a matter of being calm and disciplined enough to let your brain see the parts.

Application Practice

The first step of learning the nine rules involved recognizing when a rule applied. Before you try some first-step exercises with these new rules, let's look at some

FIGURE 10-1

The Nineteen Rules

Rules of Inference

Modus Ponens (M.P.)	Modus Tollens (M.T.)
$p \supset q$	$p \supset q$
p /∴ $\boxed{q}$	$\sim q$ /∴ $\sim p$

Hypothetical Syllogism (H.S.)	Disjunctive Syllogism (D.S.)
$p \supset q$	$p \lor q$
$q \supset r$ /∴ $p \supset r$	$\sim p$ /∴ q

Constructive Dilemma (C.D.)	Absorption (Abs.)
$(p \supset q) \cdot (r \supset s)$	$p \supset q$
$p \lor r$ /∴ $q \lor s$	/∴ $p \supset (p \cdot q)$

Simplification (Simp.)	Conjunction (Conj.)	Addition (Add.)
$p \cdot q$	p	p
/∴ p	q /∴ $p \cdot q$	/∴ $p \lor q$

Rules of Replacement

De Morgan's Theorems: $\sim (p \cdot q) \equiv (\sim p \lor \sim q)$
(De M.) $\sim (p \lor q) \equiv (\sim p \cdot \sim q)$

Commutation: $(p \lor q) \equiv (q \lor p)$
(Com.) $(p \cdot q) \equiv (q \cdot p)$

Association: $[\, p \lor (q \lor r)\,] \equiv [\,(p \lor q) \lor r\,]$
(Assoc.) $[\, p \cdot (q \cdot r)\,] \equiv [\,(p \cdot q) \cdot r\,]$

Distribution: $[\, p \cdot (q \lor r)\,] \equiv [\,(p \cdot q) \lor (p \cdot r)\,]$
(Dist.) $[\, p \lor (q \cdot r)\,] \equiv [\,(p \lor q) \cdot (p \lor r)\,]$

Double Negation (D.N.): $p \equiv \sim\sim p$

Transposition (Trans.): $(p \supset q) \equiv (\sim q \supset \sim p)$

Implication: $(\sim p \supset q) \equiv (p \lor q)$
(Impl.)

Equivalence: $(p \equiv q) \equiv [\,(p \supset q) \cdot (q \supset p)\,]$
(Equiv.) $(p \equiv q) \equiv [\,(p \cdot q) \lor (\sim p \cdot \sim q)\,]$

Exportation (Exp.) $[\,(p \cdot q) \supset r] \equiv [\, p \supset (q \supset r)\,]$

Repetition (Rep.) $p \equiv (p \lor p)$
 $p \equiv (p \cdot p)$

strategies and examples of pattern recognition with these new rules. Suppose we had the argument,

$$(A \supset B) \cdot (C \supset D)$$
$$/ \therefore \quad (A \supset B) \cdot (\sim D \supset \sim C)$$

Your task is to identify which one of the new rules justifies this inference as valid. To focus, first ask, "Has the whole line been changed in the conclusion or only a part?" In this case, only part of the premise has been changed in the conclusion. The right side of the conjunction, $(C \supset D)$, has been replaced with $(\sim D \supset \sim C)$. Essentially an *if . . . then* statement has been turned around and negations added to both the antecedent and the consequent. Scanning the new rules of replacement we see that the transposition rule, $(p \supset q) \equiv (\sim q \supset \sim p)$, justifies this move. Because $(p \supset q) \equiv (\sim q \supset \sim p)$ is a logical equivalence, $(C \supset D)$ can be replaced with $(\sim D \supset \sim C)$. So, the answer to the above problem would be "Trans."

Let's try another one, this time a whole line manipulation. Suppose we had the following:

$$(A \supset B) \cdot (B \supset A)$$
$$/ \therefore \quad A \equiv B$$

Clearly, in this case the whole line has been changed, because the major connective of the premise is a conjunction and the major connective of the conclusion is a biconditional. A change in the major connective tells us that the whole line has been changed. In such situations we must scan the new rules, looking for a rule that has a conjunction on one side of the logical equivalence and a biconditional on the other, a rule with $(\cdot) \equiv (\equiv)$. Only the equivalence rule meets this condition. But before we conclude that this is the answer, we must be sure the entire pattern fits. The first version of equivalence, $(p \equiv q) \equiv [(p \supset q) \cdot (q \supset p)]$, justifies the changing of two conditional statements connected by a conjunction to a biconditional, provided the two conditional statements have the special relationship $(p \supset q)$, $(q \supset p)$. The above premise does have two conditional statements in this special relationship, $(A \supset B)$, $(B \supset A)$, so "Equiv." is the answer to this problem.

A whole line can also be manipulated wherein the major connective stays the same. In such cases, both sides of the major connective must be changed. For instance,

$$(A \cdot B) \supset (D \cdot F)$$
$$/ \therefore \quad \sim(D \cdot F) \supset \sim(A \cdot B)$$

has the same major connective in the premise and the conclusion, but the antecedent and the consequent have been reversed and a negation added to both sides. Again the transposition rule, $(p \supset q) \equiv (\sim q \supset \sim p)$, allows for this manipulation.

In applying these new rules you must be capable of the same level of abstraction as with the nine rules. As you know by now, the following are both examples of modus ponens:

A ⊃ B   [(A ∨ B) ⊃ (D • F)] ⊃ (X ∨ ~P)
A   /∴ B   (A ∨ B) ⊃ (D • F)   /∴   (X ∨ ~P)

Similarly, both of the following are examples of transposition:

A ⊃ B   [(A ∨ B) ⊃ (D • F)] ⊃ (X ∨ ~P)
/∴   ~B ⊃ ~A   /∴   ~(X ∨ ~P) ⊃ ~[(A ∨ B) ⊃ (D • F)]

STEP 4 EXERCISES   **Rules of Replacement:** For each of the following arguments state the rule of replacement that is being used. Indicate with an X
those problems that do not have a rule of replacement justification.

1. ~A ⊃ B
   /∴   A ∨ B

2. (C ⊃ D) ∨ ~(X • Y)
   /∴   (C ⊃ D) ∨ (~X ∨ ~Y)

3. (J • K) ⊃ (D ⊃ I)
   /∴   J ⊃ [K ⊃ (D ⊃ I)]

*4. D ⊃ B
   /∴   ~D ⊃ ~B

5. O ⊃ [(X ⊃ Y) • (Y ⊃ X)]
   /∴   O ⊃ (X ≡ Y)

6. ~(R ∨ S) ⊃ (~R ∨ ~S)
   /∴   ~(S ∨ R) ⊃ (~R ∨ ~S)

7. ~(D ∨ Y)
   /∴   ~D • ~Y

8. (X ∨ Y) • (~X ∨ ~Y)
   /∴   [(X ∨ Y) • ~X] ∨ [(X ∨ Y) • ~Y]

*9. (A ∨ B) ⊃ (X ⊃ Y)
   /∴   B ∨ A

10. [C • (D • E)] ∨ (C ∨ E)
   /∴   [(C • D) • E] ∨ (C ∨ E)

11. ~~(A • D) ⊃ ~X
   /∴   (A • D) ⊃ ~X

12. (H ⊃ ~I) ⊃ (~I ⊃ ~J)
   /∴   (H ⊃ ~I) ⊃ (J ⊃ I)

13. (~K ∨ L) ⊃ (~M ∨ ~M)
   /∴   (~K ∨ L) ⊃ ~M

14. [(~O ∨ P) ∨ ~Q] ∨ ~(P ∨ ~Q)
   /∴   [~O ∨ (P ∨ ~Q)] ∨ ~(P ∨ ~Q)

15. (R ∨ ~S) ∨ (T ⊃ ~P)
   /∴   (T ⊃ ~P) ∨ (R ∨ ~S)

16. [H • (I ∨ J)] ∨ [H • (K ⊃ ~L)]
   /∴   H • [(I ∨ J) ∨ (K ⊃ ~L)]

17. [(~A • B) • (C ∨ D)] ∨ [~(~A • B) • ~(C ∨ D)]
    /∴   (~A • B) ≡ (C ∨ D)

*18. ~~A ⊃ B
   /∴   ~A ∨ B

19. ~C ⊃ (D • X)
   /∴   ~(D • X) ⊃ ~~C

20. (~M ∨ ~~N) • (O ≡ ~~P)
    /∴   (~M ∨ N) • (O ≡ ~~P)

Commonsense Origins

Before we try some second step exercises with these new rules, let's attempt to deflate some of the sensory overload you may be feeling by looking at some of these rules and their commonsense origins.

De Morgan's Theorem $\sim(p \cdot q) \equiv (\sim p \vee \sim q)$
$\sim(p \vee q) \equiv (\sim p \cdot \sim q)$

We have already seen that a little reflection shows that the statement, "Alice and Barbara are not both going to the party," $\sim(A \cdot B)$, is the same as saying that "Either Alice is not going or Barbara is not going to the party," $(\sim A \vee \sim B)$. And, asserting a *neither-nor* is the same as stating a *both-not*, as in "Neither Alice nor Barbara are going to the party," $\sim(A \vee B)$, and "Alice and Barbara are both not going to the party," $(\sim A \cdot \sim B)$. So, there are two versions of De Morgan's theorem. Remember also that a *not-both* statement is not the same as a *both-not* statement, and a *neither-nor* statement is not the same as an *either-not-or-not* statement. Thus, these versions must always be applied separately. A major mistake that students often make is to conclude $(\sim p \cdot \sim q)$, a part of the second version, from $\sim(p \cdot q)$, a part of the first version. If we knew that Alice and Barbara were not both coming to the party, we could not be certain from this alone that Alice was not coming for sure and that Barbara was not coming for sure. Similarly, it would be a mistake to infer from, "Either John did not pass the final or he did not pass the course," $(\sim F \vee \sim C)$, that "John passed neither the final exam nor the course," $\sim(F \vee C)$. As shown in Chapter 7,

not both = either not. . .or not. . . = $\sim(\underline{\quad} \cdot \underline{\quad}) = \sim\underline{\quad} \vee \sim \underline{\quad}$
both not = neither. . . nor. . .   = $\sim\underline{\quad} \cdot \sim\underline{\quad} = \sim(\underline{\quad} \vee \underline{\quad})$

But

not both ≠ both not
$\sim(\underline{\quad} \cdot \underline{\quad}) \neq \sim\underline{\quad} \cdot \sim\underline{\quad}$
neither. . . nor. . . ≠ either not. . .or not. . .
$\sim(\underline{\quad} \vee \underline{\quad}) \neq \sim\underline{\quad} \vee \sim\underline{\quad}$

Commutation: $(p \vee q) \equiv (q \vee p)$
$(p \cdot q) \equiv (q \cdot p)$

This rule simply reflects the commonsense notion that "Either Alice or Barbara is going to the party," $(A \vee B)$, is the same as "Either Barbara or Alice is going to the party," $(B \vee A)$, and "Alice and Barbara are going to the party," $(A \cdot B)$, is the same as "Barbara and Alice are going to the party," $(B \cdot A)$.[1] Note, however, that this rule applies only to statements with the connectives *or* and *and*; it allows only statements with these connectives to be "turned around," or commuted, in this

---

[1] In grammar school, we learned a similar commutation rule in mathematics, $3 + 2 = 2 + 3$ and $3 \times 2 = 2 \times 3$. In algebra these commonsense notions were generalized to $X + Y = Y + X$ and $X \times Y = Y \times X$.

manner. It is invalid to commute conditionals: $(p \supset q)$ is not logically equivalent to $(q \supset p)$. Given the premise, "If the police department does a good job, then crime will decrease," $(P \supset C)$, it does not follow that "If crime decreases, then the police department must have done a good job," $(C \supset P)$.[1]

Association:
$$[p \lor (q \lor r)] \equiv [(p \lor q) \lor r]$$
$$[p \bullet (q \bullet r)] \equiv [(p \bullet q) \bullet r]$$

This rule reflects the commonsense notion that "Either Alice, or Barbara, or Carol are going to the party," $[A \lor (B \lor C)]$, is the same as "Alice or Barbara are going to the party, or Carol is going to the party," $[(A \lor B) \lor C]$. And, "Alice is going to the party, and Barbara and Carol are also going to the party," $[A \bullet (B \bullet C)]$ is the same as "Alice and Barbara are going to the party, and Carol is going as well," $[(A \bullet B) \bullet C]$. Essentially the association rule states that when the logical connectives within a statement are all conjunctions or disjunctions, it does not matter where the parentheses go. They can be manipulated at will.[2]

However, note that association does not work with conditionals: $[p \supset (q \supset r)]$ is not logically equivalent to $[(p \supset q) \supset r]$. The statement, "If Weng has a quiz average of more than 90 percent, then if he also gets at least 88 percent on the final exam then he will receive an A for the course," $[Q \supset (F \supset A)]$ is not saying the same thing as, "If Weng has a quiz average of more than 90 percent only if he gets at least 88 percent on the final exam, then he will receive an A for the course," $[(Q \supset F) \supset A]$. The first statement refers to two conditions that together are sufficient for an A in the course, a high quiz average and a relatively high grade on the final exam, a likely state of affairs in many college courses; whereas, the second statement makes the absurd claim that a sufficient condition for an A in the course is that doing well on the final, which normally happens after the taking of quizzes, is a necessary condition for doing well on the quizzes! In other words, it would be necessary to get at least 88 percent on the final exam before averaging better than 90 percent on his quizzes, and this strange set of circumstances would then be sufficient for receiving an A in the course. So, unlike the association rule for conjunctions and disjunctions the parentheses cannot be moved at will for conditionals.[3]

Distribution:
$$[p \bullet (q \lor r)] \equiv [(p \bullet q) \lor (p \bullet r)]$$
$$[p \lor (q \bullet r)] \equiv [(p \lor q) \bullet (p \lor r)]$$

For most students the distribution rule is the most difficult to apply. But the commonsense origin of its two versions is not difficult to see. The statement, "Alice

---

[1] However, a biconditional statement can be commuted, $p \equiv q$ is logically equivalent to $q \equiv p$. The statement, "You can go out tonight if and only if you clean your room," $(O \equiv R)$, is logically equivalent to "Cleaning your room is a necessary and sufficient condition for going out tonight," $(R \equiv O)$. This will not be one of our rules though; we will prove this equivalence as a formal proof later.

[2] Similarly, in mathematics we learn that $3 \times (5 \times 2)$ is the same as $(3 \times 2) \times 5$ and hence, $X \times (Y \times Z) = (X \times Y) \times Z$.

[3] We will see, however, that association is valid for biconditionals: $[(p \equiv q) \equiv r]$ is logically equivalent to $[p \equiv (q \equiv r)]$. Rather than have this as a rule in our system, we will save this for a very entertaining proof.

is going to the party, and Barbara or Carol is going," [A • (B ∨ C)], is the same as, "Either Alice and Barbara are going to the party or Alice and Carol are going to the party," [(A • B) ∨ (A • C)]. And, the statement, "Either Rios passes the final exam or he will not receive the scholarship and not be eligible for the football team," [R ∨ (~S • ~T)] is the same as, "Rios passes the final or he will not receive the scholarship, and Rios passes the final or he will not be eligible for the football team," [(R ∨ ~S) • (R ∨ ~T)].

The important perspective to see for application purposes is that distribution always mixes conjunctions and disjunctions. It either "distributes" a conjunction through a disjunction (first version)—what was a conjunction becomes a disjunction, or it distributes a disjunction through a conjunction (second version)—what was a disjunction becomes a conjunction.

Double Negation: $p \equiv \sim\sim p$

When we say that something is not impossible we mean that it is possible. When a bureaucrat says that an action was not unauthorized we know that it was authorized. Or, when Oliver North told us that he did not believe that he went before Congress not believing that he was going to tell the truth, we know, after shaking our head a lot, that North was claiming that he went before Congress believing that he was going to tell the truth. In understanding these transformations, we are applying the double negation rule, that is, two negations cancel to make a positive statement. However, as intuitively obvious as this rule is, students often do not realize the full flexibility and power of its possible applications in terms of generating steps in a proof. Here is a trick question: "How many times can double negation be applied to $A \supset (B \supset C)$?" Most students will answer three times. Some, a little more aware of the full meaning of a variable, will reply four times, and a very few even more aware will reply five times. The answer, however, is an infinite number of times! To understand how, consider that applying double negation to $\sim\sim A$ could result in either just A or $\sim\sim\sim\sim A$.

The important point to remember is that the double negation rule says that we can either put two negations on a statement or take two negations off a statement. In formal proofs, some students will mistakingly think that we can conclude $\sim A$ from $\sim\sim A$, or that $\sim A \supset \sim B$ follows from $A \supset B$. The first application removes only one negation and hence would introduce a contradiction into a formal proof. The second application assumes that as long as the negations add up to two, we can take them from wherever we want! Both applications are invalid.

Transposition: $(p \supset q) \equiv (\sim q \supset \sim p)$

Clearly, from the statement, "If John passed the final exam, then he passed the course," (F ⊃ C), we would be able to infer, "If John did not pass the course, then he did not pass the final exam," (~C ⊃ ~F), and vice versa. If a sufficient condition for passing the course is to pass the final exam, then if the course was not passed we know that the sufficient condition was not fulfilled. However, as noted above, it is a mistake to infer (~p ⊃ ~q) from (p ⊃ q). Although a necessary condition for being pregnant is to be female, (P ⊃ F), it would surely not follow from this that if a person is not pregnant that person is not female, (~P ⊃ ~F). The bottom line in applying transposition is to always remember that the consequent and the antecedent must be switched when negations are added to both. Also remember that rules of replacement can be applied in either direction of the ≡ symbol. So, given (~A ⊃ ~B), we could infer either (B ⊃ A) or (~~B ⊃ ~~A). The latter application might be useless, silly, or at least inelegant,[1] but it would be valid.

Implication: $\qquad\qquad (\sim p \supset q) \equiv (p \vee q)$

If either Alice or Barbara are going to the party, (A ∨ B), then if Alice is not going, Barbara will go, (~A ⊃ B). And, as noted at the beginning of this chapter an inference such as the following is valid.

Either the Start II treaty is ratified or this landmark treaty will not be worth the paper it is written on. Therefore, if the Start II treaty is not ratified, this landmark treaty will not be worth the paper it is written on.

$$R \vee \sim W$$
$$/ \therefore \sim R \supset \sim W$$

It would also follow from, "If Leina does not pass the final, she does not pass the course," (~F ⊃ ~C), that, "Either Leina passes the final or she does not pass the course," (F ∨ ~C).

The important application point to remember about implication is that this rule allows us to move back and forth between disjunctions and conditionals, provided that a negation is either added (to the antecedent in moving to a conditional) or subtracted (from the antecedent in moving to a disjunction) in making the transformation.

Equivalence: $\qquad (p \equiv q) \equiv [(p \supset q) \cdot (q \supset p)]$
$\qquad\qquad\qquad\quad (p \equiv q) \equiv [(p \cdot q) \vee (\sim p \cdot \sim q)]$

As you recall, in Chapter 7 we discussed the meaning of statements such as, "You go out tonight with your friends if and only if you clean your room," (O ≡ R). There we noted that the use of *if and only if* is a way of specifying a necessary and sufficient condition, [(O ⊃ C) • (C ⊃ O)]. We can now add that this statement is also equivalent to, "Either you go out and clean your room, or you don't go out and don't clean your room," [(O • C) ∨ (~O • ~C)].

Note that the important application point for equivalence is that although all the rules make use of the biconditional symbol (≡), only equivalence has an application result that produces this connective, (p ≡ q). In looking at 17 in the

---

[1] It would be inelegant because, with two applications of double negation, ~~B ⊃ ~~A would become B ⊃ A. Why conclude something in three steps (transposition and two double negations), when we can infer the same thing in one step?

above Step 4 exercises, the only possible replacement rule would be equivalence, because no other rule allows us to transform a premise into a biconditional statement. Noting this, your original hypothesis should be that 17 is an equivalence, and then you would focus on whether all the parts fit.

It is also worth noting that, because there are two versions of equivalence, and both versions have $p \equiv q$ as a possible result, unlike a rule such as De Morgan in which it is invalid to conclude from a part of one version a part of the other version—we can't conclude $\sim p \cdot \sim q$ from $\sim(p \cdot q)$—with equivalence we can move validly from $[(p \supset q) \cdot (q \supset p)]$ to $[(p \cdot q) \vee (\sim p \cdot \sim q)]$, as we will see. But it takes two steps. Can you see the steps?

Exportation: $\quad\quad\quad [(p \cdot q) \supset r] \equiv [p \supset (q \supset r)]$

The intuitive validity of exportation is seen by noting the logical equivalence of statements such as, "If we make the car payment and also pay our medical bills this month, then we will not have enough money for basic necessities," $[(C \cdot M) \supset \sim B]$, and "If we make the car payment, then if we also pay our medical bills we will not have enough money for basic necessities," $[C \supset (M \supset \sim B)]$.

An important application point to note is that exportation cannot be applied to any statement with the form $[p \supset (q \cdot r)]$. In formal proofs, a common student mistake is to conclude statements with the form $[p \supset (q \supset r)]$ or $[(p \supset q) \supset r)]$ from $[p \supset (q \cdot r)]$. But the statement (1), "If we make the car payment this month, then we will also pay our medical bills and not have enough money for basic necessities," $[C \supset (M \cdot \sim B)]$, clearly has a different meaning from (2), "If we make the car payment, then if we also pay our medical bills we will not have enough money for basic necessities," $[C \supset (M \supset \sim B)]$, and (3), "If we make the car payment only if we also pay our medical bills, then we will not have enough money for basic necessities," $[(C \supset M) \supset \sim B)]$. Statement (1) asserts that making the car payment is a sufficient condition for also paying the medical bills, and statement (3) has as a sufficient condition for not having enough money for basic necessities the claim that a necessary condition for making the car payment is paying the medical bills! Whereas, all that is intended by (2) is that if both the car payment and the medical bills are paid, then there will not be enough money for basic necessities, $[(C \cdot M) \supset \sim B] \equiv [C \supset (M \supset \sim B)]$.

Repetition: $\quad\quad\quad p \equiv (p \vee p)$
$\quad\quad\quad\quad\quad\quad\quad\quad p \equiv (p \cdot p)$

Our last logical equivalence is so obvious that it may seem silly even to have to state it. Surely, if we know that John passed the final exam, (F), then we know that he either passed the final exam or he passed the final exam, (F ∨ F), and he passed the final exam and he passed the final exam, (F • F). However, this rule is needed to make our rules complete—to have a rule system that will enable us to construct a formal proof of validity for any argument that can be translated into propositional logic. Also, by comparison with $p \equiv (p \supset p)$ and $p \equiv (p \equiv p)$, which are both *invalid*, we see that we must be careful with even the simplest of inferences.

Strategies for Pattern Recognition Revisited

Recall that this is the major point behind all logical analysis: We get straight on the simplest components of reasoning and then use these mini-steps to analyze or produce much more complicated reasoning trails and claims. With this little excursion into the commonsense background to the rules of replacement now complete, we are ready to continue the steps necessary for constructing your own formal proofs using all the rules. As aspiring painters must first learn to mimic the styles of the great masters, it is time for you to attempt to mimic completed reasoning trails by figuring out how the trails were constructed. But let's do one problem together first.

As in Chapter 9, Step 2, the following Step 5 exercises require that you justify each line by citing the premise or premises and the rule used. Any one of the nineteen rules may now be involved. At first, this will involve a lot of staring and possible sensory overload. Recall also that in the Step 2 exercises two strategies allowed us to focus and test specific rule application possibilities, rather than just randomly stare at the rules. These same strategies will of course continue to work with lines that involve application of the first nine rules. We can now add, however, a third strategy that we were unconsciously using in Step 4 exercises. Because any application of the ten rules of replacement will necessarily involve a modification of a premise—either part of a line or a whole line will be changed—we can use this to our advantage as a strategy: we can examine whether the line we need to justify seems to be a modification of any previous line. For instance, consider the following proof:

1. $A \supset \sim(B \lor C)$
2. $(\sim B \bullet \sim C) \supset D$
3. $D \supset A$
4. $\underline{(A \equiv D) \supset X} \qquad / \therefore \quad X$
5. $A \supset (\sim B \bullet \sim C)$
6. $A \supset D$
7. $(A \supset D) \bullet (D \supset A)$
8. $A \equiv D$
9. X

The first line to be justified is line 5. In comparing line 5 with the premises, we see that it is very similar to line 1. The same statements are involved (As, Bs, and Cs), and these lines have the same antecedent (A) and the same conditional major connective. The only difference is that the consequent in line 1 is $\sim(B \lor C)$ and the consequent in line 5 is $(\sim B \bullet \sim C)$. In looking at the rules of replacement we know that one of the versions of the De Morgan rule, $\sim(p \lor q) \equiv (\sim p \bullet \sim q)$, justifies this modification. So, we would now have,

1. $A \supset \sim(B \lor C)$
2. $(\sim B \bullet \sim C) \supset D$

3. D ⊃ A
4. (A ≡ D) ⊃ X   /∴   X
5. A ⊃ (~B • ~C)   (1) De M.
6. A ⊃ D
7. (A ⊃ D) • (D ⊃ A)
8. A ≡ D
9. X

This strategy of looking for a modification of a line does not appear to work for line 6. Although line 3 has the same letters as 6, looking at the rules of replacement we see that no rule allows us to turn a conditional statement around. The closest we have to this would be transposition, but this rule requires negations to be added to both the antecedent and the consequent. The first strategy used in Step 2 exercises of seeing whether line 6 is a part of a previous line also fails. Hence, we are back to the strategy of focusing on the connective, and in this case, then, examining all the rules with ⊃ as the major connective of the conclusion.

With the addition of the ten new rules, this is now a potentially very time-consuming strategy. There are five rules that could be involved, because in five rules ⊃ either is always the connective of the conclusion (HS and Abs.) or could be the connective of the conclusion (Trans., Impl., and Exp.). For instance, line 6 could be justified by transposition if we had a previous line of ~D ⊃ ~A, since (A ⊃ D) ≡ (~D ⊃ ~A). Thus, the use of this strategy requires that you understand the rules very well and can eliminate possibilities quickly. You would need to be able to see quickly that we don't have ~D ⊃ ~A, so transposition is ruled out. Similarly, implication and exportation can be ruled out quickly. An application of implication must always result in at least one negation on the antecedent, and line 6 does not have one, and exportation always involves at least three statement parts, and line 6 only has two. Absorption is also eliminated because the consequent of the conclusion always involves two statement parts, and line 6 only has one.

Thus, we are left with hypothetical syllogism as the only viable hypothesis. This, then, means we would be looking for either,

1. 1. A ⊃ ~(B ∨ C)   or   5. A ⊃ (~B • ~C)
? ~(B ∨ C) ⊃ D /∴   A ⊃ D   ? (~B • ~C) ⊃ D   /∴ A ⊃ D

because either would produce the desired conclusion by hypothetical syllogism. Because we don't have ~(B ∨ C) ⊃ D as a premise, and do have (~B • ~C) ⊃ D in line 2, the second alternative works and we now have,

1. A ⊃ ~(B ∨ C)
2. (~B • ~C) ⊃ D
3. D ⊃ A
4. (A ≡ D) ⊃ X   /∴   X
5. A ⊃ (~B • ~C)   (1) De M.
6. A ⊃ D   (5) (2) HS

7. $(A \supset D) \cdot (D \supset A)$
8. $A \equiv D$
9. X

Next, line 7. Line 7 is definitely not a part of a previous line, nor does it appear to be a modification of a previous line. So, let's try focusing on the connective. As a conjunction for the major connective, potentially we have the conjunction rule for the first nine rules, and De Morgan, commutation, association, distribution, equivalence, and repetition as potential rules of replacement applications. In the latter case, we say "potential" because these rules of replacement all involve a conjunction as the major connective on one side of the logical equivalence. This does not mean that we must take them all equally seriously as possible applications, especially if you know the rules well. For instance, De Morgan has a conjunction as a possible result, $\sim p \cdot \sim q$, but because it involves negations, we know that line 7 cannot be the result of this rule. Similar conclusions of impossibility apply to association, distribution, and repetition. Association requires that a minor connective be a conjunction as well, and distribution requires a disjunction as the minor connective, but line 7 has conditionals as minor connectives. Repetition requires that the two statement parts be identical, $(p \cdot p)$, and the two parts in line 7 are not identical.

Both commutation and equivalence are possibilities, but because we have already noted that line 7 does not appear to be a modification of a previous line, it is not likely that any of the rules of replacement will be involved. For this reason we should try conjunction. In doing so you should see,

6. $A \supset D$
3. $D \supset A$ /∴ $(A \supset D) \cdot (D \supset A)$

So, we now have,

1. $A \supset \sim(B \lor C)$
2. $(\sim B \cdot \sim C) \supset D$
3. $D \supset A$
4 $(A \equiv D) \supset X$ /∴   X
5. $A \supset (\sim B \cdot \sim C)$ (1) De M.
6. $A \supset D$ (5) (2) HS
7. $(A \supset D) \cdot (D \supset A)$ (6) (3) Conj.
8. $A \equiv D$
9. X

Note that line 7 could have followed from either $(D \supset A) \cdot (A \supset D)$ by commutation or $A \equiv D$ by equivalence. However, neither of these lines exist *above* line 7. The statement $A \equiv D$ exists in line 8, but a line below the line we are on can never be used to justify the line we are on. However, because there is the relationship of equivalence between line 7 and 8, we have our answer for line 8. So, we now have,

1. $A \supset \sim(B \lor C)$
2. $(\sim B \cdot \sim C) \supset D$

3. D ⊃ A
4. (A ≡ D) ⊃ X   /∴   X
5. A ⊃ (~B • ~C)   (1) De M.
6. A ⊃ D   (5) (2) HS
7. (A ⊃ D) • (D ⊃ A)   (6) (3) Conj.
8. A ≡ D   (7) Equiv.
9. X

Sometimes you will just see the justification without going through any strategies consciously. Remember that often the line above is used to justify the line you are on, and by looking at this line you may just see a connection with a rule immediately. Note, however, that another way to figure out line 8 would be to focus on the connective, (≡). Because equivalence is the only rule that produces this connective, in such situations its application is always a likely possibility. Its application is an absolute guarantee in the present case, because the only other way that we can have a conclusion with ≡ would be if it occurs as a minor connective in modus ponens, disjunctive syllogism, or simplification.

Because line 9 does not have any connectives, it is most likely that strategy 1 will work, that the X is concluded as part of a line above and either modus ponens, disjunctive syllogism, or simplification is involved.[1] Viewed from this perspective, we see that line 4 has the X in the right position for a modus ponens, and line 8 matches the antecedent of line 4, giving us,

(4) (A ≡ D) ⊃ X
(8) A ≡ D /∴   X

So, we now have the completed proof,

1. A ⊃ ~(B ∨ C)
2. (~B • ~C) ⊃ D
3. D ⊃ A
4. (A ≡ D) ⊃ X   /∴   X
5. A ⊃ (~B • ~C)   (1) De M.
6. A ⊃ D   (5) (2) HS
7. (A ⊃ D) • (D ⊃ A)   (6) (3) Conj.
8. A ≡ D   (7) Equiv.
9. X   (4) (8) MP

What emerges from this practice problem is that the application of strategies is most efficient in the following order.

Strategy 1   See if the line that you are trying to justify is a part of a premise, and then try to match with either modus ponens, disjunctive syllogism, or simplification.

---

[1] In such situations, the only other possibilities would be double negation, X ≡ ~~X, or repetition, X ≡ (X • X) or X ≡ (X ∨ X). But we don't have ~~X, or (X • X), or (X ∨ X) in any of the lines above.

Strategy 2   See if the line that you are trying to justify is a modification of a previous line, and then try to match with a rule of replacement.

Strategy 3   Focus on the connective of the conclusion, and then try to match with an appropriate rule.

This order is most efficient, because strategy 3 is often the most time consuming, unless you can rule out as unlikely that any of the rules of replacement are involved. The important point to remember in attempting the following exercises is that, provided that you have mastered Steps 1 and 4,[1] these strategies will give you a way of staying focused, disciplined, and calm in developing hypotheses on possible rule applications, and testing each one in a trial-and-error process. If you find yourself so overwhelmed that you are only staring randomly at the list of rules, you will create a psychological slippery-slope process of anxiety and sensory overload, and you will not give your brain a chance to see the simplest rule applications.

STEP 5 EXERCISES: State the justification for each line that is not a premise. There are no incorrect lines, and any one of the nineteen rules could be involved.

1.
1. A ⊃ B
2. C ⊃ ~B   /∴   A ⊃ ~C
3. ~~B ⊃ ~C
4. B ⊃ ~C
5. A ⊃ ~C

2.
1. (D • E) ⊃ F
2. (D ⊃ F) ⊃ (~G ∨ Y)
3. E   /∴ ~G ∨ Y
4. (E • D) ⊃ F
5. E ⊃ (D ⊃ F)
6. D ⊃ F
7. ~G ∨ Y

*3.
1. (A ∨ T) ⊃ [J • (K • L)]
2. T   /∴   J • K
3. T ∨ A
4. A ∨ T
5. J • (K • L)
6. (J • K) • L
7. J • K

4.
1. (M ∨ N) ⊃ (C • D)
2. ~C   /∴   ~M
3. ~C ∨ ~D
4. ~(C • D)
5. ~(M ∨ N)
6. ~M • ~N
7. ~M

---

[1] If you have not mastered Steps 1 and 4, you will be wasting your time trying to do the Step 5 exercises. Go back and do the Step 1 and 4 exercises again. Often in life we have to go backward before we can go forward.

5.
1. (Q ⊃ Q) ⊃ ~R
2. ~Q ⊃ (R • ~Q)   /∴ ~R
3. Q ∨ (R • ~Q)
4. (Q ∨ R) • (Q ∨ ~Q)
5. (Q ∨ ~Q) • (Q ∨ R)
6. Q ∨ ~Q

7. ~Q ⊃ ~Q
8. Q ⊃ Q
9. ~R

6.
1. D ≡ Y
2. D ⊃ (~X ⊃ P)
3. ~X
4. ~P   /∴ ~Y
5. (D ⊃ Y) • (Y ⊃ D)
6. (Y ⊃ D) • (D ⊃ Y)
7. Y ⊃ D
8. Y ⊃ (~X ⊃ P)
9. ~X • ~P
10. ~(X ∨ P)
11. ~(~X ⊃ P)
12. ~Y

7.
1. (D • E) ⊃ F
2. ~F ∨ (G • H)
3. D ≡ E   /∴   D ⊃ G
4. (D ⊃ E) • (E ⊃ D)
5. D ⊃ E
6. D ⊃ (D • E)
7. D ⊃ F
8. (~F ∨ G) • (~F ∨ H)
9. ~F ∨ G
10. ~~F ⊃ G
11. F ⊃ G
12. D ⊃ G

8.
1. A ⊃ Z
2. Z ⊃ [A ⊃ (R ∨ S)]
3. R ≡ S
4. ~(R • S)   /∴   ~A
5. A ⊃ [A ⊃ (R ∨ S)]
6. (A • A) ⊃ (R ∨ S)
7. A ⊃ (R ∨ S)
8. (R • S) ∨ (~R • ~S)
9. ~R • ~S
10. ~(R ∨ S)
11. ~A

9.
1. A ⊃ B
2. B ⊃ C
3. C ⊃ A
4. A ⊃ ~C   /∴   ~A • ~C
5. A ⊃ C
6. (A ⊃ C) • (C ⊃ A)
7. A ≡ C
8. (A • C) ∨ (~A • ~C)
9. ~~A ⊃ ~C
10. ~A ∨ ~C
11. ~(A • C)
12. ~A • ~C

10.
1. T • (U ∨ V)
2. T ⊃ [U ⊃ (W • X)]
3. (T • V) ⊃ ~(W ∨ X)   /∴ W ≡ X
4. (T • U) ⊃ (W • X)
5. (T • V) ⊃ (~W • ~X)
6. [(T • U) ⊃ (W • X)] • [(T • V) ⊃ (~W • ~X)]
7. (T • U) ∨ (T • V)
8. (W • X) ∨ (~W • ~X)
9. W ≡ X

Subroutines

As you applied the strategies to the above lines, it is normal to focus only line by line. As a preparation for our final step, it is worth taking some time to examine the strategies and blocks of steps in the above exercises. In the final step you will be asked to start with some premises and create a symbolic trail to the desired conclusion. With the full nineteen rules, you now have an enormous flexibility in creating reasoning trails. As in life, you have many, many paths to choose from. How do you make the right choice? How do you find the right reasoning trail when there are so many trails to follow?

Part of the secret in creating successful formal proofs is recognizing what we will call **subroutines** and being proficient with the rules such that you begin to see blocks of steps in looking at one or two premises. ***In being able to quickly see where a trail will go by seeing blocks of steps, you can experiment faster.*** For instance, consider the following situation.

1. A • B
2.
3. B ⊃ (D • F)
4.
5. F ⊃ ~X

Someone proficient with the nineteen rules will be able to see ahead of actually constructing the proof that ~X will be available in this proof if needed. By the end of this chapter, you will also be able to do this. The trick is being able to recognize particular repetitive steps, or subroutines, that are used often in various formal proofs. Because line 1 is a conjunction, the following subroutine is available.

1. A • B

....

6. A   (1) Simp.
7. B • A (1) Com.
8. B   (7) Simp.

This subroutine shows that for future use whenever a conjunction is the major connective, we can now conclude both components of the conjunction, and this is true regardless of the complexity of each component. (Notice that lines 6 and 7 of problem 6 in the Step 5 exercises involved this subroutine.)

Thus, in this problem we know that both A and B are available. But if B is available, then D • F is available because B will work with modus ponens and line 3. If D • F is available, then both D and F are available due to the applicability of the same subroutine that gave us A and B. If F is available, then ~X is available because of line 5. Hence, in doing this problem, someone proficient in doing formal proofs might write down on scratch paper, A, B, D, F, and ~X—indicating that they are available—before doing even a single step.

Another common subroutine was used in problem 1 of the above Step 5 exercises. Often you may be faced with,

1. ~A ⊃ X	or	1. X ⊃ ~C
2. A ⊃ Y	or	2. Y ⊃ C

where it appears that there is little you can do to get lines 1 and 2 to work together. Often progress is made in proofs by using two premises together, and only the first nine rules allow us to use two premises. For this reason, the second ten rules often are used to set up a situation in which one of the nine rules will work. In the above situations, conditional statements have either the antecedents or the consequents opposite each other. As shown in problem 1, the trick here is to use transposition to set up hypothetical syllogism. So, in the above situation we could do,

1. ~A ⊃ X		or	1. X ⊃ ~C	
2. A ⊃ Y			2. Y ⊃ C	
3. ~Y ⊃ ~A	(2) Trans.		3. ~C ⊃ ~Y	(2) Trans.
4. ~Y ⊃ X	(3) (1) HS		4. X ⊃ ~Y	(1) (3) HS

Implementing such a subroutine does not guarantee progress or success, but it is something to try whenever we find two conditional statements in a proof and either the antecedents or consequents are opposite each other.

Part of another helpful subroutine is shown in problem 2. In the future you may be faced with,

1. A ⊃ (B ⊃ C)
2. B

and because the first nine rules can be applied to only whole lines, B in line 2 cannot be used to implement a modus ponens with line 1. That is, B cannot be used with the consequent of line 1 to get C. However, by using exportation, commutation, and exportation again, B in line 1 can be moved to work with line 2 as a modus ponens.

1. A ⊃ (B ⊃ C)	
2. B	
3. (A • B) ⊃ C	(1) Exp.
4. (B • A) ⊃ C	(3) Com.
5. B ⊃ (A ⊃ C)	(4) Exp.
6. A ⊃ C	(5) (2) MP

Often subroutines can be combined to obtain lots of steps. By combining the equivalence rule with the previous subroutine that used commutation and simplification, four steps can be generated given any biconditional statement as follows:

1. A ≡ B	
2. (A ⊃ B) • (B ⊃ A)	(1) Equiv.
3. A ⊃ B	(2) Simp.
4. (B ⊃ A) • (A ⊃ B)	(2) Com.
5. B ⊃ A	(4) Simp.

De Morgan's theorem can also be used with commutation and simplification, as follows,

1. ~(A ∨ B)
2. ~A • ~B   (1) De M.
3. ~A   (2) Simp.
4. ~B • ~A   (2) Com.
5. ~B   (4) Simp.

For future use, when you see a negated disjunction in a proof, automatically do the above subroutine at least on scratch paper to see what pieces are available.

A powerful subroutine that also combines commutation and simplification was used in lines 2–7, problem 5. Often you may be faced with this type of conditional statement,

1. ~A ⊃ (B • C)

and though we know from common sense that ~A ⊃ B and ~A ⊃ C individually follow from this statement, we cannot deduce either of these statements in one step by simplification.[1] (If Alice's not going to the party causes both Barbara and Carol to go, then clearly if Alice does not go, Barbara will go.) Because we claim that our nineteen rules give us a complete system—that a formal proof exists for any valid argument that can be translated into propositional logic—some chain of reasoning must exist that uses our system of rules and will enable us to derive ~A ⊃ B and ~A ⊃ C. Here's how:

1. ~A ⊃ (B • C)
2. A ∨ (B • C)   (1) Impl.
3. (A ∨ B) • (A ∨ C)   (2) Dist.
4. A ∨ B   (3) Simp.
5. ~A ⊃ B   (4) Impl.
6. (A ∨ C) • (A ∨ B)   (3) Com.
7. A ∨ C   (6) Simp.
8. ~A ⊃ C   (7) Impl.

This implication-distribution subroutine not only solves the problem of deriving ~A ⊃ B and ~A ⊃ C in our system, but shows the power and flexibility of the nineteen rules. From a single line, seven steps are derived. For future reference, this sequence of steps will always work for conditional statements that have a conjunction as the connective for the consequent.

A particularly elegant subroutine is shown in problem 4. Here is a similar series of steps simplified somewhat:

1. A ⊃ (B • C)
2. ~B
3. ~B ∨ ~C   (2) Add.
4. ~(B • C)   (3) De M.
5. ~A   (1)(4) MT

---

[1] Remember that only when the major connective is a conjunction can simplification be used. The major connective in this line is a conditional.

Note that with the addition of commutation, this same routine would work if line 2 were ~C instead of ~B. We would simply add ~B to ~C, then commute for ~B ∨ ~C. This shows that whenever you have a conditional statement with a conjunction as the connective in the consequent (1), and also have a negation in another premise (2) of one of the components of the conjunction consequent, adding the negation of the other component (as in line 3) starts you on your way to get these lines to work together.

There are many more subroutines than this, and most you will have to discover for yourself as you become more experienced in attempting formal proofs. You will probably even create your own, because it is possible to adopt a particular style in doing proofs. What you need is practice and the challenge of facing nothing but empty space under the premises and the thought that that space can be filled with an infinite number of possible symbolic trails. Shortly, you will be given considerable practice.

However, before we leave the subject of subroutines, note that due to the particularities of our system, often certain "bookkeeping" steps are needed to implement the rules properly. Problem 9 in the above Step 5 exercises (lines 4, 9, and 10) shows what is needed to implement implication when the antecedent of the conditional statement is not negated. Because our implication rule, $(\sim p \supset q) \equiv (p \vee q)$, states that the antecedent of the conditional must have a negation before we can move to a disjunction, to implement this rule when there is no negation, we have to "trick" our system a little by using double negation. Here's how:

1. A ⊃ (B • C)
2. ~~A ⊃ (B • C)   (1) DN
3. ~A ∨ (B • C)   (2) Impl.

The implication rule directs us to remove one negation from the antecedent of the conditional, change the connective to a disjunction, and not change what was the consequent. So, if there is no negation on the antecedent, we can't start the rule. But if there are two negations, we can implement the rule by removing one!

A similar trick can be used in situations such as the following:

1.		**2.**		**3.**	
1. A ⊃ ~Y		1. (B • ~C)		1. X ∨ ~Y	
2. Y		2. (~~B • ~C) (1) DN		2. Y	
3. ~~Y (2) DN		3. ~(~B ∨ C) (2) De M.		3. ~Y ∨ X (1) Com	
4. ~A (1) (3) MT				4. ~~Y (2) DN	
				5. X (3) (4) DS	

The first example (1) shows what we must do to implement modus tollens when originally we lack a second premise that negates the consequent of a conditional statement. Technically, Y is not the negation of ~Y, but because ~~Y is, we simply use double negation as a little bookkeeping subroutine. The second example (2) shows how to implement De Morgan's theorem when a conjunction does not have both components negated. Again double negation is used. One version of the De Morgan rule directs us to remove a negation from each side of a conjunction, change the connective from a conjunction to a disjunction, and then place a negation around the entire disjunction. Following these steps will leave us with a ~B—we remove as directed one negation from ~~B. The third example (3) shows how to implement the disjunction rule by first using commutation and double negation as bookkeeping steps.[1]

As you become more proficient with the rules, you may become frustrated with the time needed for bookkeeping steps. You will begin to have insights and be able to think much faster than you can write down each formal step. At that stage, it is appropriate to use a shortcut technique of notation that combines bookkeeping steps with other rules. Here are some typical shortcut moves:

1.
1. $A \cdot B$
2. B (1) Com. + Simp.

2.
1. $\sim X \lor Y$
2. X
3. Y (2) (1) DN + DS

3.
1. $D \supset Z$
2. $\sim D \lor Z$ (1) DN + Impl.

4.
1. $\sim(B \lor \sim C)$
2. $\sim B \cdot C$ (1) De M. + DN

5.
1. $(A \cdot B) \cdot C$
2. A (1) Ass. + Simp.

6.
1. $P \supset \sim S$
2. $S \supset \sim P$ (1) Trans. + DN

It is important to note, however, that these examples are shortcut *notation* techniques. Steps are not skipped. In working with formal systems, as with computers, every step must be rigorously specified.

Direction, Strategies, and Working Backward

As in life, one way to cut down the number of possible choices that confront us is to have a definite goal in mind. In constructing formal proofs our goal is to derive the conclusion. Often in looking at the conclusion you will not see every step ahead of time, but a little common sense in comparing the premises with the conclusion will constrain what you should try first. For instance, consider the following problem:

1. $A \supset \sim B$
2. $C \supset B$
3. $\sim C \supset X$ /$\therefore$ $\sim X \supset \sim A$

Because the premises are all conditionals and the conclusion is also a conditional, it would be wise to try the strategy that some combination of rules that involve conditionals will work in solving this problem. So, we might try using some combination of hypothetical syllogism and transposition.

---

[1] Note that a common student mistake is the following misapplication of DS: $\sim Y \lor X$, $\sim Y$ /$\therefore$ X. As in Example 3, the second premise needs to be $\sim\sim Y$.

Sometimes, however, the conclusion does not have a clear connection with the premises. In such cases, a useful strategy is to work backward on the conclusion using only the rules of replacement. Because implementing a rule of replacement produces a statement that is logically equivalent, using this strategy will reveal other possible goals. For instance, in the following problem,

1. A ⊃ ~B
2. A
3. H ⊃ ~D
4. D   /∴   ~(B ∨ H)

it is difficult to see how we would arrive at a negated disjunction as a conclusion from the given premises. But by applying De Morgan's theorem to the conclusion on scratch paper, we see that a possible goal for this proof is to find a ~B and a ~H separately, then combine them by conjunction—because ~(B ∨ H) is logically equivalent to ~B • ~H, and this statement can be produced by combining a ~B and a ~H by conjunction. If this were not done, we might be tempted to change by implication the conditional statements in lines 1 and 3 to disjunctions, because we see that the conclusion involves a disjunction. Although valid applications of a rule, this would take us on a long, messy trail. After seeing what the conclusion is logically equivalent to, we see that it would be better to just do a modus ponens with lines 1 and 2, deriving ~B, one half of what we are looking for. Often, playing a little with the conclusion on scratch paper and using rules of replacement can produce alternative statements that we might then blunder across in going forward in the proof. A common strategy in difficult proofs is to work backward a little, then blunder forward and hope that the two trails meet!

The following exercises are designed to eventually push you to a state of formal discipline that mimics the process of working with symbolic manipulation that is behind much of our technology. The first problems require only a few steps, but the last problem, which I have labeled Super Challenging, requires approximately fifty steps! You will probably not be able to solve every problem.[1] However, your goal is not to solve every problem, but rather to experiment, try strategies, fail, start again, and above all, stay disciplined and create only valid trails. The biggest mistake you can make is to be impatient and derive the conclusion using a trail that contains misapplications of the rules. Here is a typical example of a student proof with incorrect lines marked with **xx**s,

1. (X • Y) ⊃ Z
2. ~Z
3. (T ∨ P) ⊃ X
4. Y   /∴   ~T ⊃ P
5. ~(X • Y)   (1)(2) MT
6. ~X   (5) Simp. **xx**

---

[1] As I write this, in the past ten years, only six students have been able to solve the Super Challenging problem, and four of these students have been women, much to the dismay of the macho men in my classes.

7. ~T ∨ P	(3)(6) MT **xx**
8. ~~T ⊃ P	(7) Impl.
9. T ⊃ P	(8) DN

This impatient proof has only five steps, but my proof for this problem has eleven steps. The student who did this proof either did not check his or her steps or was so concerned with producing the conclusion that steps 6 and 7 were "hallucinated" as valid. Often I will give back a quiz like this and a fairly good student will immediately see the mistake or mistakes and usually remark, "Why didn't I see this?" Laziness, inattention to detail, not understanding the rules are possible explanations. However, my theory is that in many cases the student failed to see the mistakes because we are all so much a results-oriented society, so concerned with success and the competitive accomplishment of goals, that we have not been encouraged to stop and enjoy the process of accomplishing our goals. So, my advice to you in attacking the following exercises is to go slow, check and recheck your steps every so often, learn to enjoy the process of trying a trail, experimenting, failing, trying another trail, and worry more about creating a valid trail than about always getting the conclusion. This attitude should be especially useful for problems 9 (relatively difficult for this stage) and 10 (a little seven-step killer).

STEP 6 EXERCISES: the following arguments.   Construct a formal proof of validity for each of

1.
1. A ⊃ ~B
2. C ⊃ B   /∴   A ⊃ ~C

2.
1. D ⊃ (E ∨ F)
2. ~E • ~F   /∴   ~D

3.
1. (J ∨ K) ⊃ ~L
2. L   /∴   ~J

4.
1. X ≡ ~P   /∴   ~P ≡ X

5.
1. A ⊃ (B • ~C)
2. H ∨ (D • A)
3. ~H
4. T ⊃ C   /∴   T ⊃ X

6.
1. ~B ∨ [(C ⊃ D) • (E ⊃ D)]
2. B • (C ∨ E)   /∴   D

7.
1. ~A ≡ (B • C)
2. (A ∨ B) ⊃ T   /∴   T

*8.
1. (X • Y) ⊃ Z
2. ~Z
3. (T ∨ P) ⊃ X
4. Y   /∴ T ⊃ P

9.
1. X ⊃ (T • P)
2. T ⊃ X
3. ~(X • T)   /∴   ~X

10.
1. B   /∴   A ∨ ~A

Brief Truth Tables Revisited and Decision Strategies

In some of the above proofs you may have recognized subroutines that we have covered and were able to cut down on producing rambling trails. The transposition-hypothetical syllogism routine can be used in problem 1, some double negation bookkeeping can be used in problem 3, the commutation-simplification routine can be used in problems 5 and 6, the equivalence routine in problem 7, and the addition-De Morgan or the implication-distribution routine used in problem 9. As noted above, the recognition of subroutines allows us to experiment faster.

Yet in more difficult problems a situation is often produced where you have created so many trails that you begin to suffer sensory overload. Because you can take any piece of a trail and try to combine it with a piece of another trail, the possible combinations and choices become overwhelming. Sometimes, however, you may have a hunch that combining a part of one trail with a part of another trail will be fruitful, but because you can't be sure it will lead anywhere, you may be hesitant to try it because you might be wasting your time. It would be nice if a decision procedure existed that would let you know ahead of actually producing a proof that a trail will lead to the conclusion using the parts you have in mind.

There is such a procedure. Although this decision procedure will not construct a proof for you, it will give you the assurance that a proof exists. Actually, we have already studied this procedure in Chapter 8 in the section "Brief Truth Tables." Consider how a brief truth table could be used in the following situation.

8.
9.
10. A ⊃ ~X
11.
12.
13.
14.
15. B ⊃ ~X
16.

Suppose the conclusion you are attempting to derive is B ⊃ A, and in looking at lines 10 and 15 you have a hunch that it can be derived from these lines. Essentially, we are asking if the following argument is valid.

1. A ⊃ ~X
2. B ⊃ ~X   /∴   B ⊃ A

If the answer is yes, then we know that a proof exists for deriving the conclusion from these premises alone, and we need not concern ourselves with any other lines or combination of lines. If the answer is no, then we know that it would be a waste of time to attempt a derivation of B ⊃ A from these steps alone. Following the steps listed on page 280 of Chapter 8, we see that this argument is invalid.

```
T   1. A ⊃ ~X
       F    T                      A = F
                                   B = T
T   2. B ⊃ ~X   /∴   B ⊃ A         X = F
       T    T        T   F       Invalid
                       F
```

So, it would be a waste of time to attempt to derive the conclusion from steps 10 and 15 alone. These steps might be involved, but we now know for sure that at least some other steps must also be involved.

On the other hand, consider this situation where our hunch is that a conclusion ~A ∨ ~B can be derived using only lines 10, 13, and 15:

```
 8.
 9.
10. A ⊃ ~X
11.
12.
13. Y ⊃ X
14.
15. ~Y ⊃ ~B
16.
```

Here is what a brief truth table would look like.

```
T  1. A ⊃ ~X                       A = T
      T    T                       B = T
                                   X = F
F  2.  Y ⊃ X                       Y = T
       T   F                     Valid


T  3. ~Y ⊃ ~B   /∴   ~A ∨ ~B
      F     F        F    F
                  F
```

Because it is not possible to make the conclusion false and all the premises true,[1] we know that ~A ∨ ~B can be derived validly from only these premises. Hence, in the above proof we would then concentrate on only lines 10, 13, and 15 and ignore all the other lines and possible combinations with those lines.

Because brief truth tables can be done quickly on scratch paper, you can experiment with various combinations of lines in difficult proofs prior to attempting any derivations. You'll avoid a great deal of frustration if you know ahead of trying that a particular trail will not work. In problem 10 in the above Step 6 exercises, students often spend considerable time attempting to derive a ~B based on the following hypothesis as to how this problem is done.

---

[1] If we make X in premise 2 true to make this premise true, then premise 1 will become false.

1. B   /∴   A ∨ ~A
2. B ∨ A   (1) Add
3. ~B   ?????
4. A   (2) (3) DS
5. A ∨ ~A   (4) Add

Although it is admirable that they have obtained a level of proficiency to create such a hypothesis, a little reflection on the nature of a valid argument, put into the form of the simplest brief truth table, reveals that it is impossible to derive a ~B and that they are wasting their time even trying. Because we have only the premise B, deriving a ~B would be claiming that

1. B   /∴   ~B

is a valid argument. But if ~B is false, then B is true, and we have an argument with a true premise and a false conclusion, the simplest and most obvious case of an invalid argument. Unless a contradiction is already hidden in the premises, it will not be possible to derive one.

TRANSLATION AND FORMAL PROOF EXERCISES:

Translate the following valid arguments and when checked by your instructor do formal proofs for each one.

1. If we buy the new furniture, then we can't purchase the new refrigerator provided that we buy the new carpet. If we buy the new furniture, we will buy the new carpet. We are going to buy the new furniture. So, we can't purchase the new refrigerator. (F, R, C)

*2. The new administration will appoint Shalayla to a cabinet post for the following reasons. Jordan and Adams will be appointed. Furthermore, Jordan being appointed is sufficient for Shalayla to be appointed, provided that Adams is also appointed. (S, J, A)

3. Argument for why prohibition measures don't work in stopping drug traffic:

Reduction by police of the supply of illegal drugs is a sufficient condition for increasing the price of these drugs. Potential income for drug traffickers will increase if prices rise, and if potential income for drug traffickers increases, it will produce an encouraging environment for drug trafficking. Thus, if the police reduce the supply of illegal drugs, it will produce an encouraging environment for drug trafficking. (S, P, I, E)

S = Police reduce the supply of illegal drugs.
P = The price of illegal drugs increases.
I = Potential income for drug traffickers will increase.
E = An encouraging environment for drug traffickers will be produced.

4. If the president implements his tax programs, then the deficit will not continue to increase only if the economy turns around soon. The president will

implement his tax programs. However, the economy will not turn around soon. Therefore, the deficit will continue to increase. (T, D, E)

5. Treat the first premise of the following argument as an *exclusive* disjunction.

   We will hire either Aweau or Kaneshiro for the new electronics position. We will hire Aweau. So, we will not hire Kaneshiro. (A, K)

6. The burglars entered the house from either the front or the rear. Had they entered from the rear, the dog would have barked. But the dog did not bark. If the burglars entered the house from the front, then the maid must have let them in (because the alarm did not go off). This proves that the maid must have let them in. (F, R, D, M)

*7. The following statement is claimed to be true: "This statement is false." If the statement claims it is true, and is true, then it is not true. However, if the statement claims that it is true but it is not true, then it is true. Hence, the statement is true if and only if it is not true. (C, T) Translate the first statement as C.

   C = The statement, "This statement is false" is claimed to be true.
   T = The statement, "This statement is false" is true.

8. Why backup lights on cars are wired through the transmission:

   Backup lights are needed only if the car is in reverse. Because they are connected to the transmission, they are on if and only if the car is in reverse. If they are on only if the car is in reverse, then they use very little power. Thus, the backup lights are on if needed, yet they use very little power. (N, R, O, L)

   N = Backup lights are needed.
   R = The car is in reverse.
   O = Backup lights are on.
   L = Backup lights use very little power.

9. Counselor to a student named John: "The probation policy is clear, if you don't achieve at least a 2.0 GPA or you don't complete at least 50 percent of all credits attempted this semester, you will be on probation next semester." Suppose the following semester John is not on probation. Using the probation policy as a premise and the fact that John is not on probation, construct a proof for the conclusion that John achieved at least a 2.0 GPA and completed 50 percent of all his credits attempted.

   A = John achieved at least a 2.0 GPA.
   C = John completed at least 50 percent of all credits attempted.
   P = John is on probation.

10. Lee did not pass the final, but she did pass the course. So, it is not true that her passing the final was a necessary condition for her passing the course. (F, C)

*11. Famous argument attempting to prove that we are more than just a body of physical matter; that each of us has a nonphysical mind as well.

If brain processes and mental occurrences are identical—that is, if all our thoughts are just physiological processes in the brain—then the former having spatial location constitutes a necessary and sufficient condition for the latter having spatial location. No one disputes that brain processes are spatially located. However, if mental processes are spatially locatable, then it will be meaningful to assign spatial location to thought. It is not meaningful to do this. So, it is not true that brain processes and mental occurrences are identical. (I, B, M, T)

I = Brain processes and mental occurrences are identical.
B = Brain processes are spatially locatable (or have spatial location).
M = Mental processes are spatially locatable (or have spatial location).
T = It is meaningful to assign spatial location to thought.

12. An important issue in theology is known as the problem of evil. If God exists and is good, then why do humans suffer so much? Below is a valid argument that expresses this puzzle. Construct a formal proof.

If God is good, then God must wish to abolish all evil; and if God is all-powerful, God must be able to abolish evil. If God wishes to abolish evil and is able to do so, then evil should not exist. But evil does exist. Therefore, God cannot be both omnipotent and good. (G, W, P, A, E)

G = God is good.
W = God must wish to abolish all evil.
P = God is all-powerful (omnipotent).
A = God is able to abolish all evil.
E = Evil does exist.

If a fourth premise is added to this argument that "God is good," prove the conclusion, "God is not able to abolish all evil and is not omnipotent."

13. Unless people outside Korea apply pressure, the Japanese government will not listen to the women's ("comfort" women survivors) demands. If people outside Korea put pressure on the Japanese government, then the history books used in Japanese schools will be rewritten. Thus, either people outside Korea put pressure on the Japanese government and the history books used in Japanese schools are rewritten or the Japanese government will not listen to the women's demands. (P, L, H)

P = People outside Korea put pressure on the Japanese government.
L = The Japanese government will listen to the comfort women's demands.
H = The history books used in Japanese schools will be rewritten.

14. Below is an example of the Kevorkian logic of suicide for a patient with inoperable, terminal cancer. (Please note that although this is a valid argument, it contains very controversial premises and assumptions. Do you have a right to take your own life? Many religions say you do not. Do not miracle cancer remissions sometimes occur? Is not the patient's attitude or outlook on life important for the possibility of recovery?)

If you continue to take the cancer treatment, you will still die a slow, painful death or at least be a slave to pain the rest of your life. If you lie to yourself, the truth of your situation will not change; and, if the truth of your situation does not change, then you will still die a slow, painful death or at least be a slave to pain the rest of your life. You can either continue to take the cancer treatment or lie to yourself, or commit suicide. Thus, committing suicide is the only way for you to avoid a slow, painful death or being a slave to pain the rest of your life. (C, D, P, L, T, S)

C = You continue to take the cancer treatment.
D = You will still die a slow, painful death.
P = You will be at least a slave to pain the rest of your life.
L = You lie to yourself.
T = The truth of your situation changes.
S = You commit suicide.

15. Happy-birthday card logic: "WHY WORRY???"

There are only two things to worry about: either you're healthy or you're sick. If you're healthy, there's nothing to worry about, and if you're sick, there are two things to worry about, either you'll get well or you won't. If you get well there is nothing to worry about, but if you don't, you'll have two things to worry about, either you'll go to heaven or to hell. If you go to heaven, you have nothing to worry about, and if you go to hell, you'll be so busy shaking hands with all of us that you'll have no time to worry. So, either you have nothing to worry about or you'll have no time to worry. (H, S, W, G, A, B, T)

H = You are healthy.
S = You are sick.
W = You have something to worry about.
G = You will get well.
A = You will go to heaven.
B = You will go to hell.
T = You will have time to worry.

Holiday Adventures: The following two problems will give you a lot to worry about. They are quite challenging, requiring rather long valid trails for an introductory course in logic. In working on them, you may never obtain the desired conclusion. However, you will learn a lot about discipline (avoiding invalid steps to obtain the conclusion), and you will be immersing yourself into the agony and ecstasy, trial-and-error-trail-making, hypothesis-refutation-new-hypothesis process of symbolic reasoning that is so much a part of our scientific-technological culture. If nothing else, you will become very intimate with the nineteen rules and be able to do well on any exam your instructor gives you.

Challenging: A translation of the formalized argument on page 229 of Chapter 7. (Approximately thirty steps!)

1. $\sim G \supset \sim (A \lor B)$
2. $G \supset (H \supset C)$

3. (H • E) ⊃ ~C
4. B ⊃ (H • E) /∴ ~B

Super Challenging!!!: This proof shows that association works for *if and only if.* (Approximately fifty steps!!)

C ≡ (D ≡ E) /∴ (C ≡ D) ≡ E

Clarification Exercises: Using the rules of replacement transform and then rewrite the following sentences into a simpler form.

Example:

If it is not true that marijuana is not widely used or as harmful as alcohol, then it should be decriminalized. (W, H, D) ~(~W ∨ H) ⊃ D

~(~W ∨ H) ⊃ D → (~~W • ~H) ⊃ D (DeM.) → (W • ~H) ⊃ D (DN)

If marijuana is widely used and not as harmful as alcohol, then it should be decriminalized. (W • ~H) ⊃ D

1. North believed it was not illegal not to tell Congress the truth. (L = Legal to lie to Congress.) ~~L

2. It is not true that if North did not tell Congress the truth, then he did tell Congress the truth (T) ~(~T ⊃ T). Did North tell Congress the truth or not?

3. Rain and humidity are not uncommon in Hawaii. (R, H) ~~R • ~~H

4. It's noisy here even when it is quiet. (Q) Q ⊃ ~Q

5. If it was ever the common law that by marriage (a woman) gave irrevocable consent to sexual intercourse with her husband, it is no longer the common law. (C) C ⊃ ~C

6. It is not true that you can pass the final exam and not pass the course (F, C) ~(F • ~C). Transform into and rewrite as an *if. . .then* statement.

7. The view of the arbitration board that the fine against Johnson was justified but the suspension from the team was not is mistaken. (F, S) ~(F • ~S). Transform into and rewrite as an *if. . .then* statement.

8. It is not true that the accused accomplice not being in the neighborhood on the night of the crime is a sufficient condition for not having knowledge of the crime (N, K) ~(~N ⊃ ~K). Transform into a conjunction (use *but* in the rewrite) and begin the rewrite, "It is possible that the accused accomplice . . ."

9. It is not true that having true premises is a sufficient condition for a valid argument (T, V) ~(T ⊃ V). Transform into a conjunction and begin the rewrite, "It is possible to have . . ."

10. It is not true that if the president implements his tax programs, then deficits will be reduced provided that the economy turns around (T, D, E) ~[T ⊃ (E ⊃ D)]. Transform into a series of conjunctions and use *but* and *even though* in the rewrite.

ANSWERS TO STARRED EXERCISES

STEP 4   4. **X** We can't add single negations to the parts of a conditional unless the antecedent and the consequent are reversed, as in transposition.

9. **X** Commutation can be applied to part of a line, but that part can't be detached in the process. The correct application of commutation in number 9 would be:

$(A \lor B) \supset (X \supset Y)$
$/\therefore \quad (B \lor A) \supset (X \supset Y)$

18. Implication

STEP 5   **3.**

1. $(A \lor T) \supset [J \cdot (K \cdot L)]$
2. $T \qquad /\therefore \quad J \cdot K$
3. $T \lor A$ (2) Add.
4. $A \lor T$ (3) Com.
5. $J \cdot (K \cdot L)$ (1)(4) MP
6. $(J \cdot K) \cdot L$ (5) Ass.
7. $J \cdot K$ (6) Simp.

STEP 6   **8.**

1. $(X \cdot Y) \supset Z$
2. $\sim Z$
3. $(T \lor P) \supset X$
4. $Y \qquad\qquad /\therefore \quad T \supset P$
5. $\sim(X \cdot Y)$ (1) (2) MT
6. $\sim X \lor \sim Y$ (5) De M.
7. $\sim Y \lor \sim X$ (6) Com.
8. $\sim\sim Y$ (4) DN
9. $\sim X$ (7) (8) DS
10. $\sim(T \lor P)$ (3) (9) MT
11. $\sim T \cdot \sim P$ (10) De M.
12. $\sim T$ (11) Simp.
13. $\sim T \lor P$ (12) Add.
14. $\sim\sim T \supset P$ (13) Impl.
15. $T \supset P$ (14) DN

TRANSLATIONS

2.

1. $J \cdot A$
2. $A \supset (J \supset S) /\therefore \quad S$

7.

1. C
2. $(C \cdot T) \supset \sim T$
3. $(C \cdot \sim T) \supset T \quad /\therefore \quad T \equiv \sim T$

11.

1. $I \supset (B \equiv M)$
2. B
3. $M \supset T$
4. $\sim T \quad /\therefore \quad \sim I$

Chapter 11

OTHER LOGICAL TOOLS: SYLLOGISMS AND QUANTIFICATION

Introduction

A persistent theme of this book has been the interpretation of logic as a set of practical tools. I have encouraged you to see these tools as beneficial in many contexts for keeping us on track for arriving at an objective, unforced agreement by encouraging us to test our beliefs. A very important implication of this interpretation is that like any set of tools, the application of logical principles has contextual limitations. A hammer is a very good tool for pounding nails, but not very good for the more delicate task of fixing the electronic components of a computer. Thus, the logical principles in Chapters 9 and 10, which we have distilled from our common sense, should not be seen as absolute rules that guide our thinking in all contexts.

It is easy to show the limitations of the tools we have discussed thus far. Recall how the rule of commutation fails in certain contexts. If we are restricted to only the syntactical relations of propositional logic, we are forced to interpret "Mary became pregnant and married John" as equivalent to "Mary married John and became pregnant," $(P \cdot M) \equiv (M \cdot P)$. As we noted previously, propositional logic ignores the different meanings these statements would have in different cultural contexts.

Recall also the paradox of material implication discussed in Chapter 8. Our background knowledge about how the world works conflicts with some results in propositional logic. The statement, "If I sneeze within the next five minutes, then the whole world will disappear" is true when the antecedent is false—I don't sneeze within the next five minutes.

In Chapter 8 it was also noted that although there will be limitations in capturing the rich texture of human experience, logicians have developed additional logics to capture other aspects of our commonsense intuitions and explore the nature of rationality. Consider how propositional logic fails with the following argument.

EXAMPLE 11-1   Either the accused had no knowledge that the crime took place or he is guilty of being an accessory after the fact. However, it is not true that if the accused was not in the neighborhood on the night of the crime, then he had no knowledge that the crime took place. Thus, it follows that the accused is guilty of being an accessory after the fact. (K, A, N)

We are surprised in reading the conclusion, for it obviously does not follow that, from the mere possibility of someone knowing about a crime, that person was an accessory to the crime. Yet using the symbolic tools of propositional logic we can show that this argument is valid!

Translation: 1. ~K ∨ A
                   2. ~(~N ⊃ ~K)   /∴   A

Proof: 1. ~K ∨ A
                   2. ~(~N ⊃ ~K)   /∴   A
                   3. ~(N ∨ ~K)   (2) Impl.
                   4. ~N • ~~K   (3) De M.
                   5. ~~K   (4) Com + Simp.
                   6. A   (1) (5) DS

The discrepancy between our intuition and the propositional proof is due to what logicians call the **modal interpretation** that is implied in the second premise. From the statement, "It is not true that if the accused was not in the neighborhood on the night of the crime, then he had no knowledge that the crime took place," ~(~N ⊃ ~K), it follows only that it is *possible* that the accused was not in the neighborhood but still knew about the crime. Not that it is *necessary* that the accused was not in the neighborhood but knew about the crime, ~N • K (line 4 above, plus DN). So, if we only know that it is possible that the accused knew about the crime, it does not follow that he was an accessory, given premise 1.

A similar mismatch occurs between the statements, "It is not true that if John passes the final he will pass the course," ~(F ⊃ C), and "John passed the final, but he will not pass the course," F • ~C. Even though the second statement can be derived from the first: ~(F ⊃ C) ⇒ (implies) ~(~~F ⊃ C) ⇒ ~(~F ∨ C) ⇒ (~~F • ~C) ⇒ (F • ~C) (DN + Impl. + De M. + DN). The problem again is that from "It is not true that if John passes the final he will pass the course," it should follow only that it is *possible* for John to pass the final and still not pass the course.

Because of such important nuances, logicians have developed a higher logic called **modal logic** to handle inferences that involve statements regarding necessity and possibility. In this higher logic, the inferential relationship between a negation of a conditional statement and a conjunction would look like this: $\sim(p \supset q) \equiv \Diamond(p \cdot \sim q)$. The diamond symbol, as in $\Diamond A$, is read as "It is possible that A is true." Another symbol, ($\Box$), is also introduced, such that $\Box A$ reads, "A is necessarily true." Various definitions are then given, such as

$\Diamond A$ = def. $\sim\Box\sim A$ (That "A is possibly true" is equivalent by definition to "It is not true that A is necessarily false," or "It is not true that A is impossible"),

and

$\Box A$ = def. $\sim\Diamond\sim A$ (That "A is necessarily true" is equivalent to "It is not true that not A is possible").

From these definitions, various rules are derived such that the chain of reasoning from step 2 to step 4 in the above proof is rendered invalid.

There are also logics that deal with obligation and permissibility (called deontic logic), time (tense logic), multivalued logic, which differs from standard logic in considering degrees of truth and shades of gray in place of the black and white crisp values of complete truth and falsity, and even quantum logic where key features of our common sense are violated.[1] These alternate logics are the subject of advanced courses ranging from upper division to high-level graduate courses. For the most part, however, they all use propositional logic as a basis or point of departure in the sense of either adding rules or showing contextual restrictions of propositional inferences.

Syllogisms and Quantification Logic

One group of additional logical tools often covered at the introductory level is called *quantification logic* or sometimes *predicate logic*. Recall number 4 of the Part III exercises in Chapter 1. To give a correct and thorough analysis of this exercise, essentially you had to be able to distinguish between the following two arguments:

EXAMPLE 11-2   No U.S.-manufactured car built before 1970 was equipped with safety belts at the factory.
John has a U.S.-manufactured car built in 1969.
Therefore, John's car was not equipped with safety belts at the factory.

EXAMPLE 11-3   All U.S.-manufactured cars built after 1970 were equipped with safety belts at the factory.
John has a U.S.-manufactured car built in 1969.

---

[1] The legitimacy of these logics is controversial and is discussed in professional journals and philosophy books. In this context, it is worth noting that the philosopher Robert C. Solomon has argued that it is a major philosophical mistake to equate our emotions with the irrational, that our emotions have a special logic of their own. See his *The Passions* (Garden City, New York: Anchor Press/Doubleday, 1976).

Therefore, John's car was not equipped with safety belts at the factory.

The first argument (11-2) is valid, but the second (11-3) is invalid, committing a variation of the fallacy of denying the antecedent. However, we cannot show this symbolically using propositional logic alone. The techniques of propositional logic allow us to analyze simple statements and their compounds, but the validity and invalidity of the above arguments depend on the inner logical structure of noncompound statements and the meaning of the generalizations contained in the first premise of each argument. Thus, we need a way to symbolically picture the difference between statements such as "All U.S.-manufactured cars built after 1970 were equipped with safety belts at the factory," and "No U.S.-manufactured car built before 1970 was equipped with safety belts at the factory."

Historically, a method for dealing with many noncompound statements was actually developed before modern propositional logic. The philosopher Aristotle (384–322 B.C.) is credited with being the first person to attempt to distill patterns of valid and invalid inference from our everyday arguing and judging. Called classical deduction theory, Aristotelian logic focuses on *categorical propositions* and *categorical syllogisms*. Categorical propositions make statements about the relationship of categories or classes of things. The classes are designated by *terms*. In this logic, a syllogism is a very restricted deductive argument consisting of exactly three categorical propositions, a conclusion inferred from two premises, and the categorical propositions contain in total only three terms. For instance,

EXAMPLE 11-4 No U.S.-manufactured car built before 1970 was equipped with safety belts at the factory.
Some U.S.-manufactured Mustangs on the road today have safety belts equipped at the factory.
Therefore, some U.S.-manufactured Mustangs on the road today were not built before 1970.

EXAMPLE 11-5 No U.S.-manufactured car built before 1970 was equipped with safety belts at the factory.
All U.S.-manufactured Mustangs on the road today have safety belts equipped at the factory.
Therefore, no U.S.-manufactured Mustangs on the road today were built before 1970.

are syllogisms containing the terms (categories or classes):

U.S.-manufactured cars built before 1970,

Cars equipped with safety belts at the factory,

U.S.-manufactured Mustangs on the road today.

Notice also that these two syllogisms contain four types of propositions, tradition-ally referred to as ***universal affirmative***, ***universal negative***, ***particular affirmative***, and ***particular negative***, and labeled **A**, **E**, **I**, and **O**:[1]

A	(universal affirmative)	All U.S.-manufactured Mustangs on the road to-day have safety belts equipped at the factory.
E	(universal negative)	No U.S.-manufactured car built before 1970 was equipped with safety belts at the factory.
I	(particular affirmative)	Some U.S.-manufactured Mustangs on the road today have safety belts equipped at the factory.
O	(particular negative)	Some U.S.-manufactured Mustangs on the road today were not built before 1970.

A traditional symbolic method for analyzing arguments of this type involves drawing pictures with overlapping circles, called ***Venn Diagrams***.[2] For instance, the basic claims of these four propositions can be represented as follows:

A   All M is S.   Everything in the M category of things is in the S category of things.

FIGURE 11-A

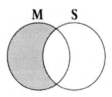

E   No B is S.   Anything that is in the B category of things is not in the S category of things.

FIGURE 11-E

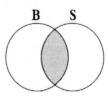

[1] The letter abbreviations most likely come from the Latin words "*Af**I**rmo*" (I affirm) and "*nE**g**O*" (I deny).
[2] After the English mathematician and logician John Venn (1834–1923) who first used this method.

I   Some M is S.   At least one thing that is in the M category of things is also in the S category of things.

FIGURE 11-I

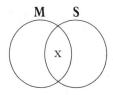

O   Some M is not B.   At least one thing that is in the M category of things is not in the B category of things.

FIGURE 11-O

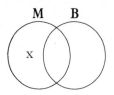

For the **A** proposition, we draw two circles for each category (M = U.S.-manufactured **Mustangs** on the road today; S = Cars equipped with **safety** belts at the factory), but shade that portion of the M circle that does not overlap with the S circle. In this way, we picture the claim of the **A** proposition that everything that is in the M category of things is also in the S category of things. For the **E** proposition, we shade the overlap area of the B (U.S.-manufactured cars **built** before 1970) and S categories, indicating that no things are in the B and S categories together. In both A and E propositions, the shading indicates emptiness. For the **I** proposition, we put a little x in the overlap area of the M and S categories, indicating that at least one thing is both an M and an S. Finally, for the **O** proposition, we put a little x in that portion of the M category that does not overlap with the B category, indicating that at least one thing is an M, but is not a B.

     We can now combine three circles, picturing the claims of each of the premises that make up the syllogisms (11-4) and (11-5), and then view the result to see if the picture produced is consistent with the conclusion. If the picture constructed is consistent with the conclusion, then the argument is valid; if it is not, then the argument is invalid. Here is what Venn diagrams of (11-4) and (11-5) would look like:

EXAMPLE 11-4   No U.S.-manufactured car built before 1970 was equipped with safety belts at the factory.
                      Some U.S.-manufactured Mustangs on the road today have safety belts equipped at the factory.

Therefore, some U.S.-manufactured Mustangs on the road today were not built before 1970.

FIGURE 11-4

No B is S.
Some M is S.
Therefore, some M is not B.

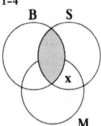

EXAMPLE 11-5   No U.S.-manufactured car built before 1970 was equipped with safety belts at the factory.
All U.S.-manufactured Mustangs on the road today have safety belts equipped at the factory.
Therefore, no U.S.-manufactured Mustangs on the road today were built before 1970.

FIGURE 11-5

No B is S.
All M is S.
Therefore, no M is B.

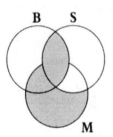

Notice that in (11-4) we first shade the overlapping area of B and S to picture the first premise that no B is S. Then we put a little x in the overlapping area of the categories M and S to picture the claim of the second premise that at least one M is an S. Once this is done, we see that the picture we have created is consistent with the conclusion that at least one M is not a B as shown by having a little x in the M circle but not in the B circle. Thus, (11-4) is valid. If the conclusion stated that "Some U.S.-manufactured Mustangs on the road today were built before 1970," the argument would be invalid, because after drawing the pictures for the premises there is no x in the overlapping area of the B and M circles as would be required to picture the conclusion.

In (11-5) we again shade the overlapping area of the B and S circles to picture the claim of the first premise that no B is S. Then we shade all of the M circle that does not overlap with the S circle to picture the claim of the second premise that all M is S. This leaves us with a picture that is consistent with the conclusion, no M is B, because the overlap area of the M and B circles is shaded. So, (11-5) is valid. Notice that if the conclusion were "All U.S.-manufactured Mustangs on the road today were built before 1970," (All M is B) then the argument would be invalid because no open unshaded area of the M circle is contained in the B circle as would be required to picture this claim. When the conclusion states more than what is contained in the premises, deductive arguments are always invalid.

Venn diagrams are fun and can be used to capture and draw out numerous nuances of commonsense reasoning. They can also be used to picture the many rules of syllogistic reasoning discovered by Aristotle and by his followers in the Middle Ages who further developed classical logic. But they cannot be used to adequately analyze the arguments (11-2) and (11-3) above. Nor do they work for arguments that contain premises such as "In the 1990s electric vehicles were both expensive and inconvenient," or complicated statements such as the quotation from Lincoln at the beginning of the first chapter, "You can fool some of the people all of the time and all of the people some of the time, but you cannot fool all of the people all of the time." Thus, by adding a few commonsense rules to propositional logic, modern logicians have developed a more powerful set of tools called **quantification logic** that combines the distinctive elements of syllogistic logic and propositional logic.

First let's see how terms in syllogistic logic would be symbolically represented in quantification logic.

U.S.-manufactured cars built before 1970.
> Bx = x is a U.S.-manufactured car *built* before 1970.

Cars equipped with safety belts at the factory.
> Sx = x is a car equipped with *safety* belts at the factory.

U.S.-manufactured Mustangs on the road today.
> Mx = x is a U.S.-manufactured *Mustang* on the road today.

Like the p, q, and r in propositional logic, the letter x in quantification logic is a **variable**. It does not represent a particular car built before 1970, a particular car equipped with safety belts, or a particular Mustang on the road today. Bx, Sx, and Mx are not propositions that are true or false; they are **propositional functions**, empty places waiting, so to speak, for something to be substituted for the variables. In quantification logic, **individual constants** represent individual cars, persons, and things and are designated by the lowercase letters *a* through *w*. For example,

Bj = John's car was built before 1970.

Sj = John's car was equipped with safety belts at the factory.

Mj = John's car is a U.S.-manufactured Mustang on the road today.

Bj, Sj, and Mj are **substitution instances** of Bx, Sx, and Mx. Just as the variable p in propositional logic could stand for any statement whatsoever, the propositional function x could stand for Annelise's car (Ba, Sa, Ma), or Cesar's car (Bc, Sc, Mc), or Tran's car (Bt, St, Mt), and so on.

Next, **quantification symbols** are introduced to capture the different meanings of generalizations (All, No) and, what we will now call existential statements (Some, Some not). Here is how our four basic categorical propositions would be translated in quantification logic:

A   (universal affirmative)   All U.S.-manufactured Mustangs on the road to-
        day have safety belts equipped at the factory.

All M is S = Given any x, if x is an M then x is also an S.
   (x)(Mx ⊃ Sx)

E   (universal negative)   No U.S.-manufactured car built before 1970 was equipped with safety belts at the factory.

No B is S = Given any x, if x is a B then x is not an S.
   (x)(Bx ⊃ ~Sx)

I   (particular affirmative)   Some U.S.-manufactured Mustangs on the road today have safety belts equipped at the factory.

Some M is S = There is at least one x such that x is M and x is S.
   (∃x)(Mx • Sx)

O   (particular negative)   Some U.S.-manufactured Mustangs on the road today were not built before 1970.

Some M is not B = There is at least one x such that x is M and x is not B.
   (∃x)(Mx • ~Bx)

Note that (x) symbolizes "Given any x," and (∃x) symbolizes "There is at least one x." We can now translate (11-2) and (11-3) as follows:

No U.S.-manufactured car built before 1970 was equipped with safety belts at the factory.
John has a U.S.-manufactured car built in 1969.
Therefore, John's car was not equipped with safety belts at the factory.

1. (x)(Bx ⊃ ~Sx)
2. Bj[1]   /∴   ~Sj

All U.S.-manufactured cars built after 1970 were equipped with safety belts at the factory.
John has a U.S.-manufactured car built in 1969.
Therefore, John's car was not equipped with safety belts at the factory.

1. (x)(~Bx ⊃ Sx)[2]
2. Bj   /∴   ~Sj

With the addition of a few rules to capture the correct inferences involving (x) and (∃x), we can show symbolically that (11-2), (11-4), and (11-5) are valid arguments. Before doing so, however, let's create a dictionary of quantification statements similar to the one in Chapter 7 for propositional logic.

Usage Dictionary

1. All humans are mortal. (Given any x, if x is an H, then x is an M.)
   (x)(Hx ⊃ Mx)

---

[1] Here we will interpret "a U.S.-manufactured car built in 1969" as "a U.S.-manufactured car built before 1970."
[2] Here we interpret "a U.S.-manufactured car built (in or) after 1970" as "a U.S.-manufactured car not built before 1970," ~Bx.

2. Only registered voters are eligible to vote. (Given any x, if x is an E, then x is an R.)

   (x)(Ex ⊃ Rx)

3. All eligible voters must be registered to vote. (Given any x, if x is an E, then x is an R.)

   (x)(Ex ⊃ Rx)

4. Socrates is human. (s is H.)

   Hs

5. If Socrates is human, then Socrates is mortal. (If s is H, then s is M.)

   Hs ⊃ Ms

6. Everything is mortal. (Given any x, x is M.)

   (x)Mx

7. No man is an island. (Given any x, if x is an M, then x is not an I.)

   (x)(Mx ⊃ ~Ix)

8. Nothing is perfect. (Given any x, x is not P.)

   (x)~Px

9. Dictators are not good leaders. (Given any x, if x is a D, then x is not a G.)

   (x)(Dx ⊃ ~Gx)

10. Some marriages are happy. (There is at least one x, such that x is M and x is H.)

    (∃x)(Mx • Hx)

11. There are happy marriages. (There is at least one x, such that x is M and x is H.)

    (∃x)(Mx • Hx)

12. Some marriages are not happy. (There is at least one x, such that x is M and x is not H.)

    (∃x)(Mx • ~Hx)

13. There are unhappy marriages. (There is at least one x, such that x is M and x is not H.)

    (∃x)(Mx • ~Hx)

14. Happiness exists. (There is at least one x, such that x is H.)

    (∃x)Hx

15. It is not true that some presidents of the United States were not rich. (It is not true that there is an x, such that x is P and x is not R. Given any x, if x is P, then x is R.)

    ~(∃x)(Px • ~Rx) or (x)(Px ⊃ Rx)

16. It is not the case that every smoker gets lung cancer. (It is not true that given any x, if x is an S, then x is an L. There is at least one x, such that x is an S and x is not an L.)

    ~(x)(Sx ⊃ Lx) or (∃x)(Sx • ~Lx)

17. Only Democrats and Republicans have a realistic opportunity for elective office in the United States (Given an x, if x is a O, then x must be either a D or a R.)

$(x)[Ox \supset (Dx \lor Rx)]$

18. Black holes exist if and only if no other explanations for what we observe regarding unusual binary systems and the center of galaxies are credible. (There is at least one x, such that x is a B if and only if given any x, if x is an E, then x is not C.)

$(\exists x)Bx \equiv (x)(Ex \supset {\sim}Cx)$

19. If some politicians lie, then some voters are cynical. (If there is at least one x, such that x is a P and a L, then there is at least one x, such that x is a V and x is a C.)

$(\exists x)(Px \cdot Lx) \supset (\exists x)(Vx \cdot Cx)$

20. In the 1990s electric vehicles were both costly and inconvenient. (Given any x, if x was an E, then x was a C and x was an I.)

$(x)[Ex \supset (Cx \cdot Ix)]$

Dictionary Elaboration

As in the case of translations in propositional logic, the goal of translating in quantification logic is to do the best we can in capturing the meaning of the statements in ordinary language. Often we will find statements that do not have the exact grammatical form of our **A**, **E**, **I**, and **O** propositions, but can be interpreted into their symbolic equivalents. For instance, number 1 in the dictionary could have been expressed in ordinary language as "Humans are mortal." Similarly, notice that number 9 is another way of saying "No dictators are good leaders," and numbers 10 and 12 are different ways of saying numbers 11 and 13 respectively.

Recall how in Chapter 1 the meaning difference between *all* and *only* easily led to mistakes in reasoning. An examination of numbers 2 and 3 reveals the secret behind translating *only* statements. As with *only if* statements in propositional logic, the term or category of things that follows the *only* will always be translated as a consequent. Thus, the secret to translating *only* statements is to always replace *only* with *all* and reverse the terms. So, the statement from Chapter 1 "Only people who believe in God are moral" becomes "All people who are moral believe in God," $(x)(Mx \supset Gx)$.

As with *only if* statements in propositional logic, *only* statements in quantification logic are interpreted as specifying a necessary condition. In numbers 2 and 3 in the dictionary, being registered to vote is specified as a necessary condition for being eligible to vote. Recall from Chapter 1 that the reason many people will object to the statement, "Only people who believe in God are moral," is that it claims that believing in the Judeo-Christian God is necessary to be a good person, which excludes Buddhists, atheists, agnostics, and secular humanists, who believe that it is possible to be moral without divine aid.

Note also some other ways of saying *only*. The statement, "None but the brave deserve the fair," is the same as "Only the brave deserve the fair." Hence, it would be translated as, (x)(Fx ⊃ Bx).

Numbers 15 and 16 show that negated existential statements can be interpreted as universal statements, and negated universal statements can be interpreted as existential statements. Consider the logic behind these interpretations. "It is not true that something exists that is perfect," ~(∃x)Px, states the same thing as number 8, "Nothing is perfect," (x)~Px. So,

$$\sim(\exists x) \equiv (x)\sim$$

thus number 15, ~(∃x)(Px • ~Rx), is equivalent to (x)~(Px • ~Rx), and

1. (x)~(Px • ~Rx)
2. (x)(~Px ∨ ~~Rx)   (1) De M.
3. (x)(~Px ∨ Rx)   (2) DN
4. (x)(~~Px ⊃ Rx)   (3) Impl.
5. (x)(Px ⊃ Rx)   (4) DN

This proves symbolically that, "It is not true that some presidents of the United States were not rich," can be stated more simply as "All presidents of the United States were rich." Similar considerations apply to number 16, because ~(x) ≡ (∃x)~. Shortly, we will formalize these equivalencies and add them as rules in quantification logic.

Numbers 17–20 show that quantification logic can capture the meaning of statements that are much more complicated than simple categorical propositions. Number 17 also shows that translating is not a rigid mechanical process, that the meaning must be understood in ordinary language first. Even though the word *and* is used in number 17, the meaning is clearly that it is necessary to be *either* a Democrat *or* a Republican to have a chance at elective office, not that one should be both.

Numbers 14 and 18 show that existence is never a term. We would not translate "Happiness exists" as (∃x)(Hx • Ex), for the (∃x) symbol already specifies an existential claim. Otherwise, we would be committing ourselves to the very strange metaphysical position that there are categories of things that exist and categories of things that do not exist. Even though our language allows us to say, "Some things don't exist," it appears reasonable to believe that it is a conceptual mistake to conclude from this that there are actual things that don't exist. We would make the same mistake that the Queen makes in *Alice in Wonderland* when Alice tells her that "Nobody is coming down the roadway," and the Queen remarks how keen Alice's eyesight is to be able to see nobody. Similarly, I can say, "Two-headed dragons do not exist," but it seems unreasonable to interpret this as meaning that there is a category for things that do not exist and a two-headed dragon is one thing in this category!

Such considerations are not mere ivory-tower distinctions. Historically, the mistake of interpreting existence as a term can be seen as an attempted major reinforcement for an entire culture. The eighteenth-century German philosopher and physicist Immanuel Kant showed that philosophers and theologians in the Middle Ages made this mistake when they argued that one could prove the existence of God by merely contemplating the definition of God as the most perfect being. Known as the **ontological argument** for the existence of God, they argued that because by definition the most perfect being had to exist, otherwise it would not be perfect, God must exist. To use quantification symbolism, they incorrectly assumed that, given any x, if x is perfect, then x exists, $(x)(Px \supset Ex)$. Then they reasoned if God is perfect, God must exist, $Pg \supset Eg$. If by definition God is perfect, Pg, it follows (modus ponens) that God must exist, Eg. Viewed this way, their argument was valid. We will see why shortly, when a new rule is discussed. But for now understand that by modern standards this proof fails because the first premise should be translated as, "There is an x, such that x is perfect," $(\exists x)Px$. This statement makes a contingent existential claim that is true or false.

Kant's criticism of the ontological argument underscores the obvious point that we cannot deduce the existence of something from a definition of that something. I can define what it means to have $100, but this does not mean that I have $100 in my wallet.

Astronomers are becoming more and more confident that black holes do exist. Observations can be made of unusual binary systems—two stars orbiting each other—where one companion is not visible but is obviously very massive. The invisible companion is also a high-energy X-ray source, an indication of matter being heated as it falls onto a very massive object. Furthermore, observations with the Hubble telescope show rapidly revolving disks of matter at the center of some galaxies. These disks are revolving much too swiftly to have a normal gravitational object at the center. Number 18 summarizes the inductive claim that it is reasonable to believe in black holes given that all other explanations for what we observe about these matters are unreasonable.

Number 18 also shows that when we translate we must assume or specify a **domain of discourse**. In discussing explanations for what we observe related to black holes, we do not mean just any explanations for any observations in general, but a particular subset of observations, that is, those related to the topic of black holes, what we observe regarding binary star systems and the center of galaxies. Recall that the arguments on Mustangs and seat belts specified the domain of U.S.-manufactured cars. Consider this argument:

EXAMPLE 11-6   No U.S.-manufactured car built before 1970 was equipped with safety belts at the factory.
All Hondas on the road today have safety belts equipped at the factory.
Therefore, no Honda on the road today was built before 1970.

Because Hondas have been manufactured in Japan and the United States, the validity of this argument cannot be established unless the domain of U.S.-manufactured cars is specified throughout the argument. The second premise is consistent with the possibility that some Hondas were built in another country before 1970 and equipped with safety belts at the factory. So, unless we specify that only U.S.-manufactured Hondas are being discussed in the second premise and the conclusion, both premises could be true and the conclusion false; it is possible that some Hondas still on the road today were equipped with safety belts in Japan before 1970. Similarly, if I tell my students, "Everyone passed the midterm," they know that I am referring to the domain of people in my course, not everyone in the world.

Number 19 shows that statements can propositionally combine quantifiers, and number 20 shows that quantified propositions can have a large number of grammatical variations. The original statement for this translation could have been, "All electric vehicles during the 1990s were both costly and inconvenient." As we have noted several times in this book, the English language is much richer than the symbolic pictures we devise. Thus, as in the case of propositional translations, there can be no mechanical procedure for translating quantified statements. In every case the meaning of the sentence, including the appropriate domain, must be understood first and then captured as best we can with logical connectives and quantifiers.

EXERCISE I   **Translations**   Translate each of the following using logical connectives and quantifiers. Use the abbreviations suggested.

1. All charitable donations are tax deductible. (Cx = x is a charitable donation; Tx = x is tax deductible.)

2. No council member voted for the bill. (Cx = x is a council member; Vx = x voted for the bill.)

*3. Every car parked on the street after 9:00 p.m. will be ticketed. (Cx = x is a car parked on the street after 9:00 p.m.; Tx = x is a car ticketed.)

4. Some new U.S.-manufactured cars are dependable. (Ux = x is a new U.S.-manufactured car; Dx = x is dependable.)

5. Japanese luxury cars are dependable. (Jx = x is a Japanese luxury car; Dx = x is dependable.)

6. Metals conduct electricity. (Mx = x is a metal; Ex = x conducts electricity.)

7. Some stocks are not good investments. (Sx = x is a stock; Gx = x is a good investment.)

*8. Not every politician makes decisions with reelection primarily in mind. (Px = x is a politician; Mx = x makes decisions with reelection primarily in mind.)

9. There are no stars in our galaxy that are not massive. (Sx = x is a star in our galaxy; Mx = x is massive.)

10. Every student who maintains a quiz average of better than 90 percent can miss one exam. (Sx = x is a student; Qx = x maintains a quiz average of better than 90 percent; Mx = x can miss one exam.)

11. All Democrats are either liberal or moderate. (Dx = x is a Democrat; Lx = x is liberal; Mx = x is moderate.)

12. Only persons older than sixty-two are eligible for Social Security benefits. (Ox = x is a person older than sixty-two; Sx = x is eligible for Social Security benefits.)

*13. Whales and humans are both mammals. (Wx = x is a whale; Hx = x is a human; Mx = x is a mammal.)

14. Quarks exist, but dragons do not exist. (Qx = x is a quark; Dx = x is a dragon.)

15. Cats are not the only animals in my house. (Cx = x is a cat; Ax = x is an animal in my house.)

16. Not every student applicant will be accepted into the medical school. (Sx = x is a student applicant; Ax = x will be accepted into the medical school.)

*17. Not only students but faculty as well attended the initiation dance. (Sx = x is a student; Fx = x is a faculty member; Ax = x attended the initiation dance.)

18. Those who ignore history are condemned to repeat it. (Ix = x ignores history; Cx = x is condemned to repeat history.)

19. Not all rocks that glitter contain gold. (Gx = x is a rock that glitters; Cx = x is a rock that contains gold.)

20. The truth is not always helpful, but deception is never helpful. (Tx = x is the truth; Hx = x is helpful; interpret "deception" as "not the truth.")

Proving Validity in Quantification Logic

We are now ready to devise a method of formal proof for arguments in quantification logic. To do so, we must add to our list of nineteen rules used in propositional logic. Consider (11-2) again:

No U.S.-manufactured car built before 1970 was equipped with safety belts at the factory.

John has a U.S.-manufactured car built in 1969.

Therefore, John's car was not equipped with safety belts at the factory.

1. (x)(Bx ⊃ ~Sx)
2. Bj   /∴   ~Sj

Even translated into the new notation of quantification logic, this argument looks suspiciously like modus ponens. What is needed prior to applying modus ponens is the intuitively obvious inference that if, given any x, if x is a B then x is not an S, then it follows that if j is a B, then j is not an S.

1. (x)(Bx ⊃ ~Sx)
    /∴   Bj ⊃ ~Sj

If all cars built before 1970 did not have safety belts equipped at the factory, then if John has a car built before 1970, he must have a car that was not equipped with safety belts at the factory. In other words, if (x)(Bx ⊃ ~Sx) is true, then so is Ba ⊃ ~Sa, Bb ⊃ ~Sb, Bc ⊃ ~Sc, and so on. We can substitute many individual constants with the domain of U.S.-manufactured cars. This is simply what it means to assert a generalization. If (x)(Bx ⊃ ~Sx) is true, then any substitution instance, Bj ⊃ ~Sj, of this proposition is true.

In quantification logic this inference is called **Universal Instantiation** (**UI**) because we are inferring "instances" from a universal statement, and it becomes our first new rule added to the nineteen rules of propositional logic. We can summarize this rule as follows:

UI (x)(Ψx)
    /∴   Ψυ

The Greek *psi* symbol (Ψ) is a variable representing an empty place for any terms whatsoever of a given domain, and the Greek *nu* symbol (υ) is also a variable representing any substituted individuals whatsoever in that domain.

Using this rule we can now provide a simple proof for (11-2).

1. (x)(Bx ⊃ ~Sx)
2. Bj   /∴ ~Sj
3. Bj ⊃ ~Sj   (1) UI
4. ~Sj   (2)(3) MP

Next, let's consider (11-5) again with its translation:

No U.S.-manufactured car built before 1970 was equipped with safety belts at the factory.
All U.S.-manufactured Mustangs on the road today have safety belts equipped at the factory.
Therefore, no U.S.-manufactured Mustang on the road today was built before 1970.

1. (x)(Bx ⊃ ~Sx)
2. (x)(Mx ⊃ Sx)   /∴   (x)(Mx ⊃ ~Bx)

This argument looks suspiciously like a subroutine we covered in Chapter 10.

1. p ⊃ q
2. r ⊃ ~q   /∴   r ⊃ ~p
3. ~q ⊃ ~p   (1) Trans.
4. r ⊃ ~p   (2)(3) HS

To use this subroutine, however, we need to employ some obvious valid inferences: (1) if all U.S.-manufactured cars built before 1970 did not have safety belts equipped at the factory, then *any arbitrarily selected individual* U.S.-manufactured car built before 1970 did not have safety belts equipped at the factory; (2) if *any arbitrarily selected individual* U.S.-manufactured Mustang on the road today was not built before 1970, then no U.S.-manufactured Mustang on the road today was built before 1970. Concerning (1), we see that our instantiation rule applies not only to individuals— a, b, c, and so on—but to any arbitrarily selected individual, which we will use the italicized *y* to designate. So, from (x)(Bx ⊃ ~Sx), By ⊃ ~Sy follows, and from (x)(Mx ⊃ Sx), My ⊃ Sy follows, and from our subroutine, My ⊃ ~By:

1. (x)(Bx ⊃ ~Sx)
2. (x)(Mx ⊃ Sx)   /∴   (x)(Mx ⊃ ~Bx)
3. By ⊃ ~Sy   (1) UI
4. My ⊃ Sy   (2) UI
5. ~~Sy ⊃ ~By   (3) Trans.
6. Sy ⊃ ~By   (5) DN
7. My ⊃ ~By   (4) (6) HS

Finally, the second obvious intuitive inference shows that we can add a new rule to our developing quantification logic. We will call this new rule *Universal Generalization* (**UG**), and it is summarized as follows:

$$\textbf{UG} \quad \Psi y$$
$$/\therefore \quad (\textbf{x})(\Psi \text{x})$$

This rule indicates that we are entitled to generalize from the special instantiation *y*, which indicates any arbitrarily selected individual. We can now complete the proof of (11-5).

1. (x)(Bx ⊃ ~Sx)
2. (x)(Mx ⊃ Sx)   /∴   (x)(Mx ⊃ ~Bx)
3. By ⊃ ~Sy   (1) UI
4. My ⊃ Sy   (2) UI
5. ~~Sy ⊃ ~By   (3) Trans.
6. Sy ⊃ ~By   (5) DN
7. My ⊃ ~By   (4) (6) HS
8. (x)(Mx ⊃ ~Bx)   (7) UG

Notice that there is nothing magical about instantiating to and then generalizing from *y*, provided that we start with universal statements. This restriction must apply otherwise we would be endorsing the fallacy of hasty conclusion as a valid rule. Clearly from the proposition, "At least one Mustang on the road today does not have safety belts equipped at the factory," $(\exists x)(Mx \cdot {\sim}Sx)$, it does not follow that "All Mustangs on the road today do not have safety belts equipped at the factory," $(x)(Mx \supset {\sim}Sx)$. What prevents this invalid inference in quantification logic is that the special instantiation to y, designating any arbitrarily selected individual from a domain, is validly inferred only from a universal statement. Thus, the following is invalid and should never be used in a proof:

$$(\exists x)\ (\Psi x)$$
$$/\therefore\quad \Psi y \quad \textbf{(incorrect)}$$

However, there are valid rules involving existential statements, and we will need these rules to complete our augmentation of propositional logic. For instance, from the statement, "John has a U.S.-manufactured Mustang on the road today with safety belts equipped at the factory," $Mj \cdot Sj$, it clearly follows that at least one U.S.-manufactured Mustang on the road today has safety belts equipped at the factory, or "Some U.S.-manufactured Mustangs on the road today have safety belts equipped at the factory," $(\exists x)(Mx \cdot Sx)$. We will call this rule *Existential Generalization* (**EG**) and summarize it as follows:

$$\textbf{EG}\quad \Psi \upsilon$$
$$/\therefore\quad (\exists x)(\Psi x)$$

where (Ψ) represents any term or terms whatsoever, and (υ) represents any individual constant whatsoever. Note that in this case an inference from Ψy to $(\exists x)\Psi x$ would be valid, for it surely follows that if any arbitrarily selected individual has a certain property, then there is some thing that has that property.

At first this may seem like a bad rule. Could we conclude from a general premise about mythical unicorns having sharp horns that unicorns actually exist? The answer is no, because although it follows from "All unicorns have sharp horns," $(x)(Ux \supset Sx)$, that any arbitrarily selected individual unicorn has a sharp horn, $Uy \supset Sy$, we cannot derive $Uy \cdot Sy$ unless we establish Uy, that an arbitrarily selected individual unicorn exists. So, if $Uy \cdot Sy$ cannot be derived from $(x)(Ux \supset Sx)$, then neither can $(\exists x)(Ux \cdot Sx)$.

There is also a version of instantiation that is valid for existential statements, but it must be used with great care. If some U.S.-manufactured Mustangs on the road today have safety belts equipped at the factory, then it follows that there is at least one U.S.-manufactured Mustang somewhere on the road today that has safety belts equipped at the factory. However, in any given context we cannot assume that we know that this refers to any specific Mustang. We cannot say that this hypothetical Mustang is Mary's, Cesar's, Kanoe's, or Tran's. This realization prevents the following argument from being valid.

EXAMPLE 11-7   Some U.S.-manufactured Mustangs on the road today have safety belts equipped at the factory.
Some cars built before 1970 had safety belts equipped at the factory.
Thus, some U.S.-manufactured Mustangs on the road today were built before 1970.

Clearly, although the conclusion may be true, it goes beyond or infers more than what is implied in the premises. If some cars built before 1970 had safety belts equipped at the factory, and some U.S.-manufactured Mustangs on the road today also have safety belts equipped at the factory, then it is possible that these U.S.-manufactured Mustangs were built before 1970. But there is no guarantee; we are not locked into this conclusion. It is just as possible that all U.S.-manufactured Mustangs on the road today were built after 1970 and equipped with safety belts at the factory, and that expensive Cadillacs were the first cars equipped with safety belts at the factory. The uncertainty of this conclusion, given the premises, can be shown by a Venn diagram.

FIGURE 11-6

1. Some M is S.
2. Some B is S.
    Therefore, some M is B.

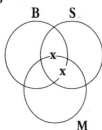

Notice that in mapping the first premise, we have a choice of where to put the little x. We could put it in that portion of the overlap of the M and S circles that also overlaps the B circle or just in that portion that overlaps S and M alone. The appropriate place to put it, then, is on the line that separates these two possibilities, thus indicating the undecidedness. A similar uncertainty results for the second premise, "Some B is S." We could put the x in that portion of overlap between just the B and S circles, or in that portion of the overlap of the B, S, and M circles. Thus, the appropriate place to put the x is on the line that separates these two possibilities. This leaves us with a Venn picture that does not explicitly contain the conclusion. Both xs are on the borders of the overlap section of the M and B circles, indicating that it is only possible that some Ms are Bs.

In quantification logic, to be consistent with this result we need to have the following important restriction when using the next quantification rule, which we will call **Existential Instantiation** (**EI**). In addition to not being able to instantiate to the special symbol y, we can instantiate existentially to an individual constant, ***provided that the individual constant occurs nowhere earlier in the context of the argument***. Here, then, is a summary of **EI**, followed by its correct use in showing (11-4) above to be valid and an incorrect use in attempting to show (11-7) to be valid.

EI (∃x)(Ψx)   (Where υ is neither γ nor any individual constant
    /∴ Ψυ   having a previous occurrence in the context of
                    the argument.)

EXAMPLE 11-4
1. (x)(Bx ⊃ ~Sx)
2. (∃x)(Mx • Sx)   /∴   (∃x)(Mx • ~Bx)
3. Ma • Sa   (2) EI
4. Ba ⊃ ~Sa   (1) UI
5. Ma   (3) Simp.
6. Sa   (3) Com. + Simp.
7. ~~Sa   (6) DN
8. ~Ba   (4) (7) MT
9. Ma • ~Ba   (5) (8) Conj.
10. (∃x)(Mx • ~Bx)   (9) EG

EXAMPLE 11-7
1. (∃x)(Mx • Sx)
2. (∃x)(Bx • Sx)   /∴   (∃x)(Mx • Bx)
3. Ma • Sa   (1) EI
4. Ba • Sa   (2) EI   **(incorrect!!)**
5. Ma   (3) Simp.
6. Ba   (4) Simp.
7. Ma • Ba   (5) (6) Conj.
8. (∃x)(Mx • Bx)   (7) EG

Note that the restriction applies to the conclusion of an argument as well. In saying that the individual constant cannot occur anywhere earlier in the context of an argument, this includes any constants that might occur on the right side of the (/∴) sign. Otherwise, we would be able to offer a proof of the obviously invalid argument:

EXAMPLE 11-8
1. Some cars built before 1970 were U.S.-manufactured Mustangs.
   Therefore, John has a U.S.-manufactured Mustang.
1. (∃x)(Bx • Mx)   /∴   Mj
2. Bj • Mj   (EI) **(incorrect!!)**
3. Mj • Bj   (2) Com.
4. Mj   (3) Simp.

One final note on these four new rules. Like the rules of inference in chapter 9, all the rules of quantification must be applied to *whole lines only*, otherwise we could prove the following argument to be valid.

EXAMPLE 11-9
If everything is a crow, then everything is black.
Some crows exist.
Therefore, everything is black.

1. $(x)Cx \supset (x)Bx$
2. $(\exists x)Cx$ $\quad$ / $\therefore$ $(x)Bx$
3. Ca $\hspace{3.5cm}$ (2) EI
4. $Ca \supset (x)Bx$ $\hspace{2cm}$ (1) UI (**incorrect!!**)
5. $Ca \supset By$ $\hspace{2.3cm}$ (4) UI (**incorrect!!**)
6. By $\hspace{3.5cm}$ (3) (5) MP[1]
7. $(x)Bx$ $\hspace{3cm}$ (6) UG

The Square of Opposition and Change of Quantifier Rules

Now we are just about ready to try some formal proofs using our new additions to the nineteen rules of propositional logic. We need to add one more rule, which is actually a series of variations of the equivalences noted in numbers 15 and 16 in the dictionary. Without this addition, we cannot prove valid arguments such as:

EXAMPLE 11-10 $\quad$ 1. It is not true that all smokers develop lung cancer.
$\hspace{3.5cm}$ 2. But all smokers are still health risks.
$\hspace{3.8cm}$ Therefore, some people who do not develop lung cancer are still health risks.

When we use the following translation:

1. $\sim(x)(Sx \supset Cx)$
2. $(x)(Sx \supset Hx)$ $\quad$ / $\therefore$ $(\exists x)(\sim Cx \cdot Hx)$

we need a rule to manipulate the first premise, and infer its simpler rendition, "Some smokers do not develop lung cancer." Before formalizing this rule, it will be helpful to review what has been traditionally called the **Square of Opposition** and compare it with its modern interpretation. Because it is so easy to confuse statements such as "All cars built after 1970 had safety belts equipped at the factory," and "No cars built before 1970 had safety belts equipped at the factory," this pictorial device was constructed by logicians to keep track of the differences and the relationships between the basic *all, no, some,* and *some not* categorical propositions. Consider the difference between these propositions:

A $\quad$ All smokers develop lung cancer.

E $\quad$ No smokers develop lung cancer.

I $\quad$ Some smokers develop lung cancer.

O $\quad$ Some smokers do not develop lung cancer.

From this we see that **O** propositions *contradict* **A** propositions. In other words, these propositions cannot both be true. If it were true that all smokers develop lung cancer, $(x)(Sx \supset Cx)$, then it would have to be false that some smokers do not

---

[1] Line 5 is unnecessary and is for illustration purposes only. We could conclude $(x)Bx$ by MP from lines 3 and 4, and then correctly instantiate to By. But the argument chain would still be broken by the invalid line 4.

develop lung cancer, ~(∃x)(Sx • ~Cx); and if it were true that some smokers do not develop lung cancer, (∃x)(Sx • ~Cx), then it would have to be false that all smokers develop lung cancer, ~(x)(Sx ⊃ Cx). Similarly, **I** propositions *contradict* **E** propositions.

In the traditional Aristotelian interpretation of categorical propositions the commonsense assumption was made that when we assert these propositions, we are making claims about actually existing things. Thus, there is a deductive relationship between **A** and **I** propositions, on the one hand, and **E** and **O** propositions on the other hand. If we are allowed to assume that at least one smoker exists, then if the **A** proposition is true, the **I** proposition must also be true; and, if the **E** proposition is true, then the **O** proposition must also be true. But notice that from a rigorous standpoint, we must make explicit the assumption that at least one smoker exists, because

1. All smokers develop lung cancer.
   Therefore, some smokers develop lung cancer.
1. $(x)(Sx \supset Cx)$ /∴ $(\exists x)(Sx \cdot Cx)$

and

1. No smokers develop lung cancer.
   Therefore, some smokers do not develop lung cancer.
1. $(x)(Sx \supset {\sim}Cx)$ /∴ $(\exists x)(Sx \cdot {\sim}Cx)$

are both technically invalid! After instantiating, there is no way to derive validly a conjunction from a conditional statement.

1. $(x)(Sx \supset Cx)$ /∴ $(\exists x)(Sx \cdot Cx)$
2. $Sa \supset Ca$ (1) UI
3. $Sa \cdot Ca$????
4. $(\exists x)(Sx \cdot Cx)$ (3) EG

1. $(x)(Sx \supset {\sim}Cx)$ /∴ $(\exists x)(Sx \cdot {\sim}Cx)$
2. $Sa \supset {\sim}Ca$ (1) UI
3. $Sa \cdot {\sim}Ca$????
4. $(\exists x)(Sx \cdot {\sim}Cx)$ (3) EG

But with the assumption that there is at least one smoker, (∃x)Sx, we can validly derive **I** propositions from **A** propositions, and **O** propositions from **E** propositions.

1. $(x)(Sx \supset Cx)$
2. $(\exists x)Sx$ /∴ /∴ $(\exists x)(Sx \cdot Cx)$
3. Sa (2) EI
4. $Sa \supset Ca$ (1) UI
5. Ca (3) (4) MP
6. $Sa \cdot Ca$ (3) (5) Conj.
4. $(\exists x)(Sx \cdot Cx)$ (6) EG

1. $(x)(Sx \supset {\sim}Cx)$
2. $(\exists x)Sx$ / $\therefore$ $(\exists x)(Sx \bullet {\sim}Cx)$
3. Sa (2) EI
4. $Sa \supset {\sim}Ca$ (1) UI
5. ${\sim}Ca$ (3) (4) MP
6. $Sa \bullet {\sim}Ca$ (3) (5) Conj.
4. $(\exists x)(Sx \bullet {\sim}Cx)$ (6) EG

In the traditional square of opposition the relationship between **A** and **I** propositions on the one hand, and between **E** and **O** propositions on the other hand, was said to be one of **subalternation**. The **A** and **E** propositions were said to be **superalterns** and the **I** and **O** **subalterns**.

Next, **A** and **E** propositions were said to be **contraries**. A contrary relationship between propositions means that **both cannot be true, but** (unlike contradictions) **both could be false**. It cannot be true both that all smokers develop lung cancer and that no smokers develop lung cancer. Thus, if it was true that all smokers develop lung cancer, then we would know that it is false that no smokers develop lung cancer, and if it was true that no smokers develop lung cancer, then we would know that it is false that all smokers develop lung cancer. Note that again we must assume $(\exists x)Sx$.

1. $(x)(Sx \supset Cx)$
2. $(\exists x)Sx$ / $\therefore$ ${\sim}(x)(Sx \supset {\sim}Cx)$

1. $(x)(Sx \supset {\sim}Cx)$
2. $(\exists x)Sx$ / $\therefore$ ${\sim}(x)(Sx \supset Cx)$

We will leave the proofs of these inferences for exercises. Note, however, that a contrary relationship is not as strong as a contradictory relationship. Because both **A** and **E** propositions could be false, if "All smokers develop lung cancer" is false, then we are uncertain about whether "No smokers develop lung cancer" is true; it could be true or false.

In the traditional square of opposition, **I** and **O** propositions were said to be **subcontraries**. A subcontrary relationship between propositions means that **both cannot be false, but both could be true**. In other words, at least one is true. So, if it is false that some smokers develop lung cancer, then it is true that some smokers do not develop lung cancer. On the other hand, if it is false that some smokers do not develop lung cancer, then it is true that some do. Again, we must assume $(\exists x)Sx$ to prove this.

1. ${\sim}(\exists x)(Sx \bullet Cx)$
2. $(\exists x)Sx$ / $\therefore$ $(\exists x)(Sx \bullet {\sim}Cx)$

1. ${\sim}(\exists x)(Sx \bullet {\sim}Cx)$
2. $(\exists x)Sx$ / $\therefore$ $(\exists x)(Sx \bullet Cx)$

We will also leave the proofs of these inferences for exercises. Note that both propositions could be true. It could be true that some smokers develop lung cancer and some do not. (In fact we believe this is true based on the reasoning discussed in Chapter 3.) So, in a subcontrary relationship, if an **I** or **O** proposition is known to be true, we cannot infer with certainty that its subcontrary is true or false. If it is true that "Some people on the football team have short hair," we cannot know deductively whether "Some people on the football team do not have short hair" is true. Nor can we know deductively whether it is false. We would have to discover the truth or falsity of this **O** proposition empirically.

We can now picture the ***Traditional Square of Opposition*** as follows:

FIGURE 11-7 THE TRADITIONAL SQUARE OF OPPOSITION.

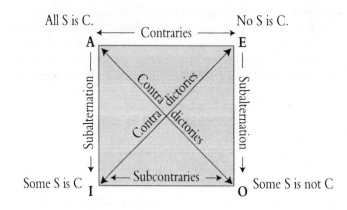

Subalternation = **I** propositions can be deduced from **A** propositions; **O** propositions can be deduced from **E** propositions.

Contradictions = opposite truth value. If **A** is true, then **O** must be false; if **O** is true, then **A** must be false. If **E** is true, then **I** must be false; if **I** is true, then **E** must be false.

Contraries = at least one is false, both cannot be true, but both could be false. If **A** is true, then **E** must be false; if **E** is true, then **A** must be false. Both **A** and **E** could be false.

Subcontraries = at least one is true, not both are false. If **I** is false, then **O** must be true; if **O** is false, then **I** must be true. But both **I** and **O** could be true.

The utility of the traditional square of opposition, as in most pictures, is that it gives us a powerful tool for summarizing many inferences. For instance, if we knew that the **A** proposition, "All smokers are health risks," is true, then we immediately know that

1. The **O** proposition, "Some smokers are not health risks," is false via the relationship of contradiction with **A**. (If **A** is true, then **O** must be false.)

2. The **I** proposition, "Some smokers are health risks," is true, because it is a subaltern of **A** and a subcontrary of **O**. (If **A** is true, then **I** is true; if **O** is false, **I** is true.)
3. The **E** proposition, "No smokers are health risks," is false because of the relationship of contradiction with **I**, and the contrary relationship with **A**. (If **I** is true, then **E** is false; if **A** is true, then **E** is false.)

As neatly as the traditional square of opposition locks in these relationships, from a modern perspective there is a fatal flaw. We obviously cannot assume in all uses of categorical propositions that the things referred to by the terms actually exist. In fact, a moment's reflection on the contents of tabloid newspaper articles while standing in your local supermarket checkout line will demonstrate that questionable assertions about existence are made all the time. In the past few years, I have seen alleged factual articles on a woman having a Martian baby, some of Hitler's men returning to Earth from an astronautical adventure that began in 1944, and even one that claimed to be able to explain the sudden reduction of the number of frogs throughout the world by an arrival of extraterrestrial creatures who were eating the frogs. That we can intelligibly, but not I believe intelligently, make statements such as, "Some Martian babies exist," and "All extraterrestrial creatures eat frogs," does not mean that such things actually exist.[1]

Thus, modern logicians claim that the Aristotelian traditional square of opposition commits an **existential fallacy** when it is applied to propositions that make assertions about things that do not actually exist. In the above discussion, notice what would be left in terms of valid deductive relations to the square of opposition if we cannot make the assumption that some smokers exist, $(\exists x)Sx$. This assumption was needed to show the validity of the subalternation, contrary, and subcontrary relationships. Thus, the only relationship that remains in what is called the **Modern Square of Opposition** is the relationship of contradiction between **A** and **O** propositions and **E** and **I** propositions.

FIGURE 11-8

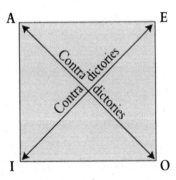

---

[1] Incidently, a much better possible scientific explanation for the disappearance of frogs is that frogs are amphibians, live partly in water, and hence have water permeable skins. Reproducing in water, their eggs are thus hypersensitive to environmental pollution and increased ultraviolet radiation due to ozone depletion. This explanation is corroborated by laboratory experiments and the fact that other species of amphibians are also disappearing at an alarming rate.

This result is not trivial. It reflects not only a major cultural difference between our age and the cozy, secure universe that Aristotle, followed by people in the Middle Ages, accepted—we have a much less certain belief environment—but this result also underscores again the importance of valid deductive analysis for testing our beliefs. What cannot be known with deductive certainty must be inductively tested. Recall from our discussion in Chapter 2 that previous philosophers in Western culture thought that they could deductively prove the foundational beliefs of their societies. Recall also from our previous discussion in this chapter that during the Middle Ages philosophers and theologians thought they could prove the existence of their version of God from words alone, that is, that from the mere contemplation of their definition of God, they could deduce His existence.

Advances in logic and science, and the general awareness of the practices and beliefs in other cultures, have epistemologically pruned our world of such certainties. To make this statement is not to imply necessarily that one cannot be rational and believe some of the same things people did in the Middle Ages, such as the existence of God. Rather, it only implies that one must work harder. From a modern perspective, there are no logical proofs of the existence of God.[1] However, as has been repeatedly emphasized in this book, this ought not to leave us with a belief-emptiness, with relativism and nihilism. Simply put, our situation is more challenging, but it is not hopeless. What logical analysis shows cannot be certain is left for the bootstrap process of inductive reasoning and the ongoing process of trying to make our inductions stronger.

The bottom line of this analysis is that we can now see clearly what must be added to our quantification rules to be able to deal with arguments such as (11-10). First, let's summarize symbolically the contradictory relationships.

$(x)(Sx \supset Cx) \equiv \sim(\exists x)(Sx \cdot \sim Cx)$	The truth of an **A** proposition is logically equivalent to the negation of an **O** proposition.
$\sim(x)(Sx \supset Cx) \equiv (\exists x)(Sx \cdot \sim Cx)$	The negation of an **A** proposition is logically equivalent to the truth of an **O** proposition.
$(\exists x)(Sx \cdot Cx) \equiv \sim(x)(Sx \supset \sim Cx)$	The truth of an **I** proposition is logically equivalent to the negation of an **E** proposition.
$\sim(\exists x)(Sx \cdot Cx) \equiv (x)(Sx \supset \sim Cx)$	The negation of an **I** proposition is logically equivalent to the truth of an **E** proposition.

From these equivalences and a little work with some of the rules of replacement we can derive the following equivalences: (We will save the work for exercises)

$$(x)(Sx \supset Cx) \equiv \sim(\exists x)\sim(Sx \supset Cx)$$
$$\sim(x)(Sx \supset Cx) \equiv (\exists x)\sim(Sx \supset Cx)$$
$$(\exists x)(Sx \cdot Cx) \equiv \sim(x)\sim(Sx \cdot Cx)$$
$$\sim(\exists x)(Sx \cdot Cx) \equiv (x)\sim(Sx \cdot Cx)$$

---

[1] According to Kant, this realization allows for the possibility of a rational faith.

Finally, we can summarize these equivalences, using our variable (Ψ), as one set of *Change of Quantifier Rules* (**CQ**). Recall that (Ψ) stands for any terms whatsoever.

$$(x)(\Psi x) \equiv \sim(\exists x)\sim(\Psi x) \quad \sim(x)(\Psi x) \equiv (\exists x)\sim(\Psi x)$$
$$(\exists x)(\Psi x) \equiv \sim(x)\sim(\Psi x) \quad \sim(\exists x)(\Psi x) \equiv (x)\sim(\Psi x)$$

Essentially, these rules formalize a set of relationships, part of which we noted above when discussing numbers 15 and 16 in the dictionary. Completing this analysis, we can now see that saying,

Everything is perfect, $(x)Px$ $\quad\equiv$ It is not true that there is something that is not perfect, $\sim(\exists x)\sim Px$

It is not true that everything is $\quad\equiv$ Some things are not perfect, $(\exists x)\sim Px$
perfect, $\sim(x)Px$

Some things are perfect, $(\exists x)Px$ $\equiv$ It is not true that nothing is perfect, $\sim(x)\sim Px$

It is not true that some things $\quad\equiv$ Nothing is perfect, $(x)\sim Px$.
are perfect, $\sim(\exists x)Px$

In using the change of quantifier rules mechanically, all you need to remember is this: If you see (x), you can change to $\sim(\exists x)\sim$ and vice versa; if you see $(\exists x)$, you can change to $\sim(x)\sim$ and vice versa; if you see $\sim(x)$, you can change to $(\exists x)\sim$ and vice versa; if you see $\sim(\exists x)$, you can change to $(x)\sim$ and vice versa.

We can now offer a proof for (11-10) as follows:

1. $\sim(x)(Sx \supset Cx)$
2. $(x)(Sx \supset Hx)$ /∴ $(\exists x)(\sim Cx \cdot Hx)$
3. $(\exists x)\sim(Sx \supset Cx)$ (1) CQ
4. $(\exists x)\sim(\sim\sim Sx \supset Cx)$ (3) DN
5. $(\exists x)\sim(\sim Sx \lor Cx)$ (4) Impl.
6. $(\exists x)(\sim\sim Sx \cdot \sim Cx)$ (5) De M.
7. $(\exists x)(Sx \cdot \sim Cx)$ (6) DN
8. $Sa \cdot \sim Ca$ (7) EI
9. $Sa \supset Ha$ (2) UI
10. Sa (8) Simp.
11. Ha (9) (10) MP
12. $\sim Ca$ (8) Com. + Simp.
13. $\sim Ca \cdot Ha$ (12) (11) Conj.
14. $(\exists x)(\sim Cx \cdot Hx)$ (13) EG

Before doing the following exercises, summarize all the quantification rules, including the change of quantifier rules, on one sheet of paper. Be sure to include notes on the restrictions.

EXERCISE   Justify each correct step in the following. Put an X for any line that
II   involves the incorrect application of a rule.

1.
1. (x)(Ax ⊃ Bx)
2. ~(∃x)(Bx • Cx) /∴ (x)(Ax ⊃ ~Cx)
3. (x)~(Bx • Cx)
4. (x)(~Bx ∨ ~Cx)
5. (x)(~~Bx ⊃ ~Cx)
6. (x)(Bx ⊃ ~Cx)
7. Ay ⊃ By
8. By ⊃ ~Cy
9. Ay ⊃ ~Cy
10. (x)(Ax ⊃ ~Cx)

2.
1. (x)(Bx ⊃ Cx)
2. (∃x)(Bx • ~Hx) /∴ (∃x)(Cx • ~Hx)
3. Bd • ~Hd
4. Bd
5. Bd ⊃ Cd
6. Cd
7. ~Hd • Bd
8. ~Hd
9. Cd • ~Hd
10. (∃x)(Cx • ~Hx)

*3.
1. ~(x)(Fx ⊃ ~Ax)
2. (∃x)(Fx • Ox) /∴ (∃x)(Ax • Ox)
3. (∃x)~(Fx ⊃ ~Ax)
4. (∃x)~(~~Fx ⊃ ~Ax)
5. (∃x)~(~Fx ∨ ~Ax)
6. (∃x)(~~Fx • ~~Ax)
7. (∃x)(Fx • ~~Ax)
8. (∃x)(Fx • Ax)
9. Fa • Aa
10. Fa • Oa
11. Aa • Fa
12. Oa • Fa
13. Aa
14. Oa
15. Aa • Oa
16. (∃x)(Ax • Ox)

4.
1. (x)[(Dx • Cx) ⊃ Ax]
2. (∃x)(Cx • ~Ax) /∴ ~(x)Dx
3. Cm • ~Am
4. (Dm • Cm) ⊃ Am
5. ~Am • Cm
6. ~Am
7. ~(Dm • Cm)
8. ~Dm ∨ ~Cm
9. ~Cm ∨ ~Dm
10. Cm
11. ~~Cm
12. ~Dm
13. (∃x)~Dx
14. ~(x)Dx

5.
1. (x)(Bx ∨ Cx)
2. (x)(Bx ⊃ Cx) /∴   (∃x)Cx
3. By ∨ Cy
4. ~By ⊃ Cy
5. By ⊃ Cy
6. ~Cy ⊃ ~By
7. ~Cy ⊃ Cy
8. Cy ∨ Cy
9. Cy
10. (x)Cx

6.
1. (∃x)~(Bx • Cx) /∴ (x)(Bx ⊃ ~Cx)
2. (∃x)(~Bx ∨ ~Cx)
3. (∃x)(~~Bx ⊃ ~Cx)
4. (∃x)(Bx ⊃ ~Cx)
5. Ba ⊃ ~Ca
6. (x)(Bx ⊃ ~Cx)

*7.
1. $(\exists x)(Ax \cdot Fx) \quad / \therefore \sim(\exists x)(Ax \supset \sim Fx)$
2. $Ay \cdot Fy$
3. $\sim\sim(Ay \cdot Fy)$
4. $\sim(\sim Ay \vee \sim Fy)$
5. $\sim(\sim\sim Ay \supset \sim Fy)$
6. $\sim(Ay \supset \sim Fy)$
7. $(x)\sim(Ax \supset \sim Fx)$
8. $\sim(\exists x)(Ax \supset \sim Fx)$

8.
1. $(\exists x)\sim(Gx \cdot Hx)$
2. $\sim(x)(Gx \cdot Hx) \supset (\exists x)(Tx \cdot \sim Px)$
3. $(x)(Cx \supset Px) \quad / \therefore \quad (\exists x)(Tx \cdot \sim Cx)$
4. $(\exists x)\sim(Gx \cdot Hx) \supset (\exists x)(Tx \cdot \sim Px)$
5. $(\exists x)(Tx \cdot \sim Px)$
6. $Ta \cdot \sim Pa$
7. $Ca \supset Pa$
8. $\sim Pa \cdot Ta$
9. $\sim Pa$
10. $\sim Ca$
11. Ta
12. $Ta \cdot \sim Ca$
13. $(\exists x)(Tx \cdot \sim Cx)$

9.
1. $\sim(\exists x)(Kx \cdot \sim Yx)$
2. $(x)[(Kx \cdot Yx) \supset Zx] / \therefore (x)(Kx \supset Zx)$
3. $(x)\sim(Kx \cdot \sim Yx)$
4. $(x)(\sim Kx \vee \sim\sim Yx)$
5. $(x)(\sim Kx \vee Yx)$
6. $(x)(\sim\sim Kx \supset Yx)$
7. $(x)(Kx \supset Yx)$
8. $Ky \supset Yy$
9. $(Ky \cdot Yy) \supset Zy$
10. $(Yy \cdot Ky) \supset Zy$
11. $Yy \supset (Ky \supset Zy)$
12. $Ky \supset (Ky \supset Zy)$
13. $(Ky \cdot Ky) \supset Zy$
14. $Ky \supset Zy$
15. $(x)(Kx \supset Zx)$

10.
1. $(x)[(Ax \vee Cx) \supset Dx]$
2. $\sim(\exists x)(\sim Ax \cdot \sim Cx)$
3. $(\exists x)(\sim Dx \vee Px) \quad / \therefore \quad (\exists x)Px$
4. $\sim Da \vee Pa$
5. $(x)\sim(\sim Ax \cdot \sim Cx)$
6. $(x)(\sim\sim Ax \vee \sim\sim Cx)$
7. $(x)(Ax \vee \sim\sim Cx)$
8. $(x)(Ax \vee Cx)$
9. $(x)(\sim Ax \supset Cx)$
10. $\sim Aa \supset Ca$
11. $(Aa \vee Ca) \supset Da$
12. $(\sim Aa \supset Ca) \supset Da$
13. Da
14. $\sim\sim Da$
15. Pa
16. $(\exists x)Px$

EXERCISE III   Write an essay, complete with a symbolic demonstration similar to (11-7) or (11-9), explaining why it is invalid to conclude "Some students are faculty" from the premise "Not only students but faculty as well attended the initiation dance." (#17, Ex. I)

EXERCISE IV   Prove the contrary and subcontrary relationships left unproved in the traditional square of opposition.

Contraries: If an **A** proposition is true, then an **E** proposition is false. If an **E** proposition is true, then an **A** proposition is false. Prove:

1. $(x)(Sx \supset Cx)$
2. $(\exists x)Sx \quad / \therefore \quad \sim(x)(Sx \supset \sim Cx)$

1. (x)(Sx ⊃ ~Cx)
2. (∃x)Sx   /∴   ~(x)(Sx ⊃ Cx)

Subcontraries: If an **I** proposition is false, then an **O** proposition is true. If an **O** proposition is false, then an **I** proposition is true. Prove:

1. ~(∃x)(Sx • Cx)
2. (∃x)Sx   /∴ (∃x)(Sx • ~Cx)
1. ~(∃x)(Sx • ~Cx)
2. (∃x)Sx   /∴ (∃x)(Sx • Cx)

EXERCISE V

Prove the equivalences on page 378 by proving the following.

*1. (x)(Sx ⊃ Cx) ≡ ~(∃x)~(Sx ⊃ Cx), because
~(∃x)~(Sx ⊃ Cx) ≡ ~(∃x)(Sx • ~Cx), so prove:

1. ~(∃x)~(Sx ⊃ Cx)   /∴   ~(∃x)(Sx • ~Cx)
    and
1. ~(∃x)(Sx • ~Cx)   /∴   ~(∃x)~(Sx ⊃ Cx)

2. ~(x)(Sx ⊃ Cx) ≡ (∃x)~(Sx ⊃ Cx), because
(∃x)~(Sx ⊃ Cx) ≡ (∃x)(Sx • ~Cx), so prove:

1. (∃x)~(Sx ⊃ Cx)   /∴   (∃x)(Sx • ~Cx)
    and
1. (∃x)(Sx • ~Cx)   /∴   (∃x)~(Sx ⊃ Cx)

3. (∃x)(Sx • Cx) ≡ ~(x)~(Sx • Cx), because
~(x)~(Sx • Cx) ≡ ~(x)(Sx ⊃ ~Cx), so prove:

1. ~(x)~(Sx • Cx)   /∴   ~(x)(Sx ⊃ ~Cx)
    and
1. ~(x)(Sx ⊃ ~Cx)   /∴   ~(x)~(Sx • Cx)

4. ~(∃x)(Sx • Cx) ≡ (x)~(Sx • Cx), because
(x)~(Sx • Cx) ≡ (x)(Sx ⊃ ~Cx), so prove:

1. (x)~(Sx • Cx)   /∴   (x)(Sx ⊃ ~Cx)
    and
1. (x)(Sx ⊃ ~Cx)   /∴   (x)~(Sx • Cx)

EXERCISE VI

Translate each of the following. Then when you are sure you have the correct translation, use the change of quantifier relationships and the rules of replacement to streamline each translation.

Eg. It is not true that there are some smokers who are not health risks. (Sx = x is a smoker; Hx = x is a health risk.)

~(∃x)(Sx • ~Hx)

1. ~(∃x)(Sx • ~Hx)
2. (x)~(Sx • ~Hx)   (1) CQ
3. (x)(~Sx ∨ ~~Hx)   (2) De M.

4. (x)(~Sx ∨ Hx)   (3) DN
5. (x)(~~Sx ⊃ Hx)   (4) Impl.
6. (x)(Sx ⊃ Hx)   (5) DN

Result as shown by line 6: All smokers are health risks.

1. It is not true that all marijuana users abuse other drugs. (Mx = x is a marijuana user; Ax = x abuses other drugs.)

*2. In the United States it is not true that some judges either do not go to law school or do not pass the bar exam. (Jx = x is a judge in the United States; Lx = x goes to law school; Bx = x passes the bar exam.)

3. Not all smokers die of lung cancer, but it is not true that some are not health risks. (Sx = x is a smoker; Dx = x dies of lung cancer; Hx = x is a health risk.)

4. It is not true that all students who passed the final did not pass the course. (Fx = students who passed the final; Cx = students who passed the course.)

5. Although it is not true that only politicians are rich, there does not exist a single politician who is not rich. (Rx = x is rich; Px = x is a politician)

EXERCISE VII

Translate each of the following arguments. When you are sure the translation is correct, provide a formal proof for each.

1. Everyone on the football team must have short hair. John does not have short hair. So, John is not on the football team. (Fx = x is on the football team; Sx = x has short hair; j = John.)

2. Only physicians can write prescriptions. Anyone who does not write prescriptions does not give business to pharmacists. Thus, only physicians give business to pharmacists. (Px = x is a physician; Wx = x writes prescriptions; Bx = x gives business to pharmacists.)

*3. Professors are either enthusiastic or they do not motivate students. Not all professors fail to motivate students. So, some professors are enthusiastic. (Px = x is a professor; Ex = x is enthusiastic; Mx = x motivates students.)

4. Those who ignore history are condemned to repeat it. Those who act for short-term gain ignore history. Some politicians act for short-term gain. Therefore, some politicians are condemned to repeat history. (Ix = x ignores history; Cx = x is condemned to repeat history; Px = x is a politician; Sx = x acts for short-term gain.)

5. All students are eligible for tuition waivers if and only if they are not freshmen and have at least a B grade point average. Kriegle

is a student with a B grade point average, but is not eligible. So, Kriegle must be a freshman. (Sx = x is a student; Ex = x is eligible for a tuition waiver; Fx = x is a freshman; Bx = x has at least a B grade point average; k = Kriegle.)

6. All supporters of abortion are either liberals or moderates. But no moderate supports abortion after the first trimester. (Assume as an obvious third premise: Anyone who supports abortion after the first trimester supports abortion.) So, only liberals support abortions after the first trimester. (Ax = x supports abortion; Lx = x is a liberal; Mx = x is a moderate; Fx = x supports abortion after the first trimester.)

7. Only persons older than sixty-two receive Social Security retirement benefits. Any person older than sixty-two is also eligible for a one-time capital gains tax reduction. But some people are eligible for a one-time capital gains tax reduction who are not older than sixty-two. Accordingly, although there is no person who receives social security who is not eligible for the one-time capital gains tax reduction, it is not true that only people older than sixty-two are eligible for the one-time capital gains tax reduction. (Ox = x is older than sixty-two; Sx = x receives Social Security retirement benefits; Ex = x is eligible for a one-time capital gains tax reduction.)

8. It is not true that every citizen in the United States owns more than one gun. The population of the United States is about 250 million. The number of guns in the United States is 600 million. If the population of the United States is about 250 million and the number of guns in the United States is 600 million, then some citizens of the Unites States own more than one gun. If some citizens of the United States do not own more than one gun and some citizens of the United States do own more than one gun, then the gun laws are flawed and should be changed. Hence, it is not true that a gun law exists that is not flawed or should not be changed. (Ux = x is a citizen of the United States; Ox = x owns more than one gun; P = the population of the United States is about 250 million; N = the number of guns in the United States is over 600 million; Gx = x is a gun law; Fx = x is flawed; Cx = x should be changed.)

*9. Not all voters make intelligent choices during elections. If some voters do not make intelligent choices during elections, then some dishonest politicians will take advantage of fallacious thinking. But some politicians are honest. If some politicians are honest and some politicians are not honest, then all voters should be critical thinkers in a democracy. It follows that it is not true that some voters should not be critical thinkers in a democracy. (Vx = x is a voter; Ix = x makes intelligent choices during elections; Hx =

x is honest; Px = x is a politician; Ax = x takes advantage of fallacious thinking; Cx = x should be a critical thinker in a democracy.)

10. No Democratic politician is both a liberal and an advocate of President Clinton's version of the trade bill. But all Democratic politicians are either liberals or advocates of President Clinton's version of the trade bill. So, it follows that no Democratic politician is a liberal if they are an advocate of President Clinton's version of the trade bill, and an advocate of President Clinton's version of the trade bill if they are a liberal; whereas, all Democratic politicians who are not liberals are advocates of President Clinton's version of the trade bill and all Democratic politicians who are not advocates of President Clinton's version of the trade bill are liberals. (Dx = x is a Democratic politician; Lx = x is a liberal; Ax = x is an advocate of President Clinton's version of the trade bill.)

Does the conclusion, "All Democratic politicians are such that they are liberals if and only if they are not advocates of President Clinton's version of the trade bill and advocates of President Clinton's version of the trade bill if and only if they are not liberals" also follow from these premises?

Final Note

The initial quantification steps we have taken are the proper place to conclude a standard introduction to logic. However, as noted at the beginning of this chapter there are many more symbolic techniques that logicians have devised in their attempt to help us weave our way through the rich and often ambiguous experience of life. Because this book began with a quotation from Abraham Lincoln, let us conclude this chapter with just a taste of the next level of quantification by showing what a translation of this famous statement would look like.

Recall that, according to Lincoln, "You can fool some of the people all of the time, and all of the people some of the time, but you cannot fool all of the people all of the time." In order to translate this statement, we must be able to translate relationships, such as John is taller than Rubin, John gained a friend, John is fooled all the time, and so on. Here is how these examples could be translated:

John is taller than Rubin = Tjr = j is T(aller) than r.

John gained a friend = $(\exists x)(Fx \cdot Gjx)$ = There is an x such that x is a F and j G(ained) x.

John is fooled all the time = $(x)(Tx \supset Fjx)$ = Given any x, if x is a T, then x is F(ooled) at x.

Accordingly, here is how the Lincoln quotation would be rendered:

"There is an x such that x is a person and given any y if y is a time then x is fooled at y, and there is a y such that y is a time and given any x if x is a person then x is fooled at y, but there is a y and there is an x such that y is a time and x is a person, and x is not fooled at y."

Hence, with Ty = y is a time, Px = x is a person, and Fxy = x is fooled at y, we have

$$\{(\exists x)[Px \bullet (y)(Ty \supset Fxy)] \bullet (\exists y)[Ty \bullet (x)(Px \supset Fxy)]\} \bullet (\exists y)(\exists x)[(Ty \bullet Px) \bullet \sim Fxy)]$$

Life gets complicated. So does symbolic logic. But let us hope that the basic logical skills covered in this book, the general analytical discipline encouraged, and the perspective of obtaining reliable beliefs in an uncertain but precious world will help decrease the number of people fooled in the twenty-first century.

ANSWERS TO STARRED EXERCISES

I   3. (x)(Cx ⊃ Tx)

8. ~(x)(Px ⊃ Mx)

13. (x)[(Wx ∨ Hx) ⊃ Mx]

17. (∃x)(Sx • Ax) • (∃x)(Fx • Ax)

II   **3.**

1. ~(x)(Fx ⊃ ~Ax)
2. (∃x)(Fx • Ox)   /∴   (∃x)(Ax • Ox)
3. (∃x)~(Fx ⊃ ~Ax)   (1) CQ
4. (∃x)~(~~Fx ⊃ ~Ax)   (3) DN
5. (∃x)~(~Fx ∨ ~Ax)   (4) Impl.
6. (∃x)(~~Fx • ~~Ax)   (5) De M.
7. (∃x)(Fx • ~~Ax)   (6) DN
8. (∃x)(Fx • Ax)   (7) DN
9. Fa • Aa   (8) EI
10. Fa • Oa   **X (EI restriction violated)**
11. Aa • Fa   (9) Com
12. Oa • Fa   (10) Com
13. Aa   (11) Simp.
14. Oa   (12) Simp.
15. Aa • Oa   (13) (14) Conj.
16. (∃x)(Ax • Ox)   (15) EG

7.

1. (∃x)(Ax • Fx)   /∴   ~(∃x) (Ax ⊃ ~Fx)
2. Ay • Fy   **X (EI restriction violated)**
3. ~~(Ay • Fy)   (2) DN

4. ~(~Ay ∨ ~Fy)   (3) De M.
5. ~(~~Ay ⊃ ~Fy)   (4) Impl.
6. ~(Ay ⊃ ~Fy)   (5) DN
7. (x)~(Ax ⊃ ~Fx)   (6) UG
8. ~(∃x)(Ax ⊃ ~Fx)   (7) CQ

V   **1.**

1. ~(∃x)~(Sx ⊃ Cx)   /∴   ~(∃x)(Sx • ~Cx)
2. ~(∃x)~(~~Sx ⊃ Cx)   (1) DN
3. ~(∃x)~(~Sx ∨ Cx)   (2) Impl.
4. ~(∃x)(~~Sx • ~Cx)   (3) De M.
5. ~(∃x)(Sx • ~Cx)   (4) DN

1. ~(∃x)(Sx • ~Cx)   /∴   ~(∃x)~(Sx ⊃ Cx)
2. ~(∃x)(~~Sx • ~Cx)   (1) DN
3. ~(∃x)~(~Sx ∨ Cx)   (2) De M.
4. ~(∃x)~(~~Sx ⊃ Cx)   (3) Impl.
5. ~(∃x)~(Sx ⊃ Cx)   (4) DN

VI   **2.** ~(∃x)[Jx • (~Lx ∨ ~Bx)]

1. ~(∃x)[Jx • (~Lx ∨ ~Bx)]
2. (x)~[Jx • (~Lx ∨ ~Bx)]   (1) CQ
3. (x)[~Jx ∨ ~(~Lx ∨ ~Bx)]   (2) De M.
4. (x)[~~Jx ⊃ ~(~Lx ∨ ~Bx)]   (3) Impl.
5. (x)[Jx ⊃ ~(~Lx ∨ ~Bx)]   (4) DN
6. (x)[Jx ⊃ (~~Lx • ~~Bx)]   (5) De M.
7. (x)[Jx ⊃ (Lx • ~~Bx)]   (6) DN
8. (x)[Jx ⊃ (Lx • Bx)]   (7) DN

Result as shown by line 8: All judges in the United States must go to law school and pass the bar exam.

VII   **3.**

1. (x)[Px ⊃ (Ex ∨ ~Mx)]
2. ~(x)(Px ⊃ ~Mx)   /∴   (∃x)(Px • Ex)
3. (∃x)~(Px ⊃ ~Mx)   (2) CQ
4. (∃x)~(~~Mx ⊃ ~Px)   (3) Trans.
5. (∃x)~(~Mx ∨ ~Px)   (4) Impl.
6. (∃x)(~~Mx • ~~Px)   (5) De M.
7. ~~Ma • ~~Pa   (6) EI

8. [Pa ⊃ (Ea ∨ ~Ma)]   (1) UI
9. ~~Pa • ~~Ma   (7) Com.
10. ~~Pa   (9) Simp.
11. Pa   (10) DN
12. Ea ∨ ~Ma   (8) (11) MP
13. ~Ma ∨ Ea   (12) Com.
14. ~~Ma   (7) Simp.
15. Ea   (13) (14) DS
16. Pa • Ea   (11) (15) Conj.
17. (∃x)(Px • Ex)   (16) EG

9.

1. ~(x)(Vx ⊃ Ix)
2. (∃x)(Vx • ~Ix) ⊃ (∃x)[(~Hx • Px) • Ax]
3. (∃x)(Px • Hx)
4. [(∃x)(Px • Hx) • (∃x)(Px • ~Hx)] ⊃ (x)(Vx ⊃ Cx)
                                 /∴ ~(∃x)(Vx • ~Cx)
5. (∃x)~(Vx ⊃ Ix)   (1) CQ
6. (∃x)~(~~Vx ⊃ Ix)   (5) DN
7. (∃x)~(~Vx ∨ Ix)   (6) Impl.
8. (∃x)(~~Vx • ~Ix)   (7) De M.
9. (∃x)(Vx • ~Ix)   (8) DN
10. (∃x)[(~Hx • Px) • Ax] (2) (9) MP
11. (~Ha • Pa) • Aa   (10) EI
12. ~Ha • Pa   (11) Simp.
13. Pa • ~Ha   (12) Com.
14. (∃x)(Px • ~Hx)   (13) EG
15. (∃x)(Px • Hx) • (∃x)(Px • ~Hx)   (3) (14) Conj.
16. (x)(Vx ⊃ Cx)   (15) (4) MP
17. (x)(~~Vx ⊃ Cx)   (16) DN
18. (x)(~Vx ∨ Cx)   (17) Impl.
19. (x)(~Vx ∨ ~~Cx)   (18) DN
20. (x)~(Vx • ~Cx)   (19) De M.
21. ~(∃x)(Vx • ~Cx)   (20) CQ

Chapter 12

FRONTIERS OF LOGIC—FUZZY LOGIC: CAN ARISTOTLE AND BUDDHA GET ALONG?

Fuzzy logic begins where Western logic ends. . . . Fuzziness begins where contradictions begin, where A and not-A holds to any degree.
—Bart Kosko, *Fuzzy Thinking: The New Science of Fuzzy Logic*

Everything must either be or not be, whether in the present or in the future.
—Aristotle, *On Interpretation*

I have not explained that the world is eternal or not eternal, I have not explained that the world is finite or infinite.
—Buddha

The fundamental idea of Buddhism is to pass beyond the world of opposites, a world built up by intellectual distinctions and emotional defilements.
—D.T. Suzuki, *The Essence of Buddhism*

Introduction

I have a complaint about the college where I teach. Complaining to my administration will not do any good, for the source of my complaint appears to be at a very deep philosophical level beyond any administrator's control. I teach in Hawaii. Outside right now, it is about 85 degrees, but as I type these words in my office, I have on a very thick jacket. The temperature in my office and adjacent classrooms averages between 55 and 60 degrees. I used to be able to control the temperature in my office and my classroom by adjusting the thermostats in each room. True, there was some minor inconvenience in the process, and at times energy was wasted when someone turned a thermostat down and then forgot to turn it back up when leaving for the day or a meeting in another building. On Monday mornings I would have to turn the thermostat way down to cool my office, which had no air-conditioning all weekend, and then turn it back up when it eventually got too cold. I would also have to adjust my classroom thermostat depending on the time of day and how many people were on my floor of the building and in the classroom. But with a little effort I could get the temperature right, because as a human being I could assess and "smoothly" control the temperature.

Now we have a so-called state-of-the-art digital computerized system, which was supposed to centrally control each room in each building and eliminate inconvenience and save taxpayers lots of money. Now we all freeze and wear jackets in Hawaii, and waste lots of energy. We don't dare complain. If our comptroller adjusts the system it will be too hot, and for a teacher, having an office and a classroom that are too hot is a fate worse than death. The mind shuts down, and suddenly even your best lecture becomes boring and students begin to fall asleep.

According to the gurus of a new logic, called *fuzzy logic*, the root of our problem is cultural and philosophical: Our air-conditioning system thinks like Aristotle rather than like Buddha. According to the proponents of this new logic, the all-or-nothing overshoot of our air-conditioning system is the technological end-product of a cultural hasty conclusion fallacy in regard to truth. Since the time of Aristotle and the ancient Greeks, Western logic has assumed that a proposition must be wholly true or false with no in-between and no shades of gray.[1] Small wonder that our computer systems are dumb, proponents of fuzzy logic say, if they are programmed on the basis of a black-and-white logic. Based on such notions of categorical truth and falsehood, on-and-off systems have no common sense; they are incapable of mimicking the simple human process of smoothly adjusting a thermostat when a room is too hot or too cold.

Perhaps it is misleading to title this chapter "frontiers" of logic. For proponents of this new logic claim that a fuzzy revolution is already taking place in Buddhist-influenced countries such as Japan, Singapore, Malaysia, South Korea, Taiwan, and China, with billions of dollars invested in fuzzy controlled cameras, camcorders, TVs, microwave ovens, washing machines, vacuum sweepers, car transmissions, and engines. Being produced are washing machines able to measure the amount of dirt in wash water and smoothly control the amount and length of water agitation to get clothes clean, fuzzy vacuum cleaners controlled by fuzzy rules that imperceptibly adjust sucking power in microseconds based on sensor readings of dirt density and carpet texture, and TV sets that instantly measure each picture frame for brightness, contrast, and color, and then smoothly adjust each, microsecond by microsecond, like a thousand little nanocreatures turning knobs with common sense.[2] Most

---

[1] In Chapter 2, Plato was blamed for the notion of categorical truth. Aristotle was a student of Plato. Aristotle is the target of proponents of fuzzy logic because Aristotle was the first to systematize many aspects of Western logic based on the generalization that statements are either wholly true or false.

[2] *Nano* means *a billionth of* as in a billionth of a second or a billionth of a meter.

important, say proponents, we will soon be able to create truly intelligent computers, *adaptive fuzzy systems*, that will think like humans, that will be able to learn, see patterns and "grow" rules based on these patterns. All of this is based on what the Japanese call "fuaji riron"—fuzzy theory.

Bivalent Logic and Paradoxes

According to the proponents of fuzzy logic, we did not have to wait for faulty air-conditioning systems to know that something was seriously wrong with the foundations of Western logic and our assumptions regarding truth. Classical Aristotelian logic is said to be founded on a *bivalent* faith, propositions are "crisply" true or false. But our experience tells us that in many areas of life a crisp categorical bivalent map oversimplifies to the point of paradox by missing important shades of gray. In short, by not recognizing that there are degrees of truth between the extremes of complete truth and complete falsehood, there is a mismatch problem between our logic and reason on the one hand, and our experience on the other hand. For instance, using Western logic founded on bivalent faith it is possible to prove the following:

1. If one-hundred thousand grains of sand make a heap of sand, and removing one grain of sand still leaves a heap, it follows that one grain of sand is still a heap.
2. If an object x is black, then any object y that is indistinguishable in color from x is also black. Consequently, any white object z is black that was produced by a sequence of gray shading such that the percentage of white increased smoothly such that each step in color was indistinguishable from the previous one.
3. If a person who is only 5 feet tall is short, then given a sequence of 999 additional persons such that starting with our 5-foot-tall person, each one is 1/32 of an inch taller than the previous person, the last person in the sequence, a little over 7 feet, 6 inches tall, is also short.

These paradoxes are known as *Sorites Paradoxes* or *paradoxes of vagueness*. Let's examine the last case in more detail. Suppose I was able to find a thousand people, such that the shortest was only five feet tall, and the tallest was a little over seven feet, six inches. Starting with our 5-foot-tall person, I was able to find a person who was 5 feet, 1/32 inch tall, then a person 5 feet, 2/32 inch tall, then 5 feet, 3/32 inch, and so on. Clearly, if the person who is only 5 feet tall is short, then the person who is 5 feet, 1/32 inch is also short. But given this, we can now create a long series of valid modus ponens steps[1] as follows:

x_1 is short. $\qquad\qquad\qquad$ ($x_1 = 5'$)
If x_1 is short, then x_2 is short. $\quad$ ($x_2 = 5'\ 1/32''$)

---

[1] Or a long series of hypothetical syllogisms and one modus ponens step. This paradox is called a sorites paradox because it can be viewed as a long series of arguments where the conclusion of each becomes a premise for the next argument.

If x_2 is short, then x_3 is short.   ($x_3 = 5'\ 2/32''$)

$$\vdots$$

$$\vdots$$

If x_{999} is short, then x_{1000} is short.
/ $\therefore$ x_{1000} is short.   ($x_{1000} \approx 7'\ 6''$)

Our common sense rebels against this conclusion. The last person in the sequence (x_{1000}) is a little over seven feet, six inches tall and is not short, but there is nothing wrong with the logic—if by logic we mean Western logic and bivalent truth values. Modus ponens is correctly applied, and, given the traditional interpretation of a conditional statement, none of the premises can be called false. There is no conditional in the sequence that can be said to have a true antecedent and a false consequent. If any given person mentioned in the antecedent is short, then a person mentioned in the consequent must be short as well, because the person mentioned in the consequent is virtually indistinguishable from the person mentioned in the antecedent. Remember that the person mentioned in the consequent is only 1/32 inch taller than the person mentioned in the antecedent. So how could it be that a person mentioned in the antecedent is short and a person mentioned in the consequent is not short? Although there is a difference in height between the persons mentioned in the antecedent and the consequent, it is too small to provide a reason to apply *short* to one person and withhold it from the other.

The problem with this logical picture is that *short* is a fuzzy concept. Membership in the class of short human beings does not have a crisp cutoff point, such that one person in our sequence is wholly short and the next in the sequence wholly not short. However, our judgment of the truth value of conditional statements demands crisp true or false antecedents and consequents. A conditional statement is false only when the antecedent is wholly true, and the consequent is wholly false.

Recall that the use of Venn diagrams covered in the previous chapter assumed a crisp division between classes of things. In Venn diagrams, the lines that mark off each class of things are sharp. Although classes can overlap—something could be a class A and a class B at the same time—something cannot both be in an A class of things and not in an A class of things at the same time. In modern logic, classes of things are called *sets,* and fuzzy logic is based on the belief that classical set theory is wrong; our experience and the categorization of our experience is based on *fuzzy sets*. There are no sharp lines, as in Venn diagrams; one fuzzy set of things can blend into another fuzzy set of things.

Consider our previous discussion of our ability at a very young age to grasp abstract concepts. Plato and Aristotle assumed that a crisp in-principle definition or conceptual apprehension of an ideal chair or chairness was possible. Fuzzy theorists reject this and argue that inclusion in a set is a matter of degree. No sharp definition is possible, because some objects may be definitely chairs, but others are "sort of" chairs. Whereas Venn diagrams picture the categorical world view of Western logic with its crisp separation of classes, the Tao yin-yang symbol of Eastern philosophy reflects the blending, shades of gray world view of fuzzy logic. (See Figure 12-1.)

FIGURE 12-1

Whereas crisp Venn circles (right) picture the categorical, either/or world of Aristotelian logic, the Tao yin-yang symbol of Eastern philosophy reflects the blending, shades-of-gray world view of fuzzy logic.

A   **Not A**

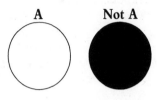

In fuzzy theory, truth is a matter of accuracy, and accuracy must be measured in degrees.

Ironically Western logicians and philosophers discovered the above paradoxes in attempting to completely systematize Western rationality, that is, logic and mathematics. The British philosopher and logician Bertrand Russell (1872–1970) once asked what we should make of a barber in need of a shave who posts a sign outside his shop that reads, "I shave all, and only, those men who do not shave themselves." Who would shave the barber? If he shaves himself, then according to his sign he does not; if he does not shave himself, then according to the sign he does. Logically, a contradiction follows from this:

(S = The barber shaves himself)

1. $(S \supset {\sim}S) \bullet ({\sim}S \supset S)$ $/\therefore$ $S \bullet {\sim}S$
2. $S \supset {\sim}S$ (1) Simp.
3. ${\sim}{\sim}S \supset {\sim}S$ (2) DN
4. ${\sim}S \vee {\sim}S$ (3) Impl.
5. ${\sim}S$ (4) Rep.
6. ${\sim}S \supset S$ (1) Com. + Simp.
7. $S \vee S$ (6) Impl.
8. S (7) Rep.
9. $S \bullet {\sim}S$ (8) (5) Conj.

But a contradiction is a disaster for Western logic, because in addition to being necessarily wholly false and, hence, making any argument that contains a contradiction in the premises unsound, any conclusion whatsoever can be derived from a contradiction.

1. $p \bullet {\sim}p$ $/\therefore$ q
2. p (1) Simp.
3. $p \vee q$ (2) Add.
4. ${\sim}p$ (1) Com. + Simp.
5. q (3) (4) D.S.

Similar paradoxes are: The man from Crete who says "All Cretans are liars"; the statement, "This statement is false"; and a bumper sticker that says, "Don't trust me." For the defenders of fuzzy logic, these paradoxes are not merely ivory-tower curiosities. They are nature's way of telling us that we have not got something right, that there is something fundamentally wrong—or incomplete to be more precise— with our Western worldview. According to the proponents of fuzzy logic, the solution to this mismatch between reason and experience, and to our overshooting air-conditioning systems is to see *Western logic as only a special case of an expanded logic*, a more general rationality that allows for, in fact demands, degrees of truth.

Fuzzy Interpretations and Degrees of Truth

Thus, the fundamental move of fuzzy logic is to generalize from classical logic by having the bivalent truth values of completely true and completely false as the extreme endpoints in a continuum of degrees of truth. If a statement is wholly true, a 1 is assigned; if a statement is wholly false, a 0 is assigned. We can then speak of a domain of discourse (people) and a set in that domain (short people), and the degree to which a particular person is in the set. Here are some typical fuzzy examples:

Person	Height	Degree of shortness
Midget	5'0"	1.00
Bea	5'1/32"	0.99
Hyon	5'6"	0.80
Taisha	6'0"	0.60
Aron	6'6"	0.40
Kuna	7'0"	0.20
Jorge	7'5"	0.03
Kareem	7'6"	0.00[1]

Thus, statements like "A is short" can now be interpreted in terms of degrees of truth, as the degree to which A participates in the set *short*. For instance, the statement "Aron is short" equals .40.

The next move in fuzzy logic is to redefine the propositional logical connectives (~), (•), and (∨). The guideline used in this redefinition is called the **Extension Principle**. The traditional Western truth values for these logical connectives are "recovered" when they connect simple statements that have the crisp values of 1 or 0, when simple statements are wholly true or wholly false. In other words, fuzzy logic is said not to be incompatible with traditional logic, but rather to be a

---

[1] Although no one assignment of degrees can be said to be absolutely correct, fuzzy logicians talk about a "reasonable assignment of degrees," given two extremes. A typical method is to assign 0 to the lowest value and 1 to the highest value, then any intermediate value will equal the original quantitative assignment (in this case height) minus the lowest value divided by the difference between the lowest and highest values. The above intermediate degrees were achieved in the following way. Given that five feet equals 1 and seven feet six inches equals 0, then any intermediate height x equals $1 - [(x-5)/2.5]$.

generalization from traditional logic. With some help from Buddha and Eastern philosophy, we are enlarging our conception of rationality, not endorsing irrationality. Fuzzy logic is a logic of fuzziness, a new set of rational guidelines that allow us to deal in a practical way with aspects of our experience that bivalent logic seems incapable of doing. The rules of fuzzy logic themselves are not fuzzy or vague. Here are the rules:

Negation (~):   The degree of truth of **not A** = 1.0 - the degree of truth of A.

                Example: "A is short" = .40 (so)
                      "A is **not** short" = .60

Conjunction (•):   **The degree of truth of A • B** = the **minimum** degree of truth of **A** and **B**.

                Example: (.40) • (.99) = .40

Disjunction (∨):   The degree of truth of **A ∨ B** = the **maximum** degree of truth of **A** and **B**.

                Example: (.40) ∨ (.90) = .90

Consider how each of these rules is a simple extension of the traditional truth table definitions. A negation gives us the opposite value of that negated ($\sim$T = F; $\sim$1 = 0; so $\sim$(.40) = .60). A traditional conjunction returned the minimum value of the parts conjoined (T • F = F; 1 • 0 = 0; so .40 • .99 = .40). A traditional disjunction returned the maximum value of its parts (T ∨ F = T; 1 ∨ 0 = 1; so .40 ∨ .99 = .99). Consider now what a fuzzy truth table would look like.

If

p = X is *not* short.

q = X is short *and* X is young.

r = X is short *or* X is young.

then we can compute the following values:

Height	Age	X is short	X is young	p	q	r
5′0″	10	1.0	1.0	0.0	1.0	1.0
5′1/32″	12	0.99	0.96	.01	.96	.99
5′6″	65	0.80	0.00	.20	0.0	.80
6′0″	55	0.60	0.18	.40	.18	.60
6′6″	18	0.40	0.85	.60	.40	.85
7′0″	25	0.20	0.73	.80	.20	.73
7′5″	45	0.03	0.36	.97	.03	.36
7′6″	29	0.00	0.65	1.0	0.0	.65

Proponents of fuzzy logic claim that the assignment of partial truth plus these new definitions of logical connectives provide us with a power of discrimination lacking in classical logic, a method for handling shades of gray that is more consistent with common sense. A person who is only 5 feet 1/32 inch tall and only 12 years old is very close to a person only 5 feet tall and only 10 years old (.96 and 1.00 respectively), but very different from a person who is 7 feet 5 inch tall and 45 years of age (.96 and .03 respectively).

Fuzzy Conditionals and Fuzzy Validity

Classical logic maintains an important relationship between valid consequence and tautologous conditionals. For instance, a truth table of the conditional [p • (p ⊃ q)] ⊃ q will show a result of all true in its final column. Note that this conditional is formed by conjoining the premises of a modus ponens argument form and making them the antecedent of a conditional with the form's conclusion as the consequent.

					*	
p	q		[p •	(p ⊃ q)]	⊃	q
T	T		T	T	**T**	T
T	F		F	F	**T**	F
F	T		F	T	**T**	T
F	F		F	T	**T**	F

p ⊃ q
p /∴ q

All of the nine rules of inference can be rendered this way, and all will produce tautologies: for modus tollens, [(p ⊃ q) • ~q] ⊃ ~p, for disjunctive syllogism, [(p ∨ q) • ~p] ⊃ q, and so on. Similarly, a truth table for the rules of replacement will show each to be a tautology. A truth table for the rule of implication, (~p ⊃ q) ≡ (p ∨ q), will show all true under the ≡ symbol.

From a fuzzy perspective, these relationships reflect the definition for a crisp notion of validity: That a valid consequence not allow for any possibility where the conclusion is false and the premises are all true. This is why a valid rule of inference reformulated into a statement will always be a tautology: there will be no row in a truth table with a true antecedent and a false consequent. As in the case of the logical connectives, in fuzzy logic the notion of validity is also generalized: *a fuzzy valid argument is one that does not allow for a loss of truth in going from the premises to the conclusion*. For example, an argument is not fuzzy valid if its premises have an overall degree of truth of .5 and the conclusion has a value of .4.

Similarly, the notion of a classical conditional statement is generalized. One way of viewing the traditional *if . . . then* statement defined in Chapter 8 is: A conditional statement is wholly true when its antecedent is no more true than its consequent, otherwise it is wholly false. In other words, when the antecedent is equal to the consequent in truth value, T ⊃ T or F ⊃ F, or has less truth value, F ⊃ T, an *if . . . then* statement is wholly true. Only when there is a loss of truth in moving from the antecedent to the consequent, T ⊃ F, is an *if . . . then* statement wholly false. In fuzzy logic, the first aspect is maintained. If there is no loss of truth, then a conditional is wholly true. However, if there is only a small loss of truth in moving from the antecedent to the consequent, such as in the statement "if .5 then .4," we want to distinguish this case from one where there is a large loss of truth, such as in the statement "if .99 then .03." A classical *if . . . then* definition loses the ability to make this discrimination by allowing only one choice, complete falsehood. Thus, the idea is to define a fuzzy conditional such that the degree of truth reflects how much truth is lost in the passage from the antecedent to the consequent. Here is the rule for fuzzy *if . . . then* statements, followed by that for biconditionals. (To reflect the important difference between the classical definitions and the fuzzy definitions, let's use the ($\rightarrow$) symbol rather than the ($\supset$) symbol, and ($\leftrightarrow$) instead of ($\equiv$).)

Conditional ($\rightarrow$): The degree of truth of $A \rightarrow B = 1-(A-B)$ if A is greater than B, otherwise 1.

      Examples:
$$.5 \rightarrow .4 \ = 1-(.5-.4) \ = .9$$
$$.99 \rightarrow .03 = 1-(.99-.03) = .04$$
$$1.0 \rightarrow 0.0 = 1-(1-0) \ = 0$$
$$0.0 \rightarrow 1.0 = 1$$
$$.5 \rightarrow .6 \ = 1$$

Biconditional ($\leftrightarrow$): The degree of truth of $A \leftrightarrow B = (A \rightarrow B) \cdot (B \rightarrow A)$.[1]

      Examples:
$$
\begin{aligned}
.5 \leftrightarrow .4 \ &= (.5 \rightarrow .4) \cdot (.4 \rightarrow .5)\\
& [1-(.5-.4)] \cdot \ 1\\
& \ .9 \quad\quad \cdot \ 1 \quad = .9
\end{aligned}
$$

$$
\begin{aligned}
.5 \leftrightarrow .6 \ &= (.5 \rightarrow .6) \cdot (.6 \rightarrow .5)\\
&= \quad 1 \ \cdot [1-(.6-.5)]\\
&= \quad 1 \cdot .9 \quad\quad = .9
\end{aligned}
$$

$$
\begin{aligned}
.2 \leftrightarrow .1 \ &= (.2 \rightarrow .1) \cdot (.1 \rightarrow .2)\\
&= [1-(.2-.1)] \cdot 1\\
&= \quad .9 \quad \cdot \ 1 \quad = .9
\end{aligned}
$$

$$
\begin{aligned}
.9 \leftrightarrow .1 \ &= (.9 \rightarrow .1) \cdot (.1 \rightarrow .9)\\
&= [1-(.9-.1)] \cdot \ 1\\
&= \quad .2 \quad \cdot \ 1 \quad = .2
\end{aligned}
$$

In this way, fuzzy conditionals are said to be capable of more discriminations, of quantifying shades of gray. The first example for the conditional definition above (.5 $\rightarrow$ .4) loses truth in moving from the antecedent to the consequent, but it loses such a little bit of truth that it is closer to being wholly true than completely false, so it receives a .9 degree of truth. A classical interpretation would force us to classify this conditional statement as all or nothing, as false because it is not wholly true. Note that the classical values are recovered when the antecedents and consequents are wholly true or false, showing again that classical logic captures only the extreme end points on a continuum of discriminations. Because a biconditional can be

---

[1] This, of course, is a derived definition showing that one version of the equivalence rule of replacement works for fuzzy logic. The definition often found in the literature on fuzzy logic is $A \leftrightarrow B = 1-|A-B|$. $|A-B|$ stands for the absolute value of A-B, so $|.5-.6|$ would equal .1, and $.5 \leftrightarrow .6 = 1-|.5-.6| = 1-.1 = .9$.

defined in terms of the fuzzy conjunction of two fuzzy conditionals, when the components of a fuzzy biconditional are close in value to each other (.5 ↔ .4, .5 ↔ .6, .2 ↔ .1), a high degree of truth will result (.9); when they are far apart (.9 ↔ .1), a low degree of truth will result (.2).

Although we will not pursue the generalization of classical logic further in this book, quantification logic is also extended in fuzzy logic such that fuzzy interpretations are given for universal and existential quantifiers. In addition, approximate reasoning terminology found in natural language is also defined so that more quantifiers are used, such as *many, few, almost,* and *usually.* In this way, fuzzy logic is said to be able to handle fuzzy syllogisms, such as:

Very old fossils are usually rare.
Rare fossils are hard to find.
Therefore, very old fossils are usually hard to find.

A fuzzy interpretation of conditional statements in conjunction with fuzzy set theory is very important for running an air-conditioning system. Typical rules in a fuzzy-controlled air conditioning system are: If the temperature is just right, then the motor speed is medium; If the temperature is warm, then the motor speed is fast. But the terms *just right* and *warm* are fuzzy sets; they do not have rigid cutoff lines as in Venn diagrams. They each cover a range of temperatures, and their ranges overlap and blend together. For instance, a temperature of 73 degrees may be interpreted as .9 degrees within the fuzzy set of just right, but .2 degrees in the set warm. In a fuzzy-controlled system, these rules are "fired" together, and an average is taken to arrive at a smooth motor speed.[1] The rules are said to fire in parallel and partially. The antecedent of the *if . . . then* rule describes to what degree the rule applies. *Warm* is not an all-or-nothing quantity. If it were interpreted in terms of just a 0 or a 1, as either being exactly warm or exactly not warm, then the motor speed would stay fast or not fast at an entire range of temperatures.

In Japan, the Sendai subway uses fifty-nine rules. They all fire constantly to some degree. Fuzzy-controlled pocket cameras use approximately ten fuzzy rules to control autofocus. Hitachi, Matsushita (Panasonic) in Japan, and Samsung in Korea have developed fuzzy washing machines using approximately thirty fuzzy rules to relate and smoothly control load size, water clarity, and water flow. Sanyo Fisher's 8mm fuzzy video FVC-880 camcorder uses nine fuzzy conditionals. Professor Michio Sugeno at the Tokyo Institute of Technology has developed a fuzzy control system for a helicopter that is said to be capable of feats that no human could match, nor any previous mathematical model. Hitachi, Matsushita, Mitsubishi, Sharp, and Samsung have developed fuzzy air-conditioner controllers that purportedly save 40 percent to 100 percent in energy. Sony has developed a palmtop computer that recognizes handwritten Kanji characters, and Sharp has developed a prototype of a refrigerator that would be capable of learning a user's pattern of usage and adjusting defrosting times and cooling times accordingly. By the early 1990s, Goldstar, Hitachi, Samsung, and Sony were all working on perfecting TV sets that would be able to

---

[1] The motor speeds of medium and fast are also interpreted as fuzzy sets.

adjust volume depending on the viewer's room location. Maruman even developed a golf diagnostic system.[1]

Resolution of Paradoxes and Implications

As our air-conditioning systems are better using fuzzy logic, so is our understanding of the sorites paradoxes. From a classical point of view, the above paradox involving height seemed technically sound. The classical valid rule of modus ponens was correctly applied, and there appeared to be no case where we could declare that one of the premises was false—no case where an antecedent was crisply true and the consequent crisply false. If person 15 is short, then person 16 has to be short because person 16 is only 1/32 inch taller, and so on.

In fuzzy logic, the resolution of the paradox comes by first realizing that we no longer need to commit ourselves to the crisp truth of all the premises. Because a degree of shortness will be assigned to each person, a lesser degree of shortness and hence truth will be assigned to each person as we move up the sequence. If the person 5 feet tall is assigned a 1, then the person 5 feet 1/32 inches will be assigned .99. Thus, the very first *if . . . then* premise under a fuzzy conditional interpretation is not wholly true: $1 \rightarrow .99 = 1-(1-.99) = .99$. Because there is a loss of truth in moving from the antecedent to the consequent, the conditional is not wholly true. Second, we note that, given our generalized notion of validity, modus ponens is no longer always valid! If we demand that a valid inference should be inconsistent with any loss of truth in moving from premises to conclusion, then in some instances modus ponens will fail this requirement. Consider the following expected results in which fuzzy sets are mixed with a classical interpretation of modus ponens:

1. $1 \rightarrow .99$
   $1 \quad / \therefore \quad .99$

2. $.99 \rightarrow .98$
   $.99 \quad / \therefore \quad .98$

The first modus ponens is fuzzy valid, but the second is fuzzy invalid! In number 1 there is no loss of truth in moving from the premises to the conclusion, but in number 2 there would be. That is, in fuzzy logic it is incorrect to say that the value of the conclusion .98 follows from the premises of number 2. Here is how this is determined in fuzzy logic:

1. $1 \rightarrow .99 = 1-(1-.99) = .99$ (The conclusion = minimum value of
   $1 \qquad\qquad\qquad =1 / \therefore .99$ the premises.)

2. $.99 \rightarrow .98 = 1-(.99-.98) = .99$ (The conclusion = minimum value of
   $.99 \qquad\qquad\qquad = .99 / \therefore .99$ the premises.)

In fuzzy logic the value of the conclusion equals the minimum value in the premises, so the correct assignment of value in number 2 is .99 for the conclusion,

---

[1] For a summary of the state of fuzzy product development by 1993, see Bart Kosko, *Fuzzy Thinking: The New Science of Fuzzy Logic* (New York: Hyperion, 1993), pp. 180–90. Kosko also notes that by the early 1990s Japanese firms held more than a thousand fuzzy patents worldwide and thirty of the thirty-eight in the United States.

not .98. The value of .98 returned by modus ponens involves a loss of truth, so the classical interpretation of number 2 is not consistent with a generalized notion of validity using fuzzy sets that demands no loss of truth occur in moving from the premises to the conclusion. At best, the classical version of modus ponens is now seen as a "weakly valid sequent"; it remains valid only at the end points of a sequence of discriminations, where the antecedent and consequent have the values one or zero or values very close to one or zero as in case number 1. In this sense, in fuzzy logic a valid argument can blend into an invalid argument. Modus ponens is valid only some of the time.[1]

We can also show that $(S \supset {\sim}S) \cdot ({\sim}S \supset S) \ / \therefore \ S \cdot {\sim}S$ is no longer valid in fuzzy logic. Consider the case where $S = .4$. The following results with a fuzzy interpretation:

$$
\begin{array}{cc}
(S \rightarrow {\sim}S) \cdot ({\sim}S \rightarrow S) & / \therefore \ S \cdot {\sim}S \\
(.4 \rightarrow .6) \cdot (.6 \rightarrow .4) & / \therefore \ .4 \cdot .6 \\
1 \quad \cdot \ [\ 1-(.6-.4)] & / \therefore \quad .4 \\
1 \quad \cdot \quad .8 & / \therefore \quad .4 \\
.8 & / \therefore \quad .4
\end{array}
$$

There is a loss of truth in moving from the premise to the conclusion. There is no loss of truth if and only if the crisp values of 1 or 0 are given to S. Furthermore, the derivation of $S \cdot {\sim}S$ fails in fuzzy logic, because the rule of implication is not fuzzy valid $({\sim}S \rightarrow S) \neq (S \vee S)$. Thus, S cannot be derived from ${\sim}S \rightarrow S$. If S is .5, we have

$$
\begin{array}{cc}
({\sim}S \rightarrow S) & / \therefore \quad S \vee S \\
(.5 \rightarrow .5) & / \therefore \quad .5 \\
1 & / \therefore \quad .5
\end{array}
$$

Only when S is a one or a zero, will there be no loss of truth from the premise to the conclusion.

Nor are contradictions a disaster for fuzzy logic. In fact, they are considered normal. Anything that can be placed in a fuzzy set also has membership in other sets partially. Remember that if the temperature is .8 just right, then it could also be .2 not just right or warm. Like a Buddhist, the fuzzy logician does not see the world as a collection of crisply separated objects, but rather as an ocean of blended drops of water, where individual drops of water may occasionally spray loose from the ocean causing an illusion of separateness. Furthermore, in fuzzy logic not everything follows from $S \cdot {\sim}S$, because $S \cdot {\sim}S$ is only wholly false when S is a one or zero. In fuzzy logic, fuzzy contradictions cannot have a value higher than .5, but they can have any value between 0 and .5. However, if S is .5, then $S \cdot {\sim}S$

---

[1] However, in fuzzy logic literature there are efforts to supply a *generalized notion of modus ponens* where arguments such as the following would be valid: "Visibility is slightly low today. If visibility is low, then flying conditions are poor. Therefore, flying conditions are slightly poor today."

equals .5, and any value less than .5 will not follow as a fuzzy valid sequent. If A is .4 and S is .5, then "S • ~S /∴ A" will have a loss of truth.

The most important statement in Aristotelian logic is A ∨ ~A. Everything must either be or not be. This principle is called the **Law of Excluded Middle** and is seen as the ultimate foundation for Western culture's pursuit of truth. We may disagree on many things, but don't we at least know for sure that something either exists or it does not exist? We may not know yet whether our solar system has a tenth planet, but surely we know ahead of time that a tenth planet either exists in reality or it does not. Hence, the little killer proof B /∴ A ∨ ~A in Chapter 10 was very important. It shows that A ∨ ~A is always true, so it can be derived from any statement.

From a fuzzy logical point of view, this allegiance to A ∨ ~A is viewed as something like a cultural questionable dilemma fallacy. As we have seen, in fuzzy logic not all contradictions are the same; some can have a degree of truth as high as .5. Similarly, the so-called law of excluded middle can have a degree of truth as low as .5. In fuzzy logic, De Morgan's theorem, double negation, and commutation still hold, so ~(A • ~A) ↔ (A ∨ ~A) regardless of the degree of truth of A. But if A equals .5, then (A • ~A) = (A ∨ ~A). This is the ultimate slap in the face for Western logic: what were formerly thought to be contradictions and tautologies dissolve into each other. In the late 1980s and early 1990s, considerable controversy focused on fuzzy interpretations of contradictions. Editors of major professional journals refused to publish articles that asserted that A • ~A could be something other than totally false.

Philosophy: What about reality?

One of the leading proponents of fuzzy logic is Bart Kosko. With degrees in philosophy, mathematics, and electrical engineering, he often makes fun of people who think that philosophy is a worthless degree. In his book, *Fuzzy Thinking: The New Science of Fuzzy Logic*, he recounts how he often turned to philosophy for guidance in discovering new mathematical and logical relationships to solve engineering problems. Repeatedly in this book we have noted the connection between philosophy and the foundations of logic and the very practical matters of technological development, belief acceptance, ethics, and meaning in life. In this chapter, we have seen how philosophy literally can be "cashed" into technology and billions of dollars in product development.

In other words, the apparent ivory-tower debates of philosophers can be of great consequence. In spite of how much the world has changed in terms of multicultural connections, Western philosophers, for the most part, still have little respect for Eastern philosophers, and vice versa. Western philosophers see themselves as rigorous, disciplined, and scientific. Eastern philosophers are thought to be wishy-washy, vague, and incoherent. Eastern philosophers see Western philosophers as dogmatic, ideologically blind, and culturally egocentric. People who try to practice what is called comparative East-West philosophy usually are scolded by both Western and Eastern philosophers for their lack of insight or weak standards of inquiry.

In this context, the whole notion of fuzzy logic has created passionate debate. Consider a few famous quotes:

> Fuzzy theory is wrong, wrong, and pernicious. What we need is more logical thinking, not less. The danger of fuzzy logic is that it will encourage the sort of imprecise thinking that has brought us so much trouble. Fuzzy logic is the cocaine of science. Professor William Kahan, University of California at Berkeley[1]

> 'Fuzzification' is a kind of scientific permissiveness. It tends to result in socially appealing slogans unaccompanied by the discipline of hard scientific work and patient observation. Professor Rudolf Kalman, University of Florida at Gainesville[2]

The influential Harvard philosopher and logician Willard Van Orman Quine once described alternatives to classical bivalent logic as "deviant," and branded some of the literature on fuzzy logic "irresponsible."[3] Furthermore, critics of fuzzy logic will claim that it is simply regular logic in disguise and that new intelligent product development is due more to improvements in sensor technology than to Buddha. There are no alternate rationalities: modus ponens is as valid in China as it is in the United States. For instance, basic rules of inference were used in developing the above presentation of fuzzy logical connectives. Take one example of the extension principle. In defining fuzzy conjunction did we not reason as follows?

If a classical conjunction returns the minimum truth value of its parts ($T \cdot F = F$), then a fuzzy conjunction will return the minimum truth value of its parts ($.8 \cdot .2 = .2$). A classical conjunction returns the minimum truth value of its parts ($T \cdot F = F$). So, a fuzzy conjunction will return the minimum truth value of its parts ($.8 \cdot .2 = .2$). Modus Ponens

Is fuzzy logic simply a gimmick? A shallow amendment to classical set theory? A surface dazzle that, when uncovered a little, shows regular logic at its core?

Fuzzy proponents will respond by reminding their critics that fuzzy logic expands logical rigor. To further the metaphor used throughout this book, fuzzy proponents claim that we are still following reasoning trails but the paths are now broader. They will also claim that they are being more scientific than the critics of fuzzy logic. Classical logic as a foundation for producing computers with artificial intelligence (AI), computers that can think intelligently and learn, has been tested and has failed. A true scientist must be empirically honest. If your pet theory fails lots of tests, then you must give it up, regardless of the effort invested. According to Kosko,

> After over thirty years of research and billions of dollars in funding, AI has so far not produced smart machines or smart products. . . . The AI crowd . . . took state funds and defense-buildup funds and set up their own classes and conferences and power networks. And they did not produce a single commercial product that you can point to or use in your home or car or office. They beat up on fuzzy logic harder than any other group because they had the most to lose from it and the fastest. Fuzzy logic

---

[1] Quoted from Lotfi Zadeh, "Making Computers Think Like People," *IEEE Spectrum*, August, 1984, page 4. Kahan made these statements in 1975.

[2] Quoted from Daniel McNeill and Paul Freiberger, *Fuzzy Logic* (New York: Simon & Schuster, 1993), pages 46–47.

[3] W. V. Quine, *Philosophy of Logic*, 2nd ed., (Cambridge, Mass.: Harvard Univ. Press, 1986, Chapter 6 and p. 85.

broke the AI monopoly on machine intelligence. Then fuzzy logic went on to work in the real world.[1]

According to Lotfi Zadeh, the University of California at Berkeley professor who first proposed the principles of fuzzy logic in the 1960s, trying to produce AI with bivalent logic is "like trying to dance the jig in a suit of armor."[2] Classical logic as a foundation for AI has failed and only dogma and inertia are keeping it around. According to Kosko, in the early days of fuzzy logic, "It was daring and novel . . . because you first had to get your university degrees in the old black-and-white school and then doubt that school and rediscover what any layman could have told you about common sense—it's vague and fuzzy and hard to pin down in words or numbers."[3] Any average human being knows what words like *many, few, almost, a little, a lot, usually,* and *quite true, very true, more or less true,* and *mostly false* mean. Computers cannot be smart unless they can compute these notions.

To conclude this book, let's end with one of the most intriguing and controversial topics of the twentieth century—the nature of reality. Even a defender of fuzzy logic could adopt a purely pragmatic stance and argue that although fuzzy logic works, we need not talk about fuzzy reality, degrees of real truth, or endorse a Buddhist ontology[4] that says that separate objects are illusions and that reality is one. To define a fuzzy set, *short,* in terms of degrees of shortness is a far cry from saying that my short friend John is one with the universe, that his apparent individuality is an illusion, and that the apparent physical world of separate objects is an illusion as well. A pragmatic defender of fuzzy logic could still argue that Aristotle was right. John is either here now or he is not here. He can't be here (in Hawaii) and simultaneously in New York.

However, one of the stimulants of fuzzy logic was the twentieth-century development of what is called quantum physics. You have no doubt heard many times the word *quantum,* as in quantum jump and quantum leap. (I have even seen beauty ads for quantum perms.) In spite of the term's wide use, few people understand just how radical is the notion of a quantum jump. Quantum physics has been very successful. It is the basic physics of how the atom and its parts work, and many of our current chemistry and electronic technologies are based on it in one way or another. But its math is very weird. The math describes electrons, for instance, as jumping around all the time. But they don't jump around like you and I can jump around. When they jump from point A to point B, they are nowhere in between. They are not discrete objects that move in a continuous space; it is not just that they jump fast, so fast that we can't detect them; they don't exist in between the points. The math, if taken literally, says that they pop in and out of existence. Worse, if the math is taken literally, a degree of each electron around every atom in the universe is a little bit everywhere. So, some of John is in Hawaii and New York at the same time.

---

[1] Kosko, 1993, pp. 159–60. Note that in addition to an inductively correct falsification point, Kosko commits an ad hominem circumstantial here.

[2] Quoted from Kosko, p. 160. A questionable analogy or simply an introduction to an argument?

[3] Ibid., p. 161.

[4] **Ontology** is a technical philosophical term that means in this context "theory of reality," a theory about what exists, about what is most fundamentally real.

In their college educations, modern physicists and engineers have been taught to ignore what the math says literally. Their educations have been heavily influenced by *logical positivism*, a philosophy popular in the early to middle decades of the twentieth century, and a philosophy of which proponents of fuzzy logic are highly critical. Logical positivists said, "Never take the math seriously in terms of what it says about reality; the job of a scientist is to adequately link experiences, not to tell us what is happening behind the scenes of those experiences." In other words, logic and experience are all that a rational person should worry about. Questions regarding the nature of reality are on par with questions about whether God exists and the meaning of existence. Objective closure on such questions is not possible. No hard evidence can ever be achieved regarding answers to such questions. The math is just a practical human tool that we use to predict results in scientific experiments and ultimately to develop technology that works.

The way this works out, in terms of the electron, is that the mathematical function that describes the electron as being "smeared out" throughout the universe is viewed as a statement of probability, as a prediction in degrees of probability of where we will find an electron when we look for it. Interpreted this way, an electron moving rapidly around an atom within an object in Hawaii has a certain probability of being in New York or the Andromeda galaxy, but these probabilities are very, very low compared to the probability that we will find the electron in Hawaii very, very close to the nucleus of its atom. The math literally says that the electron is everywhere until we interact with it; then it quantum "collapses" immediately to a specific point in Hawaii! Logical positivists said this was absurd, or more precisely, that it is absurd to believe that we have evidence that reality behaves this way. Furthermore, we need not even contemplate what the real electron is doing. They told scientists just to get on with the task of building twentieth-century technology.

In this way, the entire edifice of bivalent logic was saved from potential demolition by quantum physics. We could continue to go about our bivalent Western ways and ignore what the behavior of electrons (and all subatomic reality for that matter) seemed to say: that Aristotle was wrong, that electrons could be both here and not here. Although a few outcasts, such as the physicist David Bohm, argued that quantum physics was telling us something fundamental about reality, and that that something was closer to Eastern philosophical beliefs than Western, for the most part twentieth century physicists and engineers obediently followed their bivalent teachers, who in turn were following the logical positivists.[1]

---

[1] David Bohm was a U.S. physicist who worked with Robert Oppenheimer on the atomic bomb. When Oppenheimer was attacked during the 1950s as a possible communist sympathizer, Bohm was so disgusted with his country's behavior that he left the United States to live in Great Britain. For his philosophy and interpretation of quantum physics, see his book, *Wholeness and the Implicate Order* (London: Routledge & Kegan Paul, 1980). For a new interest in Bohm's interpretation, see "Bohm's Alternative to Quantum Mechanics," by David Z. Albert, *Scientific American*, May, 1994, pages 58–67. It is interesting to note that this same issue of *Scientific American* has an advertisement (p. 3) for a Mitsubishi Galant, hawking the new "intelligent shifting of a Fuzzy Logic transmission." This same ad could be found in other popular magazines by 1994. In the United States, ads like this in the early-to-mid-'90s represented somewhat of a marketing breakthrough. Advertising executives had been worried that the general public would view the adjective *fuzzy* negatively. Here is an example of a transitional ad, in this case for a Saturn SW2: "A 124-horsepower dual-overhead-cam engine linked to an automatic transmission utilizing fuzzy logic programming. (Huh?) It gives a Saturn the ability to adjust to different driving conditions—optimizing performance and handling. Still fuzzy on it? Any Saturn sales consultant would be more than happy to clarify things." *Scientific American*, October, 1994, page 51.

In general, supporters of fuzzy logic are very critical of probability interpretations. For instance, suppose a doctor tells her patient that there is a good chance a tumor is cancerous because it is quite large. The phrase *good chance* expresses probability and a degree of uncertainty about what is real. However, *quite large* describes reality and is not an expression of uncertainty, even though the term is not precise. Kosko does not mince words, "The ultimate fraud is the scientific atheist who believes in probability. . . . the Buddha wins . . . . The universe is deterministic but gray."[1] Supporters of fuzzy logic say that probability is a cop-out, an example of ivory-tower faith-healing, a tactic of philosophical ostriches sticking their heads in the sand afraid to face gray reality. We can't ignore that gray reality happens all the time. Probability is the timid measuring of the likelihood of something happening; fuzzy procedures measure **the degree to which it is happening**. Quantum physics is gray reality big time. The whole universe is in every part; to a degree each object is in every other object.

Western philosophers and scientists for the most part say this is absurd, that it's mysticism and opens the entire structure of scientific rationality to occult and paranormal silliness.[2] If Western logicians are bit-brains, fuzzy logicians are flip-brains (from **fuzzy logical inferences per second**). Fuzzy supporters counter that an extension of fuzzy principles will open up entirely new ways of seeing difficult ethical questions, human nature, God, and meaning in life.

We cannot follow these debates further. According to Rudyard Kipling, "Oh, East is East, and West is West, and never the twain shall meet." If Kipling is correct, Aristotle and Buddha will never get along. On the other hand, if the supporters of fuzzy logic are right, they are already getting along inside the smartest of our new computers. New computers are using chips that do both good old-fashioned bit processing and flip processing. The latter are combined with what are called **neural nets,** a process of computing that is said to mimic the way the human brain works. One thing seems certain: The twenty-first century promises to be very interesting philosophically and technologically.[3]

EXERCISE	Using the fuzzy interpretations of the logical connectives, figure out
I	the degree of truth for each of the following. Assume that A = .7, B =.3, and C = .1.

---

[1] Kosko, 1993, pages 50, 63.

[2] When the opponents of fuzzy logic really get upset, in private conversations they will fire away with ad hominem circumstantial attacks, pointing out that Kosko is from California and that fuzzy logic was first developed at Berkeley. Thus they imply that fuzzy logic should not be considered seriously because it originated in a state that is a hotbed of new-wave fads that a more disciplined person laughs at.

[3] For another introduction to fuzzy logic for the nonexpert, see *Fuzzy Logic*, Daniel McNeill and Paul Freiberger (New York: Simon & Schuster, 1993). For technical articles, see *Fuzzy Logic for the Management of Uncertainty*, Lotfi A. Zadeh and Janusz Kacprzyk, eds. (New York: John Wiley & Sons, Inc., 1992), and *IEEE International Conference on Fuzzy Systems*, March 8–12, 1992.

1. A • B   2. A ∨ C   3. A → B   4. A ↔ B

*5. (A • B) ∨ ~C   6. B → ~(A → C)   7. (A • B) ↔ C

8. ~(A • B) → C   9. A → (B • ~B)   10. ~C → ~B

11. [(A • ~B) → ~C] ↔ [~(A ∨ B) → C]

12. ~{[A → ~(B → C)] • [~C → (~A • ~B)]}

13. A → [(A → B) → B]   14. [(A → B) • (B → C)] → (~C → ~A)

15. {[(A → B) • (B → C)] • (A ∨ B)} → ~(~B • ~C)

16. ~(A ∨ C) ↔ (~A • ~C)   17. ~(A • B) ↔ (~A ∨ ~B)

18. (~A → C) ↔ (A ∨ C)   *19. (A → C) → [A → (A • C)]

20. [~A ∨ (B • C)] ↔ [(~A ∨ B) • (~A ∨ C)]

EXERCISE II
In this Chapter (p. 396) a truth table showed that modus ponens makes a tautologous conditional by conjoining the premises of modus ponens and making them be the antecedent of a conditional with the conclusion being the consequent. Repeat this procedure for the eight remaining rules of inference. Construct a classical truth table for

$$[(p \supset q) \bullet \sim q] \supset \sim p$$
$$[(p \vee q) \bullet \sim p] \supset q,^{\star} \text{ and so on.}$$

EXERCISE III
Given the values of p = .5, q = .4, r = .3, and s = .2 demonstrate which of the classical nine rules of inference return a value of 1 and which return a value of less than 1. In other words, using these values, give a fuzzy interpretation for

$$[(p \to q) \bullet p] \to q$$
$$[(p \to q) \bullet \sim q] \to \sim p$$
$$[(p \vee q) \bullet \sim p] \to q,^{\star} \text{ and so on.}$$

What does this mean? Based on this test, which rules of inference appear to be fuzzy valid? Which rules would never show a loss of truth in moving from the premises to the conclusion?

EXERCISE IV
Find a fuzzy interpretation that shows that the following classical equivalences and inferences do not always return the value 1.

1. (A ↔ B) ↔ [(A • B) ∨ (~A • ~B)]

*2. (A → B) ↔ (~A ∨ B)

3. (A → ~A) → ~A

4. [(A → B) • (A → ~B)] → ~A

5. [(A • B) → C] ↔ [A → (B → C)]

EXERCISE
V

1. Other than De M., DN, Com., and the first version of equivalence, what rules of replacement will return the value 1, given any fuzzy interpretation? Write a short essay summary explaining what degrees of truth you tested.

*2. Review the classical proof for B /∴ A ∨ ~A and explain why it fails in fuzzy logic. (What line or lines in the proof fail?)

3. In fuzzy logic what would follow from these premises? Visibility is very low today. If visibility is low then flying conditions are poor. Therefore, today flying conditions are _____.

4. As exercises, in Chapters 7 and 10 Dan Derdorf's statement made at the Minnesota Metrodome was used: "It's noisy here even when it is quiet." (Q ⊃ ~Q) In classical logic, this implies that it is never quiet at the Minnesota Metrodome, (Q ⊃ ~Q) ⊃ ~Q. Fuzzy logic, of course, would interpret the situation differently. Test fuzzy degrees of quiet (.1, .2, .3, ...9) and show that the result in degree of truth for (Q → ~Q) → ~Q will always be equal to ("fire") the degree of truth of either Q or ~Q. Then write out at least four of your tests values in English, translating Q → ~Q, using modifiers such as *slightly, almost, very, not very, moderately,* and so forth.

ANSWERS TO STARRED EXERCISES

I.

5. (A • B) ∨ ~C
   (.7 • .3) ∨ ~(.1)
       .3   ∨ .9
          .9

19. (A → C) → [A → (A • C)]
    (.7 →.1) → [.7 → (.7 • .1)]
    1-(.7-.1)→[1-(.7-.1)]
        .4 → .4
           1

II.

p	q	[(p ∨ q) • ~p] ⊃ q
T	T	T F F **T** T
T	F	T F F **T** F
F	T	T T T **T** T
F	F	F F T **T** F

III.

[(p ∨ q) • ~p] → q
[(.5 ∨ .4) • ~.5] → .4
[.5 • .5] → .4
      .5   → .4
      1-(.5-.4)
         .9

IV.

2. (A → B) ↔ (~A ∨ B)
   (.4 → .4) ↔ (.6 ∨ .4)
        1   ↔ .6
   (1 → .6) • (.6 → 1)
   [1-(1-.6)] • 1
       .6 •   1
          .6

Many values will work in showing that this relationship will not always return the value of 1. Here A and B both equal .4.

V.   1. B   /∴   A ∨ ~A
          2. B ∨ A   (1) Add
          3. A ∨ B   (2) Com.
          4. ~A → B   (3) Impl.
          5. ~A → (~A • B)   (4) Abs.
          6. A ∨ (~A • B)   (5) Impl. **X**
          7. (A ∨ ~A) • (A ∨ B)   (6) Dist.
          8. A ∨ ~A   (7) Simp.

The rule of implication will not always return the value 1. However, step 4 is fuzzy valid. Step 4 (~A → B) will never have a lower truth value than step 3 (A ∨ B). Step 6 is the problem, because the other half of the implication rule, (~p → q) → (p ∨ q), will not always return the value 1. So, step 6 can have a lower truth value than step 5.

INDEX